Energy Efficiency and Management

Energy Efficiency and Management

Edited by **Lucas Collins**

$\mathscr{CL}$ANRYE
INTERNATIONAL

New Jersey

Published by Clanrye International,
55 Van Reypen Street,
Jersey City, NJ 07306, USA
www.clanryeinternational.com

Energy Efficiency and Management
Edited by Lucas Collins

International Standard Book Number: 978-1-63240-206-6 (Hardback)

Contents

Preface

At a grass-root level, Energy Efficiency refers to using minimal energy to provide the same service. The main agenda of Energy Efficiency is to reduce the amount of energy required to provide services. However, energy conservation and energy efficiency are two different things. For instance, replacing an incandescent lamp with a compact fluorescent lamp is energy efficiency, whereas turning off a light is energy conservation. However, in contemporary times, both these concepts are extremely crucial to prevent environmental degradation. Moreover, both Energy efficiency and conservation can reduce greenhouse gas emissions.

According to a certain report by International Energy Agency, consciously practising Energy Efficiency in buildings, industrial processes and transportation could reduce the world's energy needs in 2050 by one third, and also help control emissions of greenhouse gases. Energy Efficiency has numerous advantages as well. For instance, reducing energy results in a direct decrease of carbon dioxide emissions. Also, it results in a financial cost saving to consumers by reducing energy costs.

There are a lot of ways in which Energy Efficiency can be applied. Proper placement of windows and architectural features that reflect light into a building can reduce the need for artificial lighting. Compact fluorescent lights are more energy efficient as compared to incandescent light bulbs. Electricity when generated in industries, usually creates heat as a by-product, which can be captured and used as a source of energy.

Editor

A Sustainable Energy Efficiency Solution in Power Plant by Implementation of Perform Achieve and Trade (PAT) Mechanism

Rajesh Kumar, Arun Agarwala

Instrument Design Development Centre, Indian Institute of Technology Delhi, New Delhi, India

ABSTRACT

An enhanced energy efficiency scheme, "Perform, Achieve and Trade" (PAT) is explored in relation to the existing carbon market in India, particularly the Clean Development Mechanism, Renewable Energy Certification and possible Nationally Appropriate Mitigation Actions. The PAT scheme incentivises energy-intensive large industries and facilities for Enhance Energy Efficiency, through technology upgrade and improvement in process. The PAT scheme currently identified 478 designated consumers from eight energy intensive industrial sectors namely, thermal power plants, iron and steel, cement, textiles, chlor-alkali, aluminum, fertilisers and pulp & paper. The threshold limit in thermal power plant sector to become a PAT designated consumer is 30,000 tonne of oil equivalent annual energy consumption. In the first PAT cycle, run through 2012 to 2015, total 144 designated consumers from various states have been identified with individual target. Thermal power plant sector has been categorized on the basis of their fuel input into three subsectors *i.e.* gas, oil and coal based plants. This paper reviews the state of the art in PAT mechanism design and operational features for implementation on thermal power plant sector. The possibility of implementing an Emission Trading Scheme (ETS) in India is explored from political and institutional perspectives.

Keywords: Energy Conversion; Power Plant; Perform Achieve and Trade; ESCert; Energy Certificates

1. Introduction

Low Carbon Development Strategies in the developing countries undertake domestic actions on climate mitigation vis-à-vis economic growth. An Enhance Energy Efficiency Scheme (EEES) can be one of the policy tools to promote adaptation actions in consideration to energy demand and supply scenario. The possibility of implementing an EEES in India needs to be explored from institutional perspectives in relation to the existing carbon market in India, particularly the Renewable Energy Certificate (REC), Clean Development Mechanism (CDM), Perform, Achieve and Trade scheme in line with Nationally Appropriate Mitigation Actions (NAMAs).

India is the third largest producer of greenhouse gases and its per capita basis emission is 1.4 t CO_2. Out of the eight missions initiated under National Action Plan for Climate Change, three missions focused on mitigation are conceptualised in 2008 namely, National Solar Mission, National Mission for Enhanced Energy Efficiency, National Mission for a Green India. The energy demand and supply considerations create a need for the PAT mechanism. The scheme is in the initial stages of implementation by Bureau of Energy Efficiency (BEE) as a part of the National Mission on Enhanced Energy Efficiency (NMEEE), one of the missions in National action Plan for Climate Change [1]. The PAT scheme, started by Bureau of Energy Efficiency, covers eight energy intensive Indian industrial sectors. Across these sectors, 478 of the most energy intensive installations called Designated Consumers (DC) are participating, with 144 from the power sector alone [2].

The energy saving targets have been allocated to each Designated Consumer. In case, a DC exceeds its energy saving on specified energy consumption (SEC) target as defined by PAT scheme: it is issued with energy saving certificates (ESCert) and to penalise for deficiency in target, DC is required to purchase ESCerts. Thereafter, ESCerts will be issued and can be traded between entities. The certificates traded each represent a metric tonne of oil equivalent (mtoe). BEE has targeted 6.6 million mtoe to be saved by the scheme over the first cycle (2012 to 2015) and multi fold rise in savings expected in further cycles out to 2020 [3].

The PAT programme launched by BEE is reviewed and summarised to propose an ideal model for current scenario for Indian industries. The PAT model, with more added institutions is proposed to create a balance and linkage among the existing network to extend support to DCs based on measurable performance indicators [4-7]. The PAT programme is India specific and defines its terminology to focus on highly energy consuming industries, which are explained in 3rd Section. The PAT programme implemented by the network around the various agencies from central and state government is explained in 4th Section. We then present the model of institutional network and ESCert trading in the expanded PAT environment in 5th Section. In last Section, we present the conclusion and future work.

First, we present a short recap of the energy certification across globe.

2. Review of Energy Efficiency Models

The EU Emissions Trading System (EU ETS) is a cornerstone of the European Union's policy to combat climate change and its key tool for reducing industrial greenhouse gas emissions cost-effectively. The first and biggest international scheme for the trading of greenhouse gas emission allowances, the EU ETS covers some 11,000 power stations and industrial plants in 27 EU Member States plus Iceland, Liechtenstein and Norway. It covers CO_2 emissions from installations such as power stations, combustion plants, oil refineries and iron and steel works, as well as factories making cement, glass, lime, bricks, ceramics, pulp, paper and board. The voluntary scheme Climate Change Agreements, Renewable Obligation (RO) & Emission Trading Scheme as well as mandatory schemes CRC Energy Efficiency Scheme, Tradable White Certificates, are implemented in United Kingdom [8-12].

The US RPS is a system which creates a market demand for renewable and clean energy supplies, based on the same principles as a Quota System and Connecticut adopted its RPS in 1998. US Acid Rain Programme was implemented in 1995, by the United States Environmental Protection Agency. Energy Efficiency Portfolio Standard (EEPS)-like laws are now in place in California, Colorado, Connecticut, Hawaii, Nevada, Pennsylvania, Texas, and Vermont. Chicago Climate Exchange, launched in 2003, is a voluntary, legally binding greenhouse gas reduction and trading system for emission sources and offset projects in North America and Brazil [13,14].

International Energy Agency (IEA) promoting Energy Efficiency and Demand-Side Management for global sustainable development and for business opportunities. The IEA Demand-Side Management Programme is an international collaboration of 14 countries working together to develop and promote opportunities for demand-side management (DSM). In the EU, only the UK, Italy, Poland and France have implemented White Certificate (WhC) schemes, with different targets and design characteristics

Australian federal government has introduce three energy efficiency certificate trading schemes namely; the Energy Savings Scheme (ESS) in the State of New South Wales (NSW), the Victorian Energy Efficiency Target scheme (VEET) in the State of Victoria and national cap-and-trade emissions trading scheme. Under a cap-and-trade emissions trading scheme, the Government set an annual cap on total emissions of carbon pollution covered by the scheme and issue a number of emissions permits equal to the cap [15,16].

The VEET scheme is a Victorian Government initiative promoted as the Energy Saver Incentive. It commenced on 1 January 2009 and is administered by the Essential Services Commission (ESC). Each certificate represents a tonne of greenhouse gas abated and is known as a Victorian energy efficiency certificate (VEEC). For the first three-year phase of the scheme (2009-11), the scheme target is 2.7 million VEECs per annum, increased to 5.4 million VEECs per annum during the second three-year phase, starting on 1 January 2012 [17].

The National Development Reform Commission in China launched the "Top-1000 Program," which targets energy efficiency improvements in the 1000 largest enterprises that together consume one-third of all China's primary energy [18].

India, a democratic country with states with different biodiversity, successfully implemented energy saving and RE generation with the support of central ministry. In additions, some of the schemes started by states, individually. Renewable Energy Certification (REC) programme is implemented by Central Electricity Regulatory Commission (CERC) under Jawaharlal Nehru National Solar Mission while the Bureau of energy efficiency launched the PAT programme under NMEEE mission in 2012 and the energy saving certificate (ESCert) will be issued after submission of year report by energy manager and review by BEE in 2013. PAT and REC programme are in addition to clean development mechanism (CDM) [19].

India has issued Host Country approval for 2305 projects targeting 716,590,823 Certified Emission Reduction (CER) Certificates across 13 sectors till March, 2012. The CDM approved reports for the successful issuance of CER is 459 with 36,884,665 CERs issued, and 17,759,280 CERs in 2012 [20,21]. As on June 30, 2012, China leads with 2101 registered CDM projects accounting for 48.9% followed by India (855 projects *i.e.* 19.9%) and Brazil (204 projects *i.e.* 4.7%) respectively. The national programme PAT is introduced in view of the less demand for CER in International market with reduced price varying from INR 230 ($ 4.50) to INR 300 ($6.00) in September-October, 2012

3. Perform Achieve Trade (PAT) Scheme in India

3.1. Design of Scheme

Perform, Achieve & Trade is a market based mechanism to enhance efficiency in energy intensive industries through certification of energy saving which can be traded. The market and institutional structure, as proposed by BEE would support the implementation of PAT mechanism for target achievements. The different operational models are adopted by consumers to meet the compliance targets [2,22-26] The whole process of PAT scheme includes sequential steps, namely goal setting, emission reduction, review, certification and trade.

The PAT scheme currently covers eight energy intensive Indian industrial sectors: thermal power plants, iron and steel, cement, textiles, chlor-alkali, aluminum, fertilisers and pulp and paper. Across these sectors, 478 of the most energy intensive installations. In phase I, 144 designated consumers are participating from power sector. The **Table 1** provides the details about DCs on energy reduction target in ton of oil equivalent and percentage in comparison to present consumption [3].

The thermal power sector is a large consumer of energy and the application of a PAT target should serve to provide an additional focus on energy efficiency. There is scope in the future to deepen the scheme to cover additional (smaller) installations within the relevant energy intensive sectors. The PAT scheme document targeted

selected eight sectors and fixed national energy saving target as shown in **Figure 1** [2,27].

3.1.1. Base Line and Energy Input

The net energy input calculations for the Thermal Power Plant involves the following steps:

- Identify the fuel used for input and convert the calorific values of all forms of energy sources into a single unit, namely, ton of oil equivalent using the conversion formulae.
- For E_{in}, energy input, consider all forms of energy that is, electricity, solid fuel, liquid fuel, gaseous fuel, or any other form of energy imported into the plant for consumption as energy for production of output;
- In case DC does not have disaggregated figures for energy consumed, consider the net energy consumed for calculation both in baseline year and in the target year: provided that the said designated consumer shall give adequate reasons that it was not feasible to make adjustment for energy consumed.

National Energy Saving Targets under PAT (%) (2012-15)

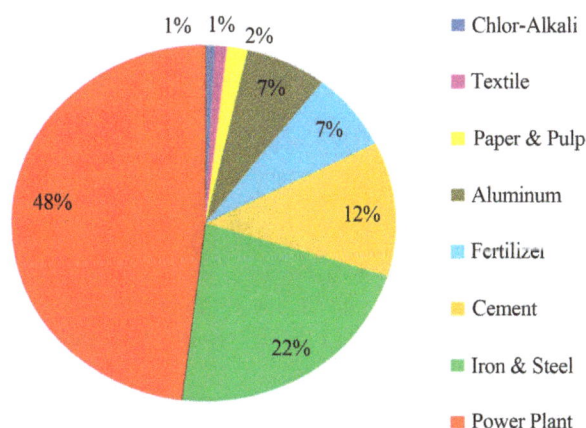

Figure 1. Energy saving by PAT-courtesy BEE.

Table 1. Breakdown of sectors and targets for PAT scheme.

Industry Sector	Annual Energy Consumption Norm to be DC (toe)	No. of identified DCs	Energy consumption million mtoe	Energy reduction (3 yrs) million mtoe	Energy reduction (3 yrs) (%)
Thermal Power	30,000	144	104	3.1	3.0
Iron & Steel	30,000	67	28	1.7	5.9
Cement	30,000	85	12	0.7	5.9
Fertilizers	30,000	29	7.9	0.5	5.8
Aluminum	7500	10	7.7	0.5	5.9
Paper & pulp	30,000	31	2.1	0.1	5.9
Textile	3000	90	1.6	0.1	6.0
Chlor-Alkali	12,000	22	0.8	0.05	5.8

- Provided further that such designated consumer shall make necessary arrangements for disaggregation of data for energy consumption to ensure that actual energy consumed for production is considered in the next cycle.

Total energy input to the designated consumers' boundary shall be estimated with the following expression:

$$E_{in} = \frac{H \times C_v}{10^7} \quad (1)$$

where, E_{in} is Energy input in toe, H is Fuel consumed quantity (kg), C_v is Gross calorific value (kCal/kg).

Specification of value of energy [24]: The value of per metric ton of oil equivalent of energy consumed shall be determined by formula;

$$P = W_c \times P_c + W_o \times P_o + W_g \times P_g + W_e \times P_e \quad (2)$$

where-

P = price of one metric ton of oil equivalent (1toe);

P_c = average price of delivered coal;

P_o = price of fuel oil;

P_g = price of gas as declared by Gas Authority of India Limited;

P_e = average price of one unit of electricity for industrial as specified by the respective State.

Electricity Regulatory Commission as on 1st April of the year.

Weightage of coal (W_c) = amount of coal consumed across all designated consumers in the baseline year (in toe)/(total energy consumption across all designated consumers in the baseline year (in toe)).

Weightage of oil (W_o) = amount of oil consumed across all designated consumers in the baseline year (in toe)/total energy consumption across all designated consumers in the baseline year (in toe).

Weightage of electricity (W_e) = amount of electricity consumed across all designated consumers in the baseline year (in toe)/total energy consumption across all designated consumers in the baseline year (in toe).

3.1.2. Specific Energy Consumption (SEC)

The overall depth of the target will be set to achieve approximately 6.6 million metric tonnes of oil equivalent (mtoe) saving during the first cycle of the scheme. Thus total energy consumption will be 6.6 m mtoe less than would have been the case had energy intensity remained constant over the period. This overall target is apportioned across energy intensive sectors using a pro-rata approach. This means that a sector that consumes more energy shall be apportioned a larger portion of the energy saving estimate [25,27-29]. The specific energy consumption (SEC) means the ratio of the net energy input into the DC boundary to the total quantity of output ex-

ported from the DC boundary, calculated as per the formula in Equation (3):

$$SEC = \frac{E_{in}}{E_{out}} \quad (3)$$

and expressed in terms of the metric ton of oil equivalent (toe)/per unit of product;

E_{in} = Net energy input into the designated consumers' boundary;

E_{out} = Total quantity of output exported from the designated consumers' boundary.

In order to calculate the SEC, the definition of product is important, given the wide range of product mixes for certain sectors, but this is much simpler in case of thermal power sector. Each installation shall be assigned independent targets expressed as a percentage reduction with respect to its estimated baseline SEC.

The targets are worked out on the basis of 'historical data' on production and energy consumption as reported by installations participating in the scheme. This approach combines elements of historical performance and comparative benchmarking to establish targets that take into account current technology employed. This approach ensures more inefficient plants are assigned higher targets or larger percentage reductions with respect to their estimated baseline consumption.

The energy consumption norms and standards for power stations shall be specified in terms of specific percentage of their present deviation of net operating heat rate, based on the average of previous three years, namely financial year 2007-08, 2008-09, 2009-10 for the first cycle and for cycles thereafter from the net design heat rate [3]. The power stations shall be grouped into various bands according to their present deviations, of operating heat rate from design heat rate and for power stations with higher deviations the energy consumption norms and standards shall be established at lower level and shall be grouped taking into account percentage deviation as under: (**Table 2**).

3.1.3. Gate-to-Gate Designated Consumer Boundary for Power Plant

As the specific energy consumption is calculated on a Gate-to-Gate concept, the plant boundary shall be selected in such a manner that the total energy input and the Electricity output, is be fully captured at the entire designated consumers' plant. Once the designated consumers' boundary has been fixed, the same boundary shall be considered for the entire cycle, and any change in the said boundary such as capacity expansion, merger of two plants, division of operation etc. shall be duly intimated to the Bureau of Energy Efficiency. Electricity generated by captive power plant (CPP), other energy purchased and consumed, electricity partially sold to grid from captive power plant is as shown in **Figure 2**.

Table 2. Reduction target for net station heat in thermal power plant.

Deviation in net station heat rate from design net heat rate	Reduction target for percentage deviation in the net station heat rate	Percentage(%) reduction target in net station heat rate
Up to 5%	10%	0.5
>5% to 10%	17%	0.75 to 1.5
>10% to 20%	21%	2.0 to 4.0
>20%	24%	5 and above

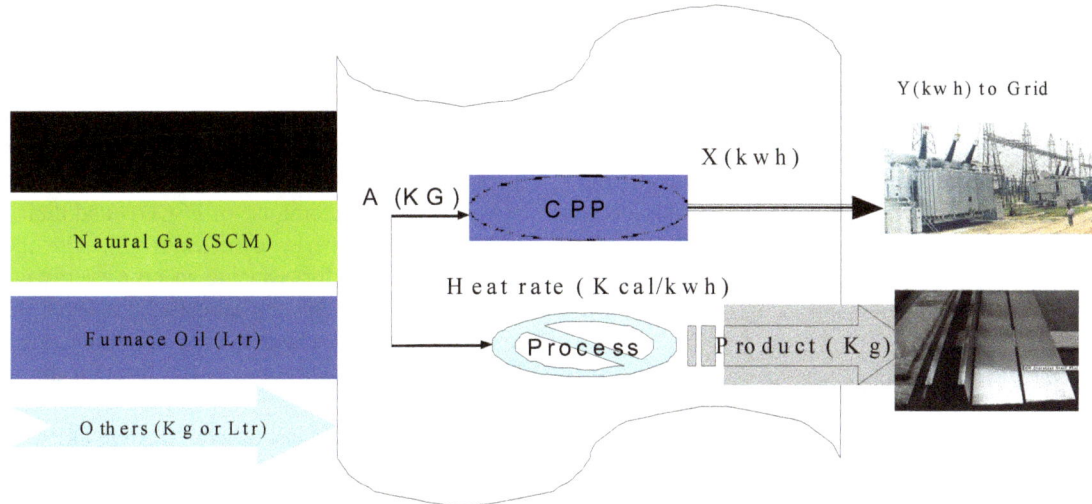

Figure 2. Captive power plant (CPP)-Electricity partially used for own process and spare supplied to grid.

3.1.4. Correction Factor Considered for Effect on Heat Rate Due to Coal Quality

Average "ash", moisture, and gross calorific value for the previous three years is considered for the calculation baseline for first cycle. For consequent cycles the value will be taken as specified for the target year shall be taken into account for the baseline year and correction factor shall be worked out based on the boiler efficiency formula as in Equation (4) [3,26]:

$$\eta = 92.5 - \frac{50 \times A + 630(M + 9H)}{GCv} \quad (4)$$

where:

η is Boiler Efficiency, A is Ash percentage in coal, M is Moisture percentage in coal, H is Hydrogen percentage in coal, G.Cv is Gross calorific value in kcal/kg and

Station heat rate (Kcal/kWh) = Turbine heat rate/ Boiler efficiency.

3.1.5. Methodology for PAT Implementation

The proposed methodology for the phase wise implementation of the PAT programme [2];

- Identification of Industry sector on study of energy consumption pattern and phase in potential designated consumer in the scheme.
- Appoint Energy Manager and expose him/her to PAT,

BEE and DC targets.

- Registration for DC with Energy audit and energy saving plan as per regulation of Bureau of Energy Efficiency Regulations 2010.
- The designated consumers shall achieve compliance with the energy consumption norms and standards within a period of three years.
- BEE establishes base year and projected energy consumption and calculating the net energy input to the plant.
- Calculate the specific energy consumption for the baseline year as well as for the target year and normalise it by taking into account the capacity utilisation, mix of grid and captive electricity, and any other factor which affects energy consumption as specified in the Schedule.
- Audit by Accredited Energy Auditor based on document review and cross checks over the claim by energy manager.
- Accredited energy auditor shall report the results of his assessment in a verification report.
- State designated agency may convey its comments, if any [2].
- The accredited energy auditor in-charge of check-verification shall submit his report with due certification to the Bureau and the concerned State Designated

- The exact number of energy savings certificates to be issued to the designated consumer and the entitlement for such energy savings certificates after determining by the following formula for thermal power plant sector:

$$\text{No. of ESCert} = \frac{(Qn - Qt)\,Eb}{10} \qquad (5)$$

No. of ESCert = number of energy savings certificates.

Qn = heat rate notified for the target year.

Qt = heat rate as achieved in the target year.

Eb = production in the baseline year in million kwh.

The total amount of energy savings certificates recommended under sub rule (2) shall be adjusted against the entitlement on conclusion of the target year as per the following formulae for thermal power plant sector:

1) Energy savings certificate to be issued after 1st year

$$\text{No. of ESCert} = \left[Qb - \frac{(Qb - Qt)}{3} - Q1 \right] \frac{80\% \times Eb}{10} \quad (6)$$

2) Energy savings certificate to be issued after 2nd year

$$\text{No. of ESCert} = \left[Qb - \frac{(Qb - Q2)2}{3} - Q2 \right] \frac{80\% \times Eb}{10} \quad (7)$$

3) Energy savings certificate to be issued in the target year

$$\text{No. of ESCert} = \frac{[Q2 - Qt]\,Eb}{10} \qquad (8)$$

Qb = heat rate in the baseline year.

Qt = heat rate notified for the target year.

Eb = production in million kwh in the baseline year.

Q1, Q2, Q3 = heat rate achieved in year 1, 2 & 3.

- The accredited energy auditor observes an unfair gain due to the deficiencies, inconsistencies, errors or misrepresentation by the designated consumer. The value of the amount payable by such designated consumer shall be as worked out in the verification report plus twenty-five per cent. Of such value because of unfair practice used by the said designated consumer for obtaining unfair advantage.

4. PAT: Institutional Network for Implementation and Trade

The PAT mechanism is designed to incentivise enhanced energy efficient technology by industry than their specified Specific Energy Consumption improvement target in a cost-effective manner. The energy efficiency improvement targets are "unit specific" and each DC is mandated to reduce its SEC by a fixed percentage based on the SEC baseline within a sectoral bandwidth. The guiding principles for developing the PAT mechanism are simplicity, accountability, transparency, predictability, consistency, and adaptability. The PAT framework includes the elements for device methodology for setting SEC for each DC and its target for SEC reduction and verification of SEC of each cluster in specified period w.r.t. base line for the issuance of Energy Savings Certificates (ESCerts). The additional certified energy savings in the form of ESCert can be traded with other designated consumers who could use these certificates to comply with their SEC reduction targets. The ESCert will be traded on special trading platforms to be created in the two power exchanges Indian Energy Exchange (IEX) and Power Exchange India (PXIL). The BEE has also set up registry and exchanges for the trading of ESCert and creation of records for creation, trading and cancellation of ESCert to enable cross-sectoral use of ESCerts and their synergy with renewable energy certificates [2]. The individual mandatory targets for reducing specific energy consumption, which have been established for over 478 DCs could be achieved either by implementing measures in their respective facilities, or offset through the purchase of EScert [27].

The flow chart for institutional linkage in PAT Mechanism in India is shown in **Figure 3**, initiated by ministry of Power. The Ministry of Power is the nodal ministry for NMEEE mission and implement PAT programme through Bureau of Energy efficiency. The institutional network is proposed for PAT process which includes identification of DCs, calculation of SEC, audit, issuance of ESCert and trading [28,30-33]. The study also proposes an additional consolidated fund needs to strengthen country research & development (R&D) capacity, technology adoption risk, PAT programme for mini, small & medium enterprises (MSME) and insurance for any natural disaster event [34].

The identified designated consumer needs to share the energy consumption and enhanced energy saving information as specified in the PAT assessment document. The BEE designed format requires critical information on technology, energy consumption and saving on technology up-gradation. The Bureau of Energy Efficiency analyses the claim submitted by the energy manager, which may require coordination with State Designated Agencies (SDAs), Designated Energy Auditors (DENA) and third party auditors. In case of doubt in energy savings claim, the BEE auditor will conduct site audits. On completion of the monitoring and verification process by BEE, the PAT registry will issue the ESCert equivalent to saving on energy for that time period. In the first phase 2012-2015, the review will be done annually and hence the ESCert will be issued after April, 2013.

The ESCerts trading mechanism is designed in this study, to enhance the response on the PAT mechanism

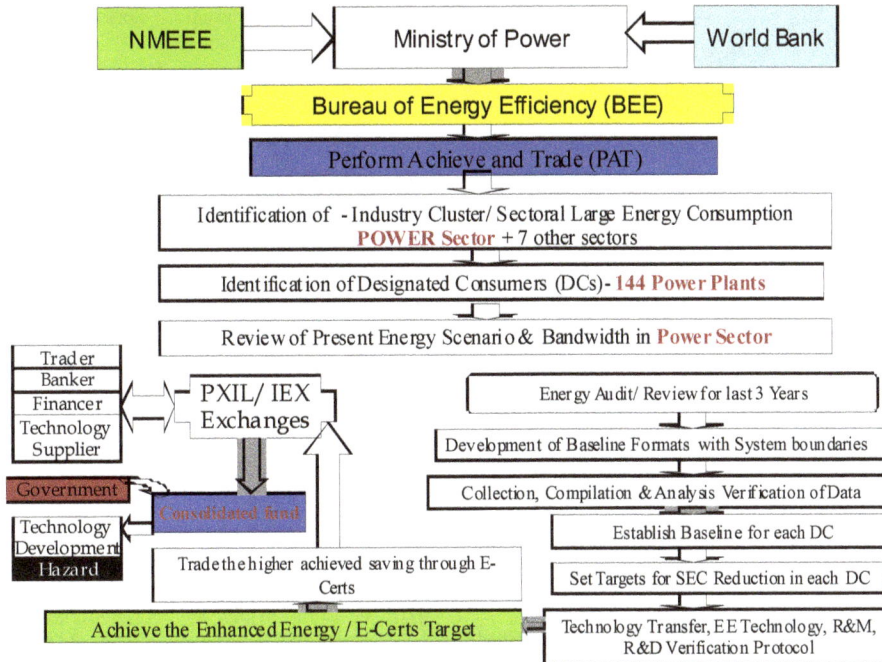

Figure 3. PAT programme implementation on power plant.

for the adoption by other energy intensive industries. This paper has identified important links for the issue and trade of ESCert between the industries, auditors and institutions from states and centre [19]. These links involve extensive information exchange between Designated Consumers, State Designated Agencies, Designated Energy Auditors, Power Exchanges, technology supplier, Bureau of Energy Efficiency and Central Registry on a regular basis. The central registry office is responsible for the creation of the information network and to timely share accurate information while maintaining confidentiality, transparency and security. The fast access and sharing of information can be achieved by adopting a suitable software platform. The Bureau of Energy Efficiency is working on developing the "**PAT-Net**", a PAT control software which is an online integrated information system for operation and data management for creation, transfer, trading and cancellation of ESCert. The main features of PAT-Net are summarized below:

1) The proposed PAT-Net is a two way communication solution for PAT operation, connecting all the stakeholders and participants. It will function as an information consortium and provide limited controlled access to all the Designated Consumers, State Designated Agencies, Designated Energy Auditors, Trading Exchanges and the Central Registry. Each one of them will be provided with a unique access depending on their category, with user rights assigned accordingly.

2) Designated consumers will access to PAT-Net, for mandatory reporting activities. Energy Manager will submit self declaration of energy savings through the PAT-Net and in response BEE will consider the issuance of ESCerts. On successful review, BEE will upload online instructions to Central Registry for issuance of ESCerts to designated consumer. During the review process, BEE will share information on credit and debit of ESCerts with designated consumers through PAT-Net. The PAT-Net software will process the data from all DC and trading exchange on issuance and trade for the generation of reports for the analysis of the PAT programme. The results will also be used to assess DC performance and provide suggestion for improvement [35].

3) The Trading Exchanges could intimate the trade details and obligations to all the participating DCs, Central Registry, SDA and DENA [19,31] and conduct constant performance monitoring of the program. The performance will be reviewed through parameters such as total ESCerts issued & traded, complying sectors or participants and market liquidity. The administrator will identify any delays in the process-chain and initiate timely action. The information on SEC targets will be issued to DC through PAT-Net and reconciliation be carried out on the issue of ESCert or penalties.

BEE will share the rich technical experience of Central Electricity Regulatory Commission (CERC), which has successfully commissioned the renewable energy certificates (REC) scheme in India. RECs are traded on two exchanges, PXIL and IEX, and BEE will plan the trading platform for ESCerts. The roles for independent sector/ trade associations in parallel with CERC and BEE should be redefined [31,36].

5. Conclusion

The PAT mechanism, with additional institutional network, will give boost to enhance energy efficiency technology driven industries. The successful implementation of PAT scheme will also provide a breakthrough in research by the industries on energy efficiency technology and process. The paper proposes to use the same platform for the trading of PAT and REC at two exchanges already started PXIL and IEX with different controlling institutions in the interactive mode with international programmes like CDM, UNFCCC and World Bank. The operational PAT mechanism focuses on large industries and this paper highlights the missing link for Mini, Small & Medium Enterprises (MSME), which has the major impact on the energy consumption and economic growth of developing country like India. To ensure liquidity and demand for ESCerts, bulk buying and bundling of ESCerts should be encouraged, with open ended time limit for trading. Some of the issues that are being still to be discussed in order to enhance liquidity are intermediate compliance timeframes to enhance market liquidity, energy allowances vs ESCerts, banking of certificates and start option for auctions and buy-back of ESCerts by the Government agency like BEE and CERC.

6. Acknowledgements

This paper is part of the ongoing research work at IIT Delhi and major initiatives from Department of Science & Technology for the promotion of "Global Technology Watch Group" under climate change programme. The authors gratefully acknowledge the guidance, inputs and feedback from Dr. Sanjay Mishra, Dr. Nisha Mendiratta and Dr. Anand Kamavisdar. The authors are also acknowledge the Bureau of Energy Efficiency and Ministry of Environment and Forest, for sharing content and views about programme.

REFERENCES

[1] National Action Plan on Climate Change Released by Prime Minister's Council on Climate Change, 2008

[2] PAT—Perform, Achieve and Trade, "Power booklet Released by Ministry of Power," Government of India, New Delhi, 2012

[3] Ministry of Power, "PAT Notification," 2012.

[4] T. Bhattacharya and R. Kapoor, "Energy Saving Instrument—ESCerts in India," *Renewable and Sustainable Energy Reviews*, Vol. 16, No. 2, 2012, pp. 1311-1316.

[5] P. Bhargava, "Project Economist Bureau of Energy Effi-

[6] ciency Presentation on Perform, Achieve & Trade (PAT) Mechanism under NMEEE". www.beeindia.in

[6] S. P. Garnaik, "Presenatation on Pre-Bid Conference on Baseline Energy Audit under Perform, Achieve & Trade (PAT) Scheme". www.beeindia.in

[7] Y. W. Wu, S. H. Lou and S. Y. Lu, "A Model for Power System Interconnection Planning under Low-Carbon Economy With CO_2 Emission Constraints," *IEEE Transactions on Sustainable Energy*, Vol. 2, No. 3, 2011, pp. 205-214.

[8] A. N. Menegaki, "Growth and Renewable Energy in Europe: A Random Effect Model with Evidence for Neutrality Hypothesis," *Energy Economics*, Vol. 33, No. 2, 2011, pp. 257-263.

[9] H. L. Raadal, E. Dotzauer, O. J. Hanssen and H. P. Kildal, "The Interaction between Electricity Disclosure and Tradable Green Certificates," *Energy Policy*, Vol. 42, 2012, pp. 419-428.

[10] P. Bertoldi and T. Huld, "Tradable Certificates for Renewable Electricity and Energy Savings," *Energy Policy*, Vol. 34, No. 2, 2006, pp. 212-222.

[11] S. Patrik, "The political economy of International Green certificate Markets," *Energy Policy*, Vol. 36, No. 6, 2008, pp. 2051-2062.

[12] V. Dinica and M. J. Arentsen, "Green Certificate Trading in the Netherlands in the Prospect of the European Electricity Market," *Energy Policy*, Vol. 31, No. 7, 2003, pp. 609-620.

[13] E. Vine and J. Hamrin, "Energy Savings Certificates: A Market-Based Tool for Reducing Greenhouse Gas Emissions," *Energy Policy*, Vol. 36, No. 1, 2008, pp. 467-476.

[14] M. Gillenwater, "Redefining RECs—Part 2: Untangling Certificates and Emission Markets," *Energy Policy*, Vol. 36, No. 6, 2008, pp. 2120-2129.

[15] G. Buckman and M. Diesendorf, "Design Limitations in Australian Renewable Electricity Policies," *Energy Policy*, Vol. 38, No. 7, 2010, pp. 3365-3376.

[16] P. Effendi and J. Courvisanos, "Political Aspects of Innovation: Examining Renewable Energy in Australia," *Renewable Energy*, Vol. 38, No. 1, 2010, pp. 245-252.

[17] G. Curran, "Contested Energy Futures: Shaping Renewable Energy Narratives in Australia," *Global Environmental Change*, Vol. 22, No. 1, 2012, pp. 236-244.

[18] S. Long, C. Michael and Y. Lin, "A Review on Sustainable Design of Renewable Energy Systems," *Renewable and Sustainable Energy Reviews*, Vol. 16, No. 1, 2012, pp. 192-207.

[19] Prabhakant and G. N. Tiwari, "Evaluation of Carbon

Credits Earned by Energy Security in India," *International Journal of Low-Carbon Technologies*, Vol. 4, No. 1, pp. 42-51.

[20] R. Kumar and A. K. Agarwala, "Energy Computing Models for Techno-Economics Feasibility," *Conference on "Computing For Nation Development*," New Delhi, 23-24 February 2012, pp. 379-384.

[21] IDBI Carbon Development News Letter, 2012.

[22] S. S. Krishnan, N. Narang, S. K. Dolly, R. King and E. Subrahmanian, "Global Mechanisms to Create Energy Efficient and Low-Carbon Infrastructures: An Indian Perspective," 2010 *3rd International Conference on Infrastructure Systems and Services: Next Generation Infrastructure Systems for Eco-Cities* (*INFRA*), Shenzhen, 11-13 November 2010, pp. 1-6.

[23] Bureau of Energy Efficiency Ministry of Power, Government of India, "PAT Consultation Document 2010-11," 2011.

[24] "PAT Rule Notification," Ministry of Power, G.S.R. 269 (E), 2012.

[25] A. K. Asthana, "Target Setting Methodology for Power Sector in Perform, Achieve & Trade (PAT) Mechanism," PAT Capacity Building, Mumbai, 2011.

[26] S. Dube, R. Awasthi and V. Dhariwal, "A Discussion Paper on India's Perform Achieve and Trade (PAT) Scheme," 2011.

[27] R. Kumar and A. K. Agarwala, "Energy Certificates REC and PAT Sustenance to Energy Model for India," *Renewable and Sustainable Energy Reviews*, Vol. 21, 2013, pp. 315-323.

[28] Ajay Mathur Energy Synergy Dialogue with Dr. Ajay Mathur, Director General Bureau of Energy Efficiency at World Energy Council, 2011. www.beeindia.in

[29] A. K. Asthana, "Can the Learning's from International Examples Make the 'Perform Achieve and Trade (PAT) Scheme Perform Better for India—An Approach for Target Setting in Perform, Achieve & Trade (PAT) Mechanism," 2011.

[30] K. Regan and N. Mehta for Verco Global Report for British High Commission (BHC) on International Finance and the PAT Scheme. http://www.vercoglobal.com/images/uploads/insights-reports/International_finance_and_the_PAT_scheme.pdf

[31] R. Kumar and A. K. Agarwala, "Renewable Energy Certificate and Perform, Achieve, Trade Mechanisms to Enhance the Energy Security for India," *Energy Policy*, Vol. 55, 2013, pp. 669-676.

[32] Minutes of Meeting, "National Level Interactive Session for Designated Consumers on PAT Awareness under NMEEE," 2011. www.beeindia.in

[33] R. Kumar and A. K. Agarwala, "RET Diffusion Model for Techno-Economics feasibility," *International Conference on "SOLARIS-2012-Energy Security Global Warming and Sustainable Climate"*, IIT Delhi and BERS, 7-9 February 2012, pp. 479-491.

[34] G. C. D. Roy, "'PAT' Scheme for Indian Industry-Some Perspectives, Stakeholder," *Meeting on "Potential India—UK Cooperation on Energy Efficiency and Trading"*, New Delhi, 18 February 2010.

[35] A. Arabali, M. Ghofrani, M. Etezadi-Amoli, M. S. Fadali and Y. Baghzouz, "Genetic-Algorithm-Based Optimization Approach for Energy Management," *IEEE Transactions on Power Delivery*, Vol. 28, No. 1, 2013, pp. 162-170.

[36] A. T. Al-Awami and M. A. El-Sharkawi, "Coordinated Trading of Wind and Thermal Energy," *IEEE Transactions on Sustainable Energy*, Vol. 2, No. 3, 2011, pp. 277-287.

2

HVAC vs Geothermal Heat Pump—Myth & Truth

Avijit Choudhury

Agm-Energy Services, Enfragy Solutions India Pvt. Ltd., New Delhi, India

ABSTRACT

In India energy auditors in some cases have recommended ground water based heat pump in place of common HVAC system to achieve better efficiency. However the users were made to understand that heat pumps are not suitable for Indian climatic condition and it produces no better result than the conventional system. This article tries to find answer to this debate thru an impartial analysis with the available data. First the concept is tested thermodynamically to compare the derived COPs and next from the experimental data on specific energy consumption.

Keywords: Heat Pump; HVAC; COP

1. Introduction

Heating ventilation & air-conditioning system is the major energy consuming part in residential or commercial buildings. In countries like India, more than 50% of input energy is consumed by HVAC system of the building. Given the next page of a typical load distribution chart of one Government office building in India (**Figure 1(a)**). For this reason, an energy auditor first looks at HVAC system when he tries to find out some meaningful energy saving potential for his client. Off late, many auditors have started recommending geothermal heat pump in place of conventional vapour compression refrigeration system and claim much better performance of the former. On the other hand, the HVAC OEMs rubbish this claim and say that the proposition is much costlier and doesn't produce any better result. Now as energy auditor when one needs to comment on this subject, he must have clear understanding of the facts & figures.

2. Look at the Basic Thermodynamics

To start with let us look at the basic law of thermodynamics. The ideal refrigeration cycle can be called as reverse Carnot's cycle (**Figure 1(b)**) where the engine draws heat from the lower temp (sink) and deliver to higher temp (source). In the above figure, the machine requires W amount of work to absorb heat Q_2 at lower temp and transfer heat Q_1 to higher temp. Hence the efficiency, in conventional term co-efficient of performance, should be defined as:

$$COP = \frac{\text{heat removed}}{\text{work input}} = \frac{Q_2}{W} = \frac{Q_2}{Q_1 - Q_2} = \frac{T_2}{T_1 - T_2}$$

So from the above equation it is clear that higher the

T_1 lower the COP. Or in other words higher the ambient temp, lesser is the COP. In countries like India where summer temperature shoots up to 42°C to 45°C, the COP or EER (energy efficiency ratio) gets affected badly for conventional HVAC system. The concept of geothermal heat pump is that—heat is discarded not into the atmosphere but at 200 ft. deep underground sump where temp is much lower and constant (10°C) throughout the year. Heat pump suppliers take this advantage of constant and lower sink temp to maximize their COP or EER.

Figures 2 and **3** are of Apollo group building at Chennai where conventional HVAC was retrofitted by geothermal heat pump [2]. Now if we look at the data carefully, the thermodynamic or Carnot COP of the above systems become

Conventional system (**Figure 2**) = 285.2/(321.8 − 285.2) = 7.792 (temp converted into Kelvin).

Retrofit system (**Figure 3**) = 284/ (313 − 284) = 9.793 (temp converted into Kelvin).

So mathematically it can be proven that the retrofit system gives much better energy performance. Now comparing the above two figures one can easily see that the retrofit system, if run exactly in identical conditions, produces air of much lower temp (lower by 5.2°C) while consuming the same power [3]. So when the air temp is controlled by a thermostat to maintain it at 17.2°C the compressor shall be on "load position" for much shorter period compared to the conventional system. This is the major area from where the savings comes. Having optimum temp difference helps to achieve heights possible suction prss at compressor inlet (or in other words compressor is made to work at lower differential prss) and thereby leading to less energy consumption.

Load Distribution

(a)

(b)

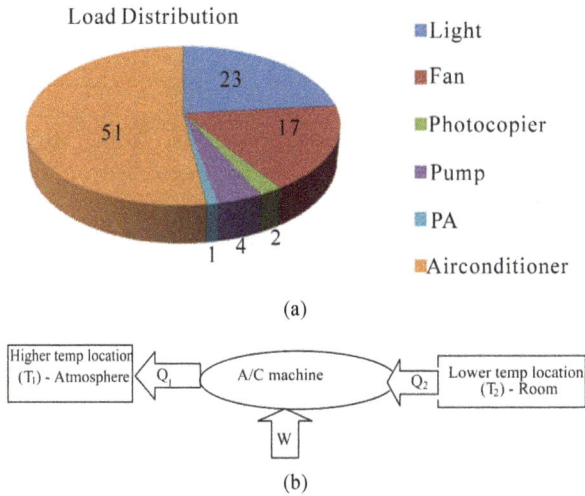

Figure 1. (a) Load distribution chart; (b) Schematic, reverse carnot cycle.

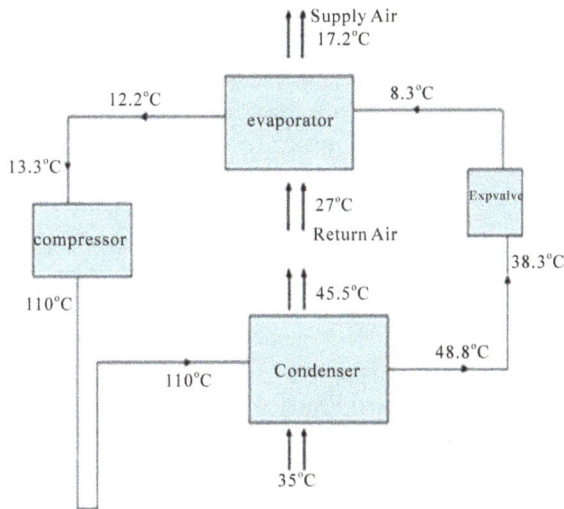

Figure 2. A conventional air conditioning system.

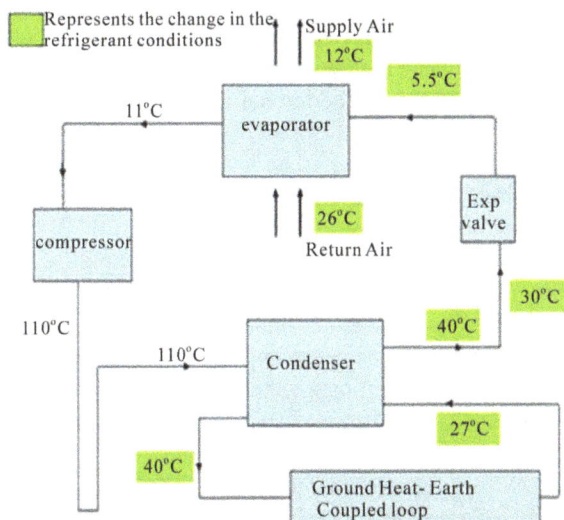

Figure 3. An earth coupled geoexchange system.

3. What Is the Basic Difference between the Two Systems?

From the above figures one may find that both the systems have much similarity in nature. The only difference is—in conventional HVAC, condenser cooling water is cooled by a cooling tower whereas for geothermal heat pump it is cooled in the underground geothermal loop. All other components are same for both the cases. Hence retrofit option for geothermal heat pump becomes easy and feasible. A study has been conducted by "Geothermal India" to compare the performance of both the systems at various ambient temps and results are given left (**Figure 4**).

From **Table 1** one may notice that EER of conventional system decreases with the rise in ambient temp whereas the EER of heat pump remains unaffected. Arguments were also given that geothermal cooling is successful only in cities with extreme climate like Delhi, Jaipur, Patna etc. To counter this argument studies and simulation were conducted for different cities across India and results **Table 2** were found quite encouraging [3]. Enfragy Solutions had conducted energy audit in more than 70 Government buildings having conventional HVAC system. The HVAC data of one such building **Table 3** is given the next page to show the energy distribution among various components of HVAC. The data suggests that even with best design and highly efficient components, chiller/compressor energy consumption remains critical for overall energy performance of HVAC system.

From **Table 3** it can be inferred that chiller unit/com-

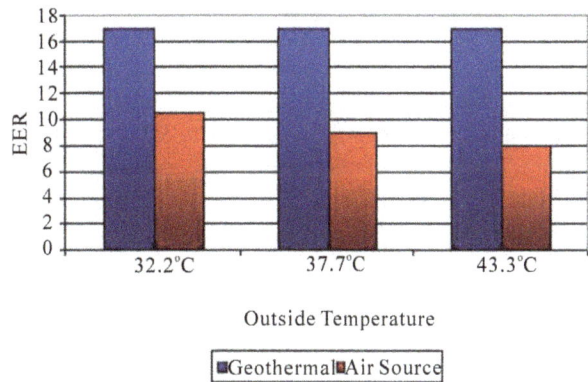

Figure 4. EER comparison.

Table 1. EER Comparison.

EER Comparison: Geothermal vs. Air Source Cooling			
Outside Temp (°C)	32.2	37.7	43.3
Geothermal EER	17	17	17
Air Source EER	10.5	9	8

Table 2. India benchmark of HVAC system 200,000 sq ft air conditioned commercial area. Geothermal heat pump rated at 0.6 kW + ground heat exchange. Chiller rated at 0.6 kW + Cooling tower. System: Space cooling + Pumps & auxiliaries + Ventilation fans.

City	Chiller (kW/ton)	Geothermal (kW/ton)	Chiller-EER	Geothermal-EER
Ahmedabad	1.15	0.48	10.46	25.01
Bangalore	1.12	0.45	10.72	26.81
Chennai	1.27	0.51	9.48	23.69
Hyderabad	1.23	0.53	9.75	22.58
Jaipur	1.11	0.43	10.85	27.61
Kolkata	1.25	0.49	10.57	24.65
Mumbai	1.16	0.46	10.34	25.85
Nagpur	1.16	0.48	10.39	25.13
NCR	1.17	0.47	10.28	25.73
Pune	1.13	0.44	10.65	27.13

Table 3. Break-up of total SEC.

HVAC Components Consuming Energy	Specific Energy Consumption (kW/ton)	
	Before Audit	After Audit Recommendation
Chiller Unit (with Compressor)	0.98	0.65
Air, Distribution System	0.33	0.18
Water Pump	0.39	0.15
Cooling Tower	0.19	0.08
Total	1.89	1.06

Table 4. Power consumption—kW/ton (centrifugal chillers) at different % of loading. % Loading.

ECWT	100%	87.50%	75%	62.50%	50%	37.50%	25%
90 DEG F	0.625	0.633	0.635	0.652	0.686	0.750	0.950
86 DEG F	0.605	0.597	0.601	0.618	0.652	0.714	0.835
82 DEG F	0.571	0.565	0.571	0.588	0.621	0.681	0.803
78 DEG F	0.539	0.536	0.544	0.561	0.593	0.652	0.775
74 DEG F	0.511	0.510	0.519	0.536	0.566	0.623	0.748
70 DEG F	0.484	0.485	0.494	0.510	0.539	0.595	0.724
66 DEG F	0.459	0.462	0.470	0.484	0.512	0.567	0.700

dit, the RH (Relative Humidity) of the conditioned air was found 60% to 69%. The data collected from Geothermal India suggests that even in city like Chennai, which is highly humid throughout the year, they are able to maintain 50% - 54% RH of the conditioned air. Since RH plays a crucial role on the comfort level, heat pump shows better results in humid areas. Experiment suggests that air with 50% RH and 24°C gives same comfort level as of air with 65% RH and 22°C. Hence lower RH generates saving of 2°C out of cooling requirement.

Other relevant observation Enfragy Audit team gathered is that most of the installed HVAC systems are not truly reversible in nature. During winter, when hot air is required in place of cold air, the same is achieved by simply turning on electric heater or in other words "heat convectors". All modern geothermal heat pumps are reversible in nature that means during heating cycle it extracts the underground heat and pump it inside the building to make the air hot. This "reversibility" makes the heat pump a better choice over conventional HVAC system.

4. Conclusions

Table 2 shows that geothermal cooling can be effective in all kind of climate zones. The heat pump can become more economical for the hotel industries as it generates hot water without any additional input energy. The only initial investment for this purpose is the installation of one de-superheater. The waste heat generated by ground source heat pump is used to heat the water. Heat pump is also beneficial to those clients who have acute shortage of raw water. Lot of water is wasted in cooling tower as drift and blow down loss. Hence for normal operation of cooling tower, substantial amount of make-up water is needed on daily basis. However this requirement can be eliminated by retrofitting geothermal cooling system.

Other than LEED rating, BEE star rating is also very popular in India for green building certification. Improving kW/ton figure through heat pump enhances the chance of getting green building certification and carbon credit. It is to be noted that ECBC (Energy Conservation

pressor consumes more than 50% of input energy in the HVAC system hence it plays a major role in controlling the overall energy consumption. Another important point in this regard to be remembered that chiller's SEC is not uniform but varies with the load condition. At full load it gives the designed value of SEC, but shows higher SEC at part load conditions **Table 4** [4].

Therefore controlling the energy consumption of chiller/compressor is the key for reducing total energy consumption. We saw that when the conventional system is retrofitted with geothermal heat pump there are changes in temperature profile which results in lesser loading of compressor *i.e.* compressor is on "load position" for shorter period and on "unload position" for longer period. Since unload power of compressor is just 1/4 of load power, retrofitting option saves substantial amount of energy. From the **Table 4** it is also clear that impact of geothermal heat pump shall be more for systems operating with part load for considerable period of time.

Another important observation can be shared here. In all most all the sites where Enfragy conducted energy au-

Building Code) is expected to get incorporated in National Building Code very soon making the statutory EPI (Energy Performance Index; kWh/sqm/year) of all commercial buildings more stringent. Heat pump shall definitely play a major role towards the compliance of ECBC. For all green field building projects, the architects/civil engineers need to take proposal for both conventional HVAC and Geothermal heat pump and do the complete life cycle cost analysis based on the guaranteed EER/COP. Our internal audit reports suggest that Geothermal heat pump option gives higher ROI, better IRR and lower payback period.

In India the apex energy efficiency body BEE should take initiative to formulate and publish the standards and norms of geothermal heat pump and bring it under national energy star leveling programme. Under such circumstances only heat pump shall become popular among the users and energy auditors.

REFERENCES

[1] Thermodynamics by P. C. Rakshit, (ISBN 0785507558, 9780785507550).

[2] Geothermal India Website.
http://www.geothermalindia.com/process.html

[3] G. Morrison and A. Parwez, "Dispelling the Myth of Geothermal HVAC by Grant Morrison & Parwez Ahmed," *Air Conditioning & Refrigeration Journal*, 2010. http://www.geothermalindia.com/data/Geothermal%20India%20Brochure/ISHRAE%201.pdf

[4] M. Nadeem, "Evaluation of Overall Chiller Performance Characteristics," *Air Conditioning & Refrigeration Journal*, Ener Save Consultants Pvt. Ltd., New Delhi, 2001. http://www.ishrae.in/journals/default.html

3

Energy-Efficiency Economics as a Resource for Energy Planning

Fabio Correa Leite, Decio Cicone Jr.,
Luiz Claudio Ribeiro Galvão, Miguel Edgar Morales Udaeta

Energy Group of Electric Energy and Automation Engineering Department,
Polytechnic School, University of São Paulo, São Paulo, Brazil

ABSTRACT

The objective of this work is a multi-criteria decision-making assessment that aims to facilitate the Energy-Efficiency Economics, introducing the Analytic Hierarchy Process (AHP) as part of power-system planning tool for an energy-efficiency application. It addresses to include qualitative aspects in the decision-making agendas of energy-efficiency projects. The manuscript details the limitations of non-rigorous financial analysis and proposes an alternative for including energy-efficiency measures in discussions pertaining to the financial opportunities available to any investor, and it presents the methodology that supports the qualitative aspects and the software package used to execute this methodology. As case study a complete example including a sensitivity analysis is presented.

Keywords: Energy Conservation; Energy Planning; Power Economics; Analytic Hierarchy Process (AHP)

1. Introduction

Energy-efficiency projects analyzed exclusively from the managerial perspective tend to make these projects less attractive once their hard benefits that are not trivial to quantify. Energy that is not spent is particularly difficult to measure and investors require a wide range of information to decide where to invest his capital and long-term benefits are not always represented in the financial indicators. "The acquisition of resources for investment is often limited, and managers tend to favor a company's core business while deteriorating secondary functions, including energy" [1].

Medium and large companies not rarely have mission statements that can be as vague as: operate safely, respect the environment and increase shareholder value. This is their way to communicate to the employees, company's owners and society what drives their goals. Meanwhile, investment decisions are still based on traditional methods and a subjective ingredient: the decision maker's own beliefs. From that perspective, how does a company decide between its environmental footprint or a more attractive IRR? It is a challenging decision because projects can fulfill regulations and industry standards and still impact the society in different ways. Regulations are not perfect, and one project is supposedly less harmless to the environment than the other. Then the question is: how to translate business major subjective goals into an organized and repeatable decision-making process?

Project managers, capital leaders and engineering managers have to make project decisions all the time. Some decisions are as comprehensive as whether to build a LEED office building or an ordinary one. This could never happen without committed leadership on the part of the organization. Or the simple choice of a high-efficiency motor rather than a standard one involves an additional capital investment, which, if not supported by top management, can lead to unpleasant disputes.

The consideration of non-monetary aspects in an investment decision through AHP is a possibility to help decision-makers choose energy-efficient and environmentally friendly projects through a standardized process. In this sense, it is proposed a way of analyzing energy-efficiency projects in terms of similar parameters to any other expansion, acquisition or fusion project, including, in addition, elements that are not part of the traditional decision-making process, such as qualitative factors used in the AHP.

This article aims to discuss certain specifics, which, when added to well-known methods of financial analysis like Net Present Value (NPV) and Internal Rate of Return (IRR), make energy-efficiency projects more attractive from the financial perspective and an alternative way to base decisions with a more comprehensive approach. Accordingly, it quantifies qualitative aspects by considering them in the decision-making process.

2. Features of Energy Efficiency Economics

Industries that make intensive use of utilities (electricity, industrial water, vapor, cooled water, etc.) are, in their majority, capital-intensive industries. The growth process of capital-intensive industries requires large investments; on the other hand, long-term bank credit for these companies is limited, and thus, companies can opt to go public as an alternative for fund raising. Therefore, low investment returns must be compensated by solid profit margins in order to attract investors.

Typical shareholders are not willing to use a part of their resources to generate a capital reserve for energy projects. Investors study any investment from the standpoint of a risk-return relationship, in any type of project. They attribute energy-efficiency applications with the same level of importance as new product projects and expansions, for instance, in decision-making processes.

Therefore, there is the need to show investors why they should invest in a project that might save some thousands of dollars in electricity compared to a project that would increase sales, the company's working capital and consequently the profits.

2.1. Economics of Project

There are several well-known methodologies for project analysis and investors are used to following only a few parameters that support comparisons between projects. The most popular methods are simple payback and discounted cash flow analysis.

To illustrate an example of the different perspective between the traditional and multivariate project decision-making processes, a US $60,000 project is presented to improve a condensate return system in the manufacturing area, including a new high-pressure condensate flash tank with controls to capture low-pressure flash steam, a new condensate return pumping system, and the stan-

dardization of steam traps.

This project could save US$30,000/year in energy due to increased condensate return and recovery of flash steam, reduce fixed-cost savings by US$3000 through the standardization of parts. However, variable maintenance costs would be increased by US$7000, due to new sophisticated equipment. Reference [2] indicated that over the next few years, the cost of energy is expected to rise faster than other costs. Inflation in Brazil is projected to keep hovering at around 4.5%/yr; accordingly, maintenance- and fixed-cost inflation was set at 4%/yr and energy inflation at 5%/yr.

According to Brazilian regulations, this kind of investment must be equally depreciated in 10 years, resulting in a US $6000/year cost with depreciation. Taking all of these considerations in account results in the following cash flow data (see **Table 1**).

The NPV (Net Present Value), calculated from the PTOI (Pre-Tax Operating Income) net savings, is 52,500 US dollars for an IRR of 47%. Since the NPV is positive and the IRR is greater than the discount rate, this means that the project creates the amount of 52,500 US dollars over 10 years, as long as we only consider the cash flow from the estimated savings and expenses. On the other hand, it is necessary to remember that the calculation of NPV from the estimated savings minus the expenses is not the actual amount created for the investor. The depreciation of the assets and the additional taxes paid have to be included. Therefore, it is necessary to include these two items in the NPV calculation. In Brazil, the tax system is complex; the total amount of taxes paid in 2008 was about 35% of Gross Domestic Product (GDP) [3]. For the purposes of simplification, in this calculation the fraction considered as tax is 30%. Therefore, the new calculation, including depreciation and taxes, gives us an NPV of 28,000 US dollars and an IRR of 36%. It means

Table 1. Cash flow aimed at the improvement of the condensate system.

Year	0	1	2	3	4	5	6	7	8	9	10
Capital Investment	−60,000										
Energy Savings		30,000	31,500	33,075	34,729	36,465	38,288	40,203	42,213	44,324	46,540
Maintenance savings		3000	3120	3245	3375	3510	3650	3796	3948	4106	4270
Additional Maintenance		−7000	−7280	−7571	−7874	−8189	−8517	−8857	−9212	−9580	−9963
Pre Tax Operating Income	−60,000	26,000	27,340	28,749	30,229	31,786	33,422	35,142	36,949	38,849	40,847
Depreciation		−6000	−6000	−6000	−6000	−6000	−6000	−6000	−6000	−6000	−6000
Gross Income	−60,000	20,000	21,340	22,749	24,229	25,786	27,422	29,142	30,949	32,849	34,847
Taxes		−6000	−6402	−6825	−7269	−7736	−8227	−8742	−9285	−9855	−10,454
Net Income	−60,000	14,000	14,938	15,924	16,961	18,050	19,195	20,399	21,664	22,995	24,393
Depreciation		6000	6000	6000	6000	6000	6000	6000	6000	6000	6000
Net Income + Depreciation	−60,000	20,000	20,938	21,924	22,961	24,050	25,195	26,399	27,664	28,995	30,393

43% lower financial attractiveness. Indeed, the financial attractiveness is not actually smaller, only more realistic from the investor's standpoint.

3. Decision Making for Sustainable Energy

Most of the time, only immediate monetary aspects are considered in energy-efficiency projects, although quailtative aspects can be quantified and considered in the decision-making process. That is very important when looking for more expressiveness in energy planning, coming from energy-efficiency as a resource for sustainability.

This brings about a number of difficulties, since the insertion of different kinds of qualitative aspects into a quantitative decision-making model is complex. It is easy to consider the costs saved per kWh, but inserting aspects like "local community support is something more difficult to figure out" [4].

One alternative method for financial-only analysis is the Cost Benefit Analysis (CBA). The CBA is a tool that attempts to assign a monetary value to all non-economic aspects of a decision. The relation between the implement cost of a project and the benefit created by it is determined by the CBA. Although the CBA, in thesis, includes the net value of non-economic impacts as a monetary value, the "CBA can be used to produce almost any result desired by the analyst to suit his own prejudices or the interests of his sponsor, since the attempt to transform every potentially significant effect into a monetary value requires an arbitrary and subjective judgment on the part of the analyst" [5]. More recently [6] used the CBA to evaluate domestic energy-efficiency programs and was able to overcome the difficulties in order to reach monetary values for non-economic aspects, as mentioned by [5]. In this way [6] used the sensitivity analysis for costs to minimize impacts of inaccurate estimations and concluded "a perfect methodology for evaluating large-scale energy-efficiency programs is not yet available" [6]. To overcome the limitations of CBA analysis, [5] suggest, in their article, the use of a multi-dimensional approach and a way to deal with this issue using the analytical hierarchy process. It uses the concept of a decision tree with paired alternative comparisons.

3.1. The Analytical Hierarchy Process Issue

The analytic hierarchy process (AHP) is a powerful and flexible method used in decision-making that supports the determining of priorities. It also identifies the best option within a number of possible alternatives considering quantitative and qualitative aspects. Through the reduction of complex decisions to decisions that can be paired and compared (paired comparison), the AHP does not only help decision makers obtain the best option, but it also provides a clear view of why that alternative is the best. The AHP is executed in 3 phases: Structuring, Judgment and Synthesis of Results [7].

The AHP has already been used in multi-criteria decisions regarding energy conservation policies by [8] and [9], among other authors, but a common aspect among them is that traditional decision-making processes usually ignore non-monetary aspects. Reference [8] used the AHP to determine what the most effective policy instruments were for promoting energy conservation in Jordan and remembered that "the hierarchical structure of AHP allows the DM (Decision Maker) to break the complex decision problem down into smaller, but related, problems in the form of goals, criteria, sub-criteria, and alternatives".

Initially, it is necessary to establish the criteria and sub-criteria to be used in the evaluation as well as the alternatives. The next step is to organize these alternatives in a hierarchical manner, as shown in **Figure 1**. This organization can have as many levels as necessary. The top level must always be the goal to be attained and the bottom level must consist of the alternatives. The intermediate levels consist of the criteria, their respective sub-criteria and, if necessary, other, lower, levels.

After the definition of the hierarchical tree, an evaluation within each level is started. This evaluation is normally carried out by comparing the pairs to each option at each level. However, it is also possible to evaluate them using absolute rates. A scale from 1 to 9 was arbitrated to compare the pairs, as proposed by [7]. This scale amplifies the distance and makes the most important choice (9) nine times more important than the choice of equal importance (1). If one were to use a 2 to 10 scale, for example, the most important choice (10) would only be five times more important than the choice of equal importance (2). Using more than 5 categories could complicate judgment, whereas less than 4 would impose limits on judgment and increase inconsistency. For this reason, 5 categories were adopted, where:

1 = equal importance;
3 = a little more important;
5 = more important;
7 = clearly or strongly more important;
9 = much more important.

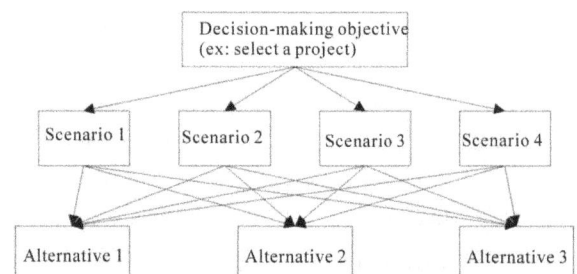

Figure 1. Hierarchical tree for the AHP methodology.

The intermediate values can be used to clearly demonstrate the evaluation. Each alternative is compared with the rest using this scale. Upon finishing the evaluation, Equation (1) can be formulated, with the dimensions n by n for the Z matrix, where n is the number of items to be compared.

$$Z = \begin{array}{c} \\ A \\ B \\ C \\ D \end{array} \begin{array}{cccc} A & B & C & D \\ \begin{bmatrix} 1 & Z_{12} & Z_{13} & Z_{14} \\ 1/Z_{12} & 1 & Z_{23} & Z_{24} \\ 1/Z_{13} & 1/Z_{23} & 1 & Z_{34} \\ 1/Z_{14} & 1/Z_{24} & 1/Z_{34} & 1 \end{bmatrix} \end{array} \quad (1)$$

In Equation (1), the Z_{ij} terms result from the paired comparison of items i and j. The Z_{ij} terms will always be equal to 1 when $i = j$ (similar to self-comparison alternatives), while Z_{ji} is the inverse of Z_{ij}.

From this matrix, the priority vector (Vector P) is found in this comparison. Then it is necessary to normalize the matrix from its columns, add the rows and divide the result by the number of elements in the row. The priority vector will be included in the comparison. When employing the AHP method, it is essential to verify data consistency.

Human beings are subjective by nature and this subjectivity means conflicting choices will sometimes be made. Such conflicts in the pairwise comparisons of the AHP method can be computed. This is possible by obtaining the Consistence Index (*CI*). Mathematical methods are used to calculate this index through the maximum matrix autovalue (λ_{max}). The *CI* can be calculated as follows:

$$CI = \frac{\lambda_{max} - n}{n - 1} \quad (2)$$

After calculating the *CI*, the consistency ratio (*CR*) then needs to be calculated. The *CR* is based on the comparison between the actual consistency index (*CI*) and what the *CI* would be if random choices had been made. In order to calculate the *CR*, it is necessary to use tables to provide a random index (*RI*) for each dimension of the matrix, as shown in the **Table 2** [7].

After finding the RI for the respective matrix dimension, the *CR* is calculated using the ratio between *CI* and *RI*. The *CI* is used to check whether the judgments are consistent among themselves. This means that if one person makes random choices, or simply fails to clearly express his or her opinions in a consistent manner, there

will be proximity between the *CI* and the *RI*. The result can be considered significant for a *CR* lower than, or equal to, 0.1. If it is greater than 0.1, the comparisons must be evaluated again in order to achieve acceptable consistency. After carrying out this process for the top criteria, the same process needs to be executed with lower-level criteria. This makes it possible to find the weight of the sub-criteria within the superior criteria. Once this procedure is complete for all sub-criteria, the alternatives can be analyzed within the bottom sub-criteria, based on the same method.

After finding the weights for each alternative within the bottom criteria, the general ranking can be constructed through the following steps:

1) Multiply the alternative priority within the sub-criteria by the sub-criteria priority;

2) Once step one has been executed for all the sub-criteria of a certain level and its criteria/sub-criteria subordinates of the next superior level, add up the results. It is necessary to multiply the result by the weight of the criteria or sub-criteria of the next level;

3) Repeat step 2 at a superior level;

4) Repeat step 3 until the top level. Upon reaching this level, the value obtained will be the total priority of the alternative.

As steps above demonstrate, the AHP is an interactive method, where the number of comparisons grows rapidly in accordance with the number of criteria. As [10] warn, "a primary criticism of the Analytic Hierarchy Process (AHP), as originally presented by [7], is the number of comparisons required to develop the judgment matrices". For large AHP problems, the number of interactions needed may become a problem when solving the problem manually or using a spreadsheet. Reference [10] suggested a method to reduce the number of comparisons needed, but there are certain software packages that implement the AHP, such as the Expert Choice®, Super Decisions and Decision Lens™, which minimize this problem with current computational capabilities. The user inputs the criteria, sub-criteria and alternatives into the software interface and the program assembles compareson sheets where the user is required to vote. After voting, the software automatically assembles the matrix, calculates the inconsistency index, the consistency ratio, the normalized weight and the ranking of alternatives. These packages also offer the possibility of using ratings or direct voting.

3.2. Decision Making for Energy Efficiency Economics

The criteria are sorted into two camps, economic and welfare-environmental. Each criteria has a few sub-criteria, such as IRR, NPV and risk for the economic criteria, as well as employee exposure, higher job creation

Table 2. Random index for the matrix dimension.

Matrix Dimension	1	2	3	4	5	6	7	8	9	10
Random Index	0	0	0.58	0.90	1.12	1.24	1.32	1.41	1.45	1.49

and lowest environmental impacts for the welfare-environmental criteria. In the following example three alternatives are considered in the decision making:

- Alternative 1 is doing nothing;
- Alternative 2 is improving the current condenser system;
- Alternative 3 is installing a brand new condenser.

Attention must be drawn to the decision-making method and not to the values alone. The decision tree is shown in the **Figure 2**.

This decision tree simulates qualitative and quantitative aspects. The top-down analysis starts by attributing importance to the criteria. In other words, choosing which criteria is the most important: economic or welfare-environmental.

Within the sub-criteria qualitative analysis, Saaty's verbal scale is used to compare pairs: IRR with NPV, NPV with Risk, IRR with Risk, employee exposure with job creation, job creation with minimum environmental impacts and employee exposure with minimum environmental impacts.

The alternatives are classified according to each sub-criteria. In this manner, there are some sub-criteria that must use quantitative grading and others that must use qualitative grading. The quantitative ones are IRR, NPV and job creation. The remaining ones are all qualitative.

Regarding qualitative sub-criteria, the alternatives are compared in pairs using the Saaty verbal scale [7]; meanwhile, the ratings concept is used in the quantitative sub-criteria. It is a relation where, for each quantitative value, another value from 0 to 1 is attributed. For example, in the IRR sub-criteria, the rating can be defined starting from 11% to 100%. In this manner, if any alternative has an IRR equal to 55%, the grade will be exactly 0.5.

The quantitative values do not need judgment, as they are numeric values, but the qualitative values do need judgment from specialists or the team involved in thedecision making, who in this case are those carrying out the work [4].

Specialist judgment infuses the decision with a subjective ingredient. The main difference for traditional subjectivism in the decision-making process is based on the fact that specialists are putting their experience to use in order to contribute to solving small issues, rather than the big decision. Using the AHP [11] analyzes the changes in the energy resources priority stack when replacing specialists with community members. Surprisingly, there were no significant changes in the priority list. Once the starting criteria are defined, the decision is much more sensitive to the criteria than to the judgments. Conversely, this particular case demonstrates only that the AHP study result is less sensitive to grading by specialists than to the definition of criteria.

In the following example, the weights of the economical and welfare-environmental criteria were changed and the response was automatically modified. By changing the weights as presented in the **Table 3**, the decision is changed to a new condenser installation project.

From the judgments of specialists, the software calculates all the priorities for each level, compared to the lower level. The software adds up the results and calculates the priority for each alternative. A table featuring the grading of the software response summary for each project alternative is presented below.

From **Table 4**, the project option with the best grade is to improve the current condenser. But if the decision maker wants to have a better understanding of the decision-building process, these software packages are able to perform a useful sensitivity analysis for each criteria. Accordingly, **Figure 3** shows how each sub-criteria affects the decision.

4. Conclusions

A more deep analysis from the investor's perspective of energy-efficiency projects demonstrates the need to consider accountable details that will be a key factor in comparing projects fairly. The popular payback and IRR and NPV methods have deficiencies known to investors.

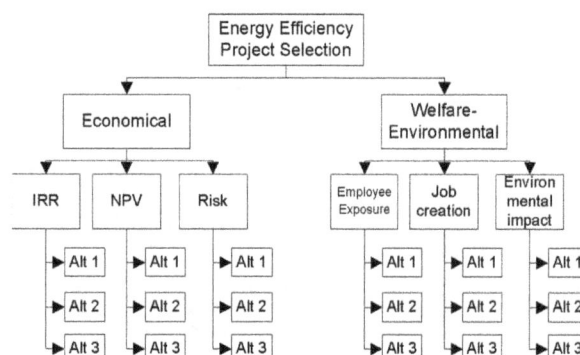

Figure 2. Decision tree for the example.

Table 3. Alternative decision change due to weight adjustment.

Criteria	Economical	Welfare Environmental	Alternative suggestion
Original weight	73.90%	26.10%	Improvement
New weight	10.00%	90.00%	New condenser

Table 4. Software response summary.

Order	Alternative	Score
1	Alternative 1 is doing nothing	0.650
2	Alternative 2 is improving the current condenser system	0.747
3	Alternative 3 is install a brand new condenser	0.370

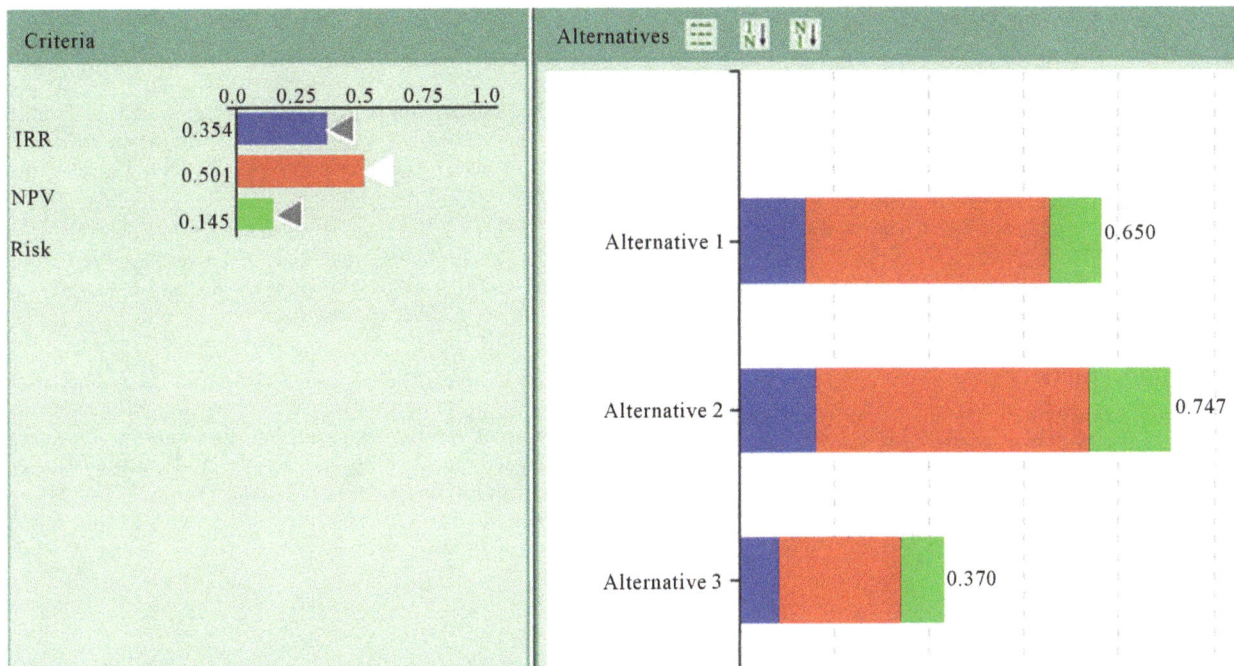

Figure 3. Economic sub-criteria sensitivity analysis.

However, they are useful if used with caution. The calculation of NPV from the cash flow savings versus expenses proved its inefficiency inasmuch as it does not bring out the actual value of the returns. In this case, it is necessary to consider depreciation and taxes. Apart from this, depreciation period is a determining factor to the investor return.

The addition of qualitative aspects in the decision-making is a very important factor when making a tho rough and reliable evaluation. Non-economic factors are often not considered due to the lack of knowledge or even difficulty in bringing them to the decision agenda.

The AHP is an alternative method for solving problems pertaining to qualitative aspects in the decision-making process, particularly in energy-efficiency economics as energy planning tool. It is also helpful to support the often-ignored decision-making development process.

From the calculated quantitative aspects using well-known methodologies such as NPV, IRR and Risk Evaluation, the AHP makes it possible to include qualitative aspects in the decision-making agenda in a standardized and repeatable manner.

Reference [5] warned about a significant limitation of CBA, AHP and other decision making methods. In the long term, they are limited to consider the interests of future generations since they transform qualitative aspects into a monetary value and execute the voting process or calculate the discount rates based on present insight.

Even though grading is able to affect the result of AHP project choices, according to [11], it is also possible to observe that the AHP study results are less sensitive to grading by specialists than to criteria definition.

From the practical standpoint, the AHP enables companies and decision-makers as whole to include qualitative aspects in the decision-making agenda that currently depend on the personal beliefs of managers. That is true even for the modern power system planning looking for sustainable development where the energy-efficiency economics inclusion is necessary.

5. Ackknowledgements

To the São Paulo Research Foundation—FAPESP (from the portuguese acronym "Fundação de Amparo à Pesquisa do Estado de São Paulo"), for the support given to the project "New Instruments for Regional Energy Planning Aiming at Sustainable Development of the western region of São Paulo" ("Novos Instrumentos de Planejamento Energético visando o Desenvolvimento Sustentável do Oeste Paulista"), code 03/06441-7.

REFERENCES

[1] C. E. M. Russell, "Strategic Industrial Energy Efficiency: Reduce Expenses, Build Revenues and Control Risk," Alliance to Save Energy Report 2003, Washington.

[2] F. Couto, "Agência Canal Energia, Negócios, 2007." http://www.canalenergia.com.br/

[3] O Estado de São Paulo, 2009. Carga tributária no Brasil bate recorde e chega a 35.8% do PIB. http://www.estadao.com.br/noticias/economia,carga-tribut

aria-no-brasil-bate-recorde-e-chega-a-35-8-do-pib,399074 ,0.htm

[4] M. E. M. Udaeta, "Planejamento Integrado de Recursos Energéticos para o Setor Elétrico—PIR (Pensando o Desenvolvimento Sustentado)," Tese de Doutorado, Escola Politécnica da Universidade de São Paulo, Brasil, 1997.

[5] D. Simpson and J. Walker, "Extending Cost-Benefit Analysis for Energy Investment Choices," *Energy Policy*, Vol. 15, No. 3, 1987, pp. 217-227.

[6] J. P. Clinch and J. D. Healy, "Cost-Benefit Analysis of Domestic Energy Efficiency," *Energy Policy*, Vol. 29, No. 2, 2001, pp. 113-124.

[7] T. L. Saaty, "The Analytic Hierarchy Process: Planning, Priority Setting Resource Allocation," McGraw-Hill, London, 1980.

[8] M. M. Kablan, "Decision Support for Energy Conserva-

tion Promotion: An Analytic Hierarchy Process Approach," *Energy Policy*, Vol. 32, No. 10, 2004, pp. 1151-1158.

[9] P. Konidari and D. Mavrakis, "A Multi-Criteria Evaluation Method for Climate Change Mitigation Policy Instruments," *Energy Policy*, Vol. 35, No. 12, 2007, pp. 6235-6257.

[10] K. H. Lim and S. R. Swenseth, "An Iterative Procedure for Reducing Problem Size in Large Scale AHP Problems," *European Journal of Operational Research*, Vol. 67, No. 1, 1993, pp. 64-74.

[11] D. Cicone, "Modelagem e Aplicação da Avaliação de Custos Completos Através do Processo Analítico Hierárquico Dentro do Planejamento Integrado de Recursos," Dissertação de Mestrado, Escola Politécnica da Universidade de São Paulo, Brasil, 2008.

4

Energy and Population Policies in Australia

Doug Hargreaves
Queensland University of Technology, Brisbane, Australia

ABSTRACT

The Australian Government is about to release Australia's first sustainable population policy. Sustainable population growth, among other things, implies sustainable energy demand. Current modelling of future energy demand both in Australia and by agencies such as the International Energy Agency sees population growth as one of the key drivers of energy demand. Simply increasing the demand for energy in response to population policy is sustainable only if there is a radical restructuring of the energy system away from energy sources associated with environmental degradation towards one more reliant on renewable fuels and less reliant on fossil fuels. Energy policy can also address the present nexus between energy consumption per person and population growth through an aggressive energy efficiency policy. This paper considers the link between population policies and energy policies and considers how the overall goal of sustainability can be achieved. The methods applied in this analysis draw on the literature of sustainable development to develop elements of an energy planning framework to support a sustainable population policy. Rather than simply accept that energy demand is a function of population increase moderated by an assumed rate of energy efficiency improvement, the focus is on considering what rate of energy efficiency improvement is necessary to significantly reduce the standard connections between population growth and growth in energy demand and what policies are necessary to achieve this situation. Energy efficiency policies can only moderate unsustainable aspects of energy demand and other policies are essential to restructure existing energy systems into on-going sustainable forms. Policies to achieve these objectives are considered. This analysis shows that energy policy, population policy and sustainable development policies are closely integrated. Present policy and planning agencies do not reflect this integration and energy and population policies in Australia have largely developed independently and whether the outcome is sustainable is largely a matter of chance. A genuinely sustainable population policy recognises the inter-dependence between population and energy policies and it is essential that this is reflected in integrated policy and planning agencies.

Keywords: Population; Energy; Energy Efficiency; Energy Intensity; Sustainability

1. Introduction

Australia is well endowed with energy resources and energy is relatively cheap and readily available to consumers and businesses. These are key factors underpinning Australian economic growth. In the last twenty years the energy sector has experienced an intense period of reform. State owned, vertically integrated, monopolies in the electricity sector were dismantled into separate generating, transmission and distribution and retail businesses operating in a competitive framework. There has been extensive privatisation of these businesses and the construction of integrated electricity grids, including one of the physically largest grids in the southern hemisphere, the NEM, National Electricity Market, covering New South Wales, Victoria, Queensland, South Australia and Tasmania facilitating trade in generating capacity.

The benefits of reform have been impressive and include applying a brake to electricity prices, rationalisation of excess generating capacity and over staffing and

the capacity to effectively manage unanticipated breakdowns and other emergencies. Almost 90% of Australia's electricity generation is fossil-fuelled, mainly black and brown coal but a growing amount of natural gas.

In the natural gas sector, private sector ownership has been the norm from the outset and government's role has been to drive microeconomic reform through regulations that underpin construction and efficient operation of pipelines by the multiple operators. Early developments were focused on domestic markets, but most recent gas fields, especially those in remote locations, have been developed specifically for export markets. The historical pattern of gas field development has meant that domestic natural gas prices have been well below global prices, encouraging the take-up of gas by industry, businesses and consumers. There are signs that this is changing, particularly in Western Australia and for new coal seam gas fields in eastern Australia. There are abundant reserves of coal and gas in Australia and policy has been based on a preoccupation with utilising them.

Australia was self-sufficient in oil until about 2002 when the main source of supply for Australian refineries, the Bass Straight oil field, peaked. While the country's oil reserves overall have not peaked assisted by the development of new oil fields, particularly in Western Australia, this event changed the course of the oil industry in Australia. Australian refineries were geared to the lighter oil from Bass Strait and the heavier oils from the new fields were unsuitable as feedstock and have been exported instead. Australian refineries are comparatively small and no new refineries have been built for some time. Initially oil imports were necessary to fill the gap between local crude production and feedstock requirements. However, demand for liquid transport fuels has increased much faster than refinery capacity and now an increasing proportion of Australia's refined petroleum products demand is imported in addition to a growing volume of crude oil. Crude oil exports from newer fields have been unable to keep up with import growth and there is a growing deficit in trade of oil and petroleum products embedded within an overall positive energy trade balance. Australia needs high value exports of fossil fuels and other commodities to balance a voracious appetite for oil and petrol.

Energy reform has ensured that Australia's energy sector is highly competitive but ignored sustainability. Australia's dependence on fossil fuels is highlighted by per capita greenhouse emissions amongst the highest in the world. There is growing awareness and concern about this and concern about future susceptibility to supply and price instability of fuel supplies. Australia's capacity to meet persistent trade deficits in liquid fuels may change, especially as the world moves to lower global greenhouse emissions. But the immediate concern is that the demand for energy continues to grow strongly under the influence of strong population growth and policy settings that reinforce present consumption habits.

2. Australian Population Policy

Until 2010, Australia operated a *de facto* rather than a formal population policy. The release of a Treasury report exploring the implications of recent population growth highlighted the downside of this approach and led to a Government decision to develop a sustainable population policy. Of course, sustainable meant different things to different people. Some saw continuation of high population growth as essential to sustaining economic growth; others argued that skills shortages in key disciplines, including engineering, meant that increased skilled migration was essential to avoid bottle-necks. Engineers Australia has taken a conventional sustainable development approach and has argued that sustainable population growth is population growth that is consistent with sus-

tainable development principles and practices and improvements in well-being of all Australians.

During the past forty years, Australia's population has grown from 12.5 million in 1970 to 22.2 million in 2010. On average population growth was 1.4% per annum, but attracted little attention because economic growth, both in terms of GDP, gross domestic product, (average 3.3% per annum) and per capita GDP (average 1.9% per annum) were relatively strong. More recently Australians have become more concerned about population growth. There has been increasing frustration about the conesquences of persistent under-investment in infrastructure, particularly in the larger cities, intensified by the realisation that recent growth population rates in Australia have exceeded rates in many developing countries, as well as, in most developed countries. Concern has also developed about the consequences for economic growth and fiscal balances of the aging of Australia's population.

Sustainable economic growth is more complex than simply enlarging the economy to achieve economies of scale. It is about the optimisation of economic, social and environmental variables and raises questions about the viability of status quo policy settings. At present energy policy presumes that a larger population implies a commensurately larger demand for energy. But Australia's greenhouse emissions show this is not sustainable and alternative policies are necessary. The challenge will be to find policies that move Australia onto a sustainable pathway while still realising the economic and social benefits derived from energy consumption. In the Australian context, this also means better understanding the role of population growth. This paper examines the links between these concepts and shows that population and energy policies require stronger integration.

3. Population, Energy and Sustainability in Australia

This paper employs decomposition methodology to link sustainability, energy supply and consumption, economic progress and population growth. This technique has been used by Turton and Hamilton to examine the contribution migration makes to Australia's greenhouse emissions [1], by ABARE, Australian Bureau of Agricultural and Resource Economics, (now ABARE-BRS, having merged with the Bureau of Rural Sciences), to examine the relative contribution of economic structure and energy efficiency to changes in energy intensity [2,3] and by the IEA, International Energy Agency, to highlight developments in the global energy system [4].

Many countries now have greenhouse gas reduction targets, sometimes expressed as reductions from a historical emissions level and sometimes as reductions from a historical level of emission intensity. Although desired objectives were not achieved at the Copenhagen and

Cancun climate change conferences, there was a significant increase in the number of countries committing, in various ways, to reducing emissions. Australia has an unequivocal medium term target to reduce emissions by 5%, relative to 2000 levels, by 2020 and has indicated a willingness to strengthen this target to 15% to 25% of 2000 levels, depending on international actions. Australia has also committed to a long term target to reduce emissions by 60%, relative to 2000 levels, by 2050.

In this paper the growth of greenhouse emissions is used as a proxy for sustainable growth in the energy system. Sustainable population growth extends beyond energy but the key links are evident in the energy system and the concepts revealed apply more generally. The relationship between energy related greenhouse emissions, energy supply, energy demand and population can be expressed as:

$$G = (G/F)*(F/PE)*(PE/C) \\ *(C/GDP)*(GDP/POP)*POP \quad (1)$$

where G is greenhouse emissions from energy use (millions tonnes); F is the consumption of fossil fuels (PJ); PE is total primary energy supply (PJ); C is the final consumption of converted energy (PJ); GDP is gross domestic product in 2008-09 prices (millions of Australian dollars) POP is the Australian population (millions)

In (1), greenhouse emissions are a proxy for sustainability and each term has a meaning conducive to policy development:

(G/F) is the emissions intensity of fossil fuel combustion and reflects the fuel mix between black and brown coal, natural gas and oil.

(F/PE) is the proportion of primary energy supply sourced from fossil fuels and reflects the diversification of primary energy supply.

(PE/C) is the ratio of primary energy supply necessary to deliver final consumption of converted energy and reflects primary energy conversion efficiency and to a lesser degree, the fuel mix.

(C/GDP) is the energy intensity of the economy and reflects both changing economic structure and end user energy efficiency.

(GDP/POP) is gross domestic product per person and is the conventional measure of economic well-being or affluence.

An approximation is used to estimate changes over time. Take the ratio of Equation (1) for time period "t" and for time period "1" and express each term as a percentage. Thus, (G_t/G_1), expressed as a per cent becomes the percentage change in emissions between periods "1" and "t". Similarly, $(G_t/F_t)/(G_1/F_1)$ expressed as a per cent becomes the percentage change in the emissions intensity of fossil fuels between period "1" and "t" and so on for the remaining terms. A more precise method is to differ-

entiate the logarithm of (1) with respect to time but the approximation is a more practical way to proceed here. Cross-effects are ignored because they are typically small, but may be reflected in some summation errors that are not significant.

Turton and Hamilton's [1] examination of the period 1989 to 1997 was repeated using their energy and greenhouse emissions data but new data for GDP in 2008-09 prices and revised actual population statistics sourced from the ABS, Australian Bureau of Statistics [5,6]. The factors contributing to growth in greenhouse emissions during this decade were compared to the later period 1997 to 2008 to examine the changes that had occurred.

Energy statistics were sourced from ABARE [7], economic and population statistics from the ABS [5,8] and [6] and greenhouse statistics from the DCC & EE, Department of Climate Change and Energy Efficiency, national greenhouse inventory [9,10]. Since the connections discussed in the paper hinge on the energy and population projections, it is pertinent to comment on their derivation.

The energy projections were generated by the ABARE E4 cast econometric model comprising 19 energy sources, 5 energy conversion sectors, 19 energy end use sectors and 7 geographic regions. This model has been developed over several years and its projections provide are inputs to the Australian Government's energy policy deliberations. Given the structure of the model, its key "drivers" for the issues dealt with here are population growth and macroeconomic growth. Population projections are from the ABS whose methodology is a standard demographic approach applied to the best population data base in Australia. These projections were used as the basis for the development of Australia's sustainable population policy. The macroeconomic assumptions used are a variant of those used by the Australian Treasury in its annual budget projections. The common feature of these sources is they each are from credible organisations used by the Australian Government in its policy deliberations. This paper is a commentary on aspects of these deliberations and it is appropriate to draw on the same sources.

Projections to 2020 were used to examine whether policy settings had learnt from the past. Prevailing policy settings are the assumptions that shape projection scenarios. ABARE has published energy projections to 2030 [7] and growth rates from this work were used to estimate the values for relevant energy variables in 2020. In its estimation model, ABARE included the following policies to reduce greenhouse emissions:

- The Renewable Energy Target that aims to ensure that by 2020, 20% of Australia's electricity comes from renewable sources.
- An emissions trading scheme that set a price on car-

bon emissions consistent with Australia's unequivocal commitment to reduce emissions by 5% on 2000 levels by 2020.

- Several State energy efficiency and emissions reduction programs that have been in place for some time.
- National energy efficiency policies and programs.

Population projections were obtained from the ABS [6] and reflect present natural growth parameters and present migration policies. GDP growth rates from the Australian Treasury were used to estimate GDP in 2020 [11] and reflects present economic policy settings.

The DCC & EE has published projected greenhouse emissions for 2020 [9] based on greenhouse abatement policies and programs common to those used by ABARE for its energy projections but with one major difference that demonstrates the fluidity of Australian politics in this area. ABARE's energy projections were published when there was a consensus that legislation establishing an emissions trading scheme would be passed by the Australian Parliament. As events turned out, this legislation was rejected by the Senate. The DCC & EE projections of greenhouse emissions were completed much later and did not include emissions trading. The dichotomy between the two sets of projections was resolved by including a stochastic term when calculating the changes in the components of Equation (1).

Figure 1 shows the contributions to increases in greenhouse emissions for the three periods discussed above.

1) 1989 to 1997

During this period, illustrated by the blue bars in **Figure 1**, Australian greenhouse emissions per capita increased from 15.8 tonnes per capita to 17.0 tonnes per capita and from 265 Mt CO_{2-e} to 315 Mt CO_{2-e} overall, an increase of about 19%. The main contributions were an increase in GDP per capita from \$A39,103 in 1989 to \$A44,890 in 1997 (both in 2008-09 prices), an increase of about 15%, and an increase in Australia's population from 16.8 million to 18.5 million, an increase of about 10%.

Australia reduced its energy intensity from 3.58 TJ/\$m

GDP to 3.4 TJ/\$m GDP, a fall in energy intensity of about 5%. Changes in energy intensity occur either as a result of structural change in the economy away from energy intensive industries towards less energy intensive industries or from improvements in end user energy efficiency. ABARE [2] explored the relative size of these effects using a time period that overlapped the one in question here and found that most of the change in energy intensity occurred during the second half of the period, with virtually no change during the first half. Averaged over the decade, about 60% of the reduction in energy intensity was the result of changed industry structure and about 40% was from improved energy efficiency. As engineers this distinction is important and shows that a focus on the improve energy efficiency of a machine, appliance or process is insufficient and the context in which change occurs is also important. Careful and comprehensive measurement and monitoring of progress is essential.

There was considerable inertia in the energy system itself. The emission intensity of fossil fuels increased marginally (about 0.1%); the proportion of fossil fuels in primary energy remained above 93% and barely changed and there was a small improvement in primary energy conversion efficiency (about 0.5%). These small changes show how strongly Australia was locked into fossil fuel consumption in electricity and in transport. The changes in the energy system and in energy efficiency were swamped by the effects of increased affluence and increased population and the result was a large and unsustainable increase in greenhouse emissions.

2) 1997 to 2008

During this period, illustrated by the red bars in **Figure 1**, Australia's per capita greenhouse emissions increased from 18.5 to 21.3 tonnes CO_{2-e}, and from 315 to 416 Mt CO_{2-e} overall, an increase of about 32%. Per capita GDP increased from \$A44,890 to \$A57,981 in 2008 (both in 2008-09 prices), an increase of 29% and the population increased from 18.5 million to 21.3 million, an increase of 15%.

Energy intensity fell to 3.0 TJ/\$million GDP, about 11%. Further research by ABARE [7] has shown that over the longer period 1989 to 2008, Australian energy intensity fell by about 23% with over half the change (about 56%) due to changes in economic structure and the rest (about 44%) due to improved end user energy efficiency. Most of the change in energy efficiency has occurred since the early 1990s and there is a suggestion that the pace of change has been relatively steady.

During this decade there was an increase in the amount of natural gas and renewable energy used in electricity generation and some energy saving in transport. These changes are shown as a reduction of about 4% in the emission intensity of the energy sector in **Figure 1**. Al-

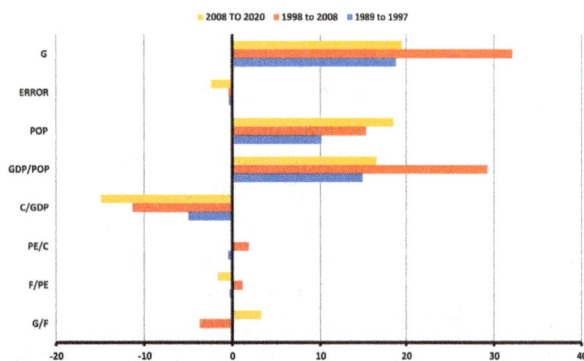

Figure 1. Contributions to Australian greenhouse emissions growth, actual 1987 to 2008 and projected 2008 to 2020.

though an improvement over the previous decade, emission intensity remains well above sustainable levels. The proportion of fossil fuels in primary energy increased to over 95% and the amount of primary energy required to supply final energy consumption increased reflecting growth in electricity consumption (the proportion of final energy consumed as electricity rose from 19.2% to 21.3%) and growth in the consumption of transport fuels. While the consumption of renewable energy also increased, this development was swamped by the large increase in consumption of fossil fuel energy resources. Conversion efficiency deteriorated reflecting aging of energy infrastructure. The energy system responded well to increases in the demand for energy but not to concerns about sustainability.

3) Comparing the last two decades

The pattern of events in Australia's energy system in each of the past two decades is very similar. Increased affluence and strong population growth resulted in strong and unsustainable increases in overall greenhouse emissions and in emissions per capita.

A range of policies and programs has been introduced in Australia designed to improve the sustainability of the energy system. At the margin, there is evidence of positive impacts from some of these policies but the Australian energy system remains highly dependent on fossil fuels, with little evidence of changes to reduce the system's contributions to greenhouse emissions. However, there is concrete evidence that end user energy efficiency policies are working and contributing energy savings that offset the impact of affluence and population on emissions growth. Research by ABARE suggests that the scale difference in impact between the two decades is probably because these programs began to realise results half way through the first decade.

4) The future to 2020

Extending this analysis into the future is not straightforward because available projections are based on different statistics. In **Figure 1** the yellow bars illustrate the pattern of events when available statistics are taken at face value.

What is immediately apparent is that the projections, based on present policy settings, show that continued economic growth and population growth will have similar unsustainable impacts on greenhouse emissions as occurred in the preceding two decades. Australia's GDP per capita is projected to increase from $A57,981 in 2008 to $A76,567 in 2020 (both in 2008-09 prices), an increase of 16.5%. The population is expected to increase from 21.3 million to 25.3 million, an increase of 18.5%. Greenhouse emissions per capita are projected to increase from 19.5 to 19.7 tonnes CO_{2-e} per capita and from 416 to 498 Mt CO_{2-e} an overall increase in emissions of 19.5%.

The energy intensity of the Australian economy is projected to continue its downwards path and by 2020 to be 2.5 TJ/$million GDP, reducing greenhouse emissions by 15%. It is important to realise that, in ABARE energy projections, energy efficiency improvements enter the model as exogenous variables assumed to reduce energy consumption by 0.5% per year for most fuels in the non-energy intensive industries and by 0.2% per year in energy intensive industries [7]. Some of this change will be contributed by current policies such as the National Strategy on Energy Efficiency but the connection between policy and assumed outcomes is far from clear.

Another issue is the acceptability of the reduced energy intensity from the standpoint of sustainable population policy. Even with a reduction of 15% in energy intensity, the net effect when growing affluence and population are considered is to contribute 20% to the growth of greenhouse emissions. The Prime Minister's Task Group on Energy Efficiency [12] believes that "Australia has not consciously or explicitly targeted world best practice in energy efficiency and, by comparison with other countries, has significant gaps in its energy efficiency policy armoury". The Task Force was asked to examine the feasibility of achieving a step-wise increase in energy efficiency in Australia and its key recommendation was that Australia should adopt an aspirational target of improving primary energy intensity by 30% by 2020. This objective is twice the reduction in energy intensity illustrated in **Figure 1**. Achieving such an objective would mean that the net addition of increased affluence, population growth and energy intensity to growth in greenhouse emissions would be limited to about 5%. My view is that this comparison shows that current policy settings are indeed inadequate and that adoption of stronger energy efficiency policies is implicit in the formulation of sustainable population policy.

Energy efficiency is an important technical matter for engineers. There are numerous examples in practically every aspect of engineering. However important, technical improvements are one thing but persuading households and businesses to adopt energy efficiency innovations is another. The simple analysis in this paper shows the importance of measuring and monitoring energy efficiency. Without significant improvement to data, analytical tools and widely available information on energy efficiency progress, energy efficiency policy will remain a hit or miss affair. The purpose of an energy efficiency target is to construct a framework to organise policy on energy efficiency to ensure the target is actually met. Other key aspects of a comprehensive energy efficiency framework are building an energy efficiency culture and overcoming non-market barriers to the adoption of energy efficiency. Many aspects of an energy efficiency framework involve considerations beyond the scope of engineering as most of us know it. My point is that to make real difference collaboration between the engineering profession and other disciplines is indispensable.

Although calculations relating to the energy system in **Figure 1** are complicated by the inconsistency in statistics described earlier, it is apparent that the inertia evident in the previous two decades is still part of the policy settings reflected in the projections. As important as it is ambitious energy efficiency is not sufficient to ensure that growth is sustainable. In 2000, energy sector greenhouse emissions were 361 Mt CO_{2-e} and a uniform 5% cut in emissions would mean that the sector emissions would be restricted to 343 Mt CO_{2-e} in emissions in 2020. This is equivalent to a 31% *reduction* in business as usual emissions and the net change from increased affluence, population growth and a 30% energy efficiency target is a 5% *increase* in emissions. In other words, the energy system will need to find ways to deliver about 36% reductions in emissions simply to meet a uniform 5% emissions reduction, relative to 2000 levels, by 2020. This can only be achieved through the structural adjustment brought about by a price on carbon emissions.

4. Policy Implications

This paper used decomposition analysis applied to reputable energy and population projections used by the Australian Government to demonstrate that population policy, sustainability and energy policy are inter-connected in such a way that sustainable development is possible only with integrated planning. The decomposition also demonstrated the critical role that energy efficiency has to play in achieving sustainable development. Finally, the decomposition demonstrates the pervasive inertia to change of Australia's fossil fuel based energy system.

Australia, like many countries has a Government energy policy, but it has at best vague connections to population policies. The Australian population policy is essentially a policy dealing with the quantum of migration and the type of immigrants that may be admitted and in the eyes of many is seen as a driver of economic growth through increasing demand for housing and other goods and services required by newly arrived migrants. What is not addressed is a key issue; although this approach to population "widens" demand by adding more of the same, it ignores "deepening" demand through innovation, improved efficiency and growing sophistication.

Certainly there is future work for engineers in this prescription, but such an approach fails to take advantage of the advances in engineering knowledge, particularly in

energy systems and in energy efficiency, and simply perpetuates past problems rather than address them. Until Governments acknowledge that population growth, for example, increases greenhouse emissions through the simple medium of more people and by using outmoded technologies, change towards sustainable development outcomes will elude.

5. Acknowledgements

This work was completed when the author was National President of Engineers Australia and the contribution of Engineers Australia is acknowledged.

REFERENCES

[1]   H. Turton and C. Hamilton, "Population Growth and Greenhouse Emissions, Sources, Trends and Projections in Australia," *Discussion Paper*, the Australia Institute, Canberra, No. 26, 1999.

[2]   Australian Bureau of Agricultural and Resource Economics, "Trends in Australian Energy Intensity, 1973-74 to 1997-98," *Research Report* 13, Canberra, 2000.

[3]   Australian Bureau of Agricultural and Resource Economics-Bureau of Rural Sciences (ABARE-BRS), "End Use Energy Intensity in the Australian Economy," *Research Report* 10.08, Canberra, 2010.

[4]   International Energy Agency (IEA), "30 Years of Energy Use in IEA Countries," Organisation for Economic Co-operation and Development, Paris, 2004.

[5]   Australian Bureau of Statistics (ABS), "Australian National Accounts, National Income, Expenditure and Product," *Cat No.* 5206.0, Canberra, 2010.

[6]   Australian Bureau of Statistics (ABS), "Australian Demographic Statistics," *Cat No.* 3101.0, Canberra2010,

[7]   Australian Bureau of Agricultural and Resource Economics (ABARE), "Australian Energy Projections to 2020-30," *Research Report* 10.02, Canberra, 2010.

[8]   Australian Bureau of Statistics (ABS), "Population Projections Australia," *Cat No.* 3222.0, Canberra, 2008.

[9]   Department of Climate Change and Energy Efficiency (DCC & EE), Australia's Emission Projections, Canberra, 2010.

[10]   Department of Climate Change and Energy Efficiency (DCC & EE), "Australian National Greenhouse Accounts, National Greenhouse Inventory," *Accounting for the Kyoto Target May* 2010, Canberra, 2010.

[11]   Australian Treasury, "Australia to 2050: Future Challenges," Canberra, 2010.

[12]   Prime Ministerial Task Group, "Report on Energy Efficiency," Canberra, 2010.

5

Test Apparatus for Performance Evaluation of Compressed Air End-Use Applications

Vishal Sardeshpande[1], Vijay Patil[2], Cherat Murali[2]

[1]A.T.E. Enterprises Private Limited, 2 Shreenivas Classic, Pune, India
[2]Birla Accucast Limited, MIDC Waluj Industrial Area, Aurangabad, India

ABSTRACT

Compressed air is an integral utility part of industrial utility systems. Any improvement in compressed air system will lead to reduction in utility cost. The effectiveness of utilization side of compressed air is usually dependent upon operator's discretion. There is no performance testing methods available for testing existing end use equipments. A test apparatus for estimation of compressed air flow based on measurement of pressure reduction in a fixed volume cylinder in a given time is developed. The test apparatus is easy to build and simple to operate in an industrial environment. This can be used for measuring performance of any pneumatic end-use equipment and for benchmarking the performance. The test apparatus was used in a foundry for quantifying the performance of the old and new blow guns.

Keywords: Compressed Air; Utilization Side; Test Apparatus; Performance Estimation; Pneumatic Equipment; Pneumatic Blow Gun

1. Introduction

Compressed air is a key utility after electricity, gas and water, and contributes to a major factor of utility cost in manufacturing plants. In compressed air systems the efficiency of conversion from electrical input-to-kinetic energy output at usage point is around 10 percent Ming [1]. It is generated centrally using compressors and then distributed for utilization in the manufacturing plant. The flow and pressure of compressed air determine the power requirement of the compressor, typically 1 kW of electricity can produce 10 m^3/hr (6 cfm) compressed air at 7 bar (g) pressure [2]. Compressed air is used in every sector of industry in the areas of process control, actuators, material conveying, blowing and moulding etc. For any compressed air application, the amount of air used should be at the minimum pressure required for the operation and used for the shortest possible duration. Compressed air usage should be constantly monitored and re-evaluated [3].

In a best practice program for compressed air James and Peter [4] suggested the target towards the efficient operation of compressor is achieved by minimizing leakages, avoiding wastages and misuse. More and more efficient compressors are being introduced to reduce the specific energy consumption (SEC) measured in terms kWh/m^3. Efficient use of compressed air system on utilization side is possible only by user's initiative. Antonio *et al.* [5] has discussed utilization side performance improvement solutions like end-use pressure reduction, volumetric flow control, demand flow control and alternative solution for compressed air usage. However, the performance of end-use equipment is not tracked in terms of flow measurement in industries as an operational practice.

Compressed air on utilization side is used for powering pneumatic tools, operation of air-actuated equipment, dust filtration shake down, blow guns and other similar applications. The manufacturers of compressed air end-use equipment generally specify the compressed air consumption at different air pressures. Amount of compressed air consumed for the end-use operation in an intended task is determined by the operating pressure and time duration of the operation. At end-use points pressure is regulated using local pressure reducing valves for achieving desired action from pneumatic equipments. Any deterioration in the pneumatic equipment performance is usually overcome by increasing the operating pressure (changing pressure reducing valve manually) which leads to higher consumption of compressed air.

The use of efficient end-use equipment on utilization side is a promising area for improving compressed air system performance. The end-use equipment performance can be measured in terms of compressed air consumption at a given pressure. The flow rate of individual end-use equipment is small (about 1% to 5%) as compared to the overall compressed air flow in the system.

There are conventional flow meters available for measurement of compressed air flow which are generally used on the main distribution lines. These flow meters are costly and difficult to adopt for measurement of small flow rates encountered for end-use equipment Koski [6].

In a compressed air system the compressor performance is measured with pump up test using existing compressor receivers. **Table 1** summarizes compressed air system performance measuring methods. There is a leakage test method for quantification of compressed air leakage in distribution system which is used effectively by industries for monitoring the leakage and initiate leakage sealing drives [2]. There is no simple quantification test method or test apparatus available for quantification of compressed air flow in an end-use equipment.

A simple and cost effective test apparatus is developed for flow measurement of the compressed air end-use equipment. The test apparatus is used in a foundry for measurement of compressed air flow for old and new blow guns and the effectiveness of the test apparatus is demonstrated.

2. Test Apparatus

The proposed test apparatus consists of a compressed air storage tank with an inlet, outlet and drain valve. The storage tank is mounted on a movable skid for testing different compressed air usage points. The test apparatus is installed with pressure gauges and a temperature gauge. The drain valve installed is used for removing the moisture from the storage tank. The construction detail of the test apparatus is provided in **Figure 1**. The reduction of pressure and the operating temperature of compressed air in the storage tank are the main parameters measured during the operation of the test apparatus. The test apparatus is used for estimation of air flow based on reduction of compressed air pressure in the storage tank (of known volume) in a given time. The principle used considers air as an ideal gas with unsteady flow from the storage tank given in Equation (1),

$$m_a = \frac{\left\{ \dfrac{\left(P_i - P_f \right) \times V}{R \times T} \right\}}{t} \quad (1)$$

where, m_a is actual mass flow rate of air in kg/s; P_i is initial pressure in kPa; P_f is final pressure in kPa; t is time in s; R is characteristic gas constant for air in kJ/kg K; T is temperature in K.

The test apparatus is connected to the end-use point through a flexible pipe after a valve V_1, shown in **Figure 1**. The air is supplied to the storage tank from a compressed air header using a valve V_2. After the storage tank is filled to the level of the compressed air header pressure monitored with the help of a pressure gauge P_1, the valve

V_2 is closed to isolate the header from the storage tank. The moisture in the storage tank is drained with help of a valve V_3. The valve V_1 is adjusted to set the required pressure of end-use point with the help of a pressure gauge P_2.

The compressed air is used for application after recording of initial pressure of the storage tank using the pressure gauge P_1 and a temperature sensor T_1. The time taken for pressure reduction in the pressure gauge P_1 is monitored with help of a stop watch. The reduction in pressure and time required for the pressure reduction is used for estimation of mass flow rate with the help of Equation (1).

Some of the advantages of the test apparatus are

- Low cost compared to compressed air mass flow meter;
- Easy and accurate measurement of small quantity of flow rates;
- Simple to operate in an industrial environment similar to compressor pump-up test;
- No need of special calibration, standard calibrated pressure and temperature sensors can be used;
- Does not create any pressure drop in the system;
- Limitations of test apparatus developed;
- Cannot be installed for online measurement;
- Size and weight of system is more as compared to mass flow meters;
- Manual intervention required for operation and observations.

The test apparatus developed can be used for the quantification of compressed air flow for an end-use equipment. The quantification of flow can be used as a benchmark of performance and any increase in the compressed air flow compared to benchmark reflects performance deterioration. The quantification of performance deterioration will be helpful for taking corrective actions to restore the performance. The test apparatus can also be used for testing effectiveness of repair and maintenance of the compressed air end-use equipment by measuring per-

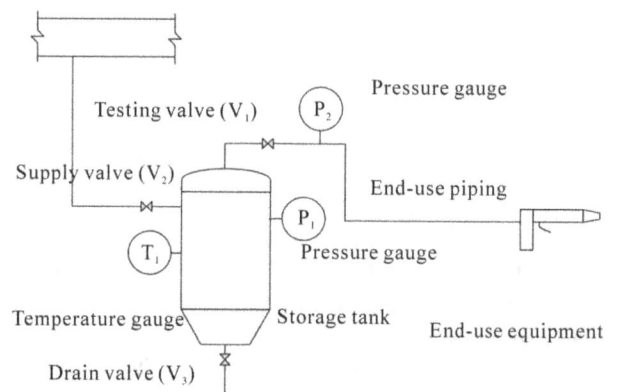

Figure 1. Experimental setup for characterization of blow gun.

Table 1. Compressed air system performance measuring methods.

Areas of compressed air system	Generation side	Distribution side	Utilization side
Performance indicator	Compressor output flow	Compressed air leakage from distribution system	Compressed air flow consumption for end-use equipment
Method of performance measurement	Existing method: Pump-up test	Existing method: Leakage test	No existing method: Test apparatus proposed
Use of performance measurement	Tracking performance deterioration of compressor Checking service effectives after overhauling of compressor	Benchmarking leakage levels Quantification of leakage plugging drives	Performance benchmarking of end-use equipment Quantification of performance and service effectiveness for end-use equipments

formance before and after the maintenance. This can also be used for selecting energy efficient end-use equipment, where the performance of the new and old end-use equipment is compared. This also can be used for validating the efficiency of end-use equipment compared to the supplier's claims.

3. Performance Measurement of a Blow Gun

Field studies to measure the performance of a blow gun using the test apparatus were carried out in a foundry located in Aurangabad, India. The foundry has three numbers of Atlas Copco make compressors installed in the utility block of the foundry and compressed air is used for various end-use applications. The compressed air system layout along with compressor specifications and end-use applications is presented in **Figure 2**.

The Shell machine (SM) and Core machine (CM) are moulding machines which pack sand and draw steel pattern to manufacture the sand shell and core (Choudhary and Choudhary [7]). The machines uses resin sand in steel patterns and are heated externally for curing. The compressed air is used mainly for sand shooter, pneumatic cylinders and cleaning of shell and core pattern by direct air blowing.

The foundry had taken up the drive for leakage minimization and eliminating major leakage by conducting periodic leakage test during scheduled maintenance. The next area identified for improvement was use of efficient blow guns for direct blowing applications. The installation of efficient blow guns (air nozzles) would meet the process requirements with significantly reduced compressed air consumption.

Compressed air flow in open end-use applications such as blow-off, part clearing, and moisture removal is auto controlled using a solenoid valve or manually controlled by equipment operators via in-line valves. Blow-off and part positioning on selected equipment is achieved with a steady stream of high velocity compressed air. Open or crimped tubing is often used as a delivery nozzle and pressures are often unregulated.

In the foundry high velocity jets created by blow guns are used for the cleaning of the resin sand burrs generated on the steel pattern during thermosetting of shell and

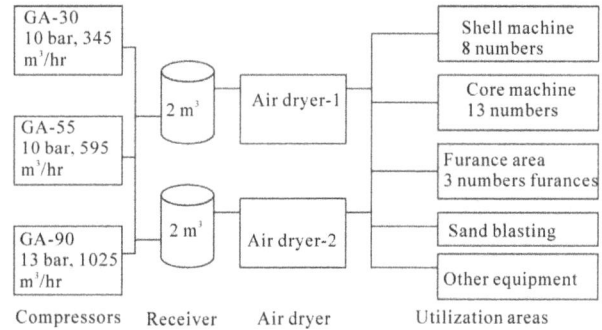

Figure 2. Compressed air system layout at the foundry.

core making operation. Cleaning of the pattern is critical for proper curing of the shell and core. The SM and CM operator does blow gun pressure setting for each blow gun independently. During the SM and CM survey the variation in local pressure setting was observed in the range of 3 - 5 bar (g) at different machines.

More efficient use of compressed air can be achieved by regulating pressures and using flow control devices for delivery such as entraining blow gun with nozzles and air amplifiers. These devices use flow dynamics and geometric design to entrain ambient air into the compressed air stream. By doing so, a greater volume of air is delivered at the point-of-use, thereby limiting the quantity of air that must be supplied directly by the compressed air system.

The foundry was interested for replacing their existing blow guns (old blow guns) with air amplifying blow guns (new blow guns). In order to adopt the new blow guns it was important to measure the air consumption of the old and new blow guns at a given operating pressure. The test apparatus is used for the new and old blow gun to measure the flow output at different operating pressures. The specifications of the test apparatus used in the foundry are provided in **Table 2**.

No performance curve of air consumption was available for the old blow guns. The new blow gun supplier provided the characteristic curve for mass flow rate at various operating pressures for a particular model of blow gun-"Ex-blow gun". The data generated from the characterization of the Ex-blow gun with the test apparatus is used for the validation with the manufacturer's data sheet. **Figure 3** presents the comparison of the performance characterization of the Ex-blow gun for measured data and the data supplied by the manufacturer. The experimental measurement follows the manufacturer's data except for the error band due to the measuring instruments.

Ex-blow gun is used for low penetration and wide flow blow application; it may be difficult to clean stubborn dirt particles in narrow passages and deep grooves. The shell and core cleaning with blow gun falls in this category. In order to cope up with such situation the manufacturer recommended a high thrust blow (HTB) nozzle mounted on a pistol shape (PSH) blow gun.

The performance characterization for PSH with HTB is not available in the manufacturer's catalogue. After successful testing of PSH gun on for shell and core blowing, a performance characterization is performed using the test apparatus. **Figure 4** indicates the variation of flow output with operating pressure for old gun and PSH gun. The flow delivered by PSH gun is more than old gun at same operating pressure. The difference in the flow delivered by the new and the old gun increases with increase in operating pressure (shown in **Figure 4**). The old blow gun was operating at 4 bar (g) and delivering about 9.8 m³/hr flow rate, the new blow gun can deliver same flow rate at 2.7 bar (g). The test apparatus was useful for the foundry to quantify the performance of the new and old blow guns before taking replacement decision. The test apparatus can be utilized for other compressed air end-use equipments for performance characterization.

The supply pressure for both blow guns is in the range 2 to 6 bar (g). Due to the operating pressures the pressure ratio is less than 0.528. Hence the compressible flow Eq. 2 is used for estimation of theoretical flow from the blow gun. The ratio of actual measured flow to that of theoretical flow (Equation (3)) provides coefficient of discharge for the old and PHS blow gun.

$$m_t = A \times \sqrt{\frac{2 \times \gamma}{\gamma - 1} \times \left(\rho \times P_{\text{in}} \right) \times \left\{ \left(\frac{P_{\text{out}}}{P_{\text{in}}} \right)^{2/\gamma} - \left(\frac{P_{\text{out}}}{P_{\text{in}}} \right)^{(\gamma+1)/\gamma} \right\}} \quad (2)$$

$$C_d = \frac{m_a}{m_t} \quad (3)$$

where, m_t is theoretical mass flow rate from blow gun kg/s; C_d is coefficient of discharge for blow gun; A is area of nozzle m²; γ is polytrophic index for air; P_{in} is upstream pressure Pa; P_{out} is down stream pressure Pa; ρ is density of air kg/m³.

The result of estimation of coefficient of discharge with experimental uncertainty for the old blow gun was 0.42 ± 0.1 and for the PSH blow gun was 0.75 ± 0.12. The test apparatus can also be used for the estimation of compressed air flow for a pneumatic cylinder used for clamping and actuating purpose. The performance of the pneumatic cylinder deteriorates with operation due to wear and tear of piston rings, scouring on cylinder etc., which leads to increase in compressed air consumption. The test apparatus can be used to measure compressed air consumption, and quantification of the present performance.

The other area of utilization of the test apparatus is to measure overall compressed air consumption in a processing or manufacturing machine, where the compressed air is supplied at a single point from compressed air header. In the machine, compressed air is distributed with internal pneumatic piping and used for different tasks like actuating, blowing, cooling, material conveying etc. The measurement of overall compressed air consumption of the machine at a regular interval with test apparatus will indicate the scenarios of leakage in internal piping

Table 2. Specifications of test apparatus.

Description	Specification
Storage tank of plain carbon steel	0.11 m³
Globe valve for inlet, outlet and drain point	25 NB
Pressure gauge	0 - 7 bar with least count of 0.1 bar
Temperature gauge	0°C - 200°C with least count of 1°C

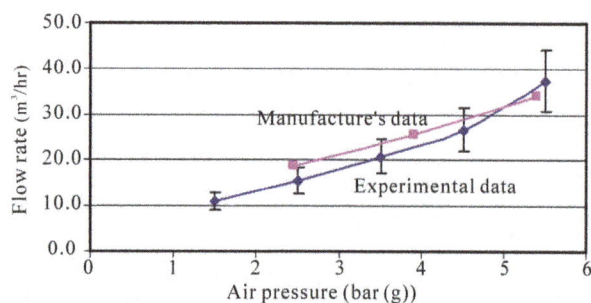

Figure 3. Comparison of experimental and manufacture's performance for Ex-blow gun [8].

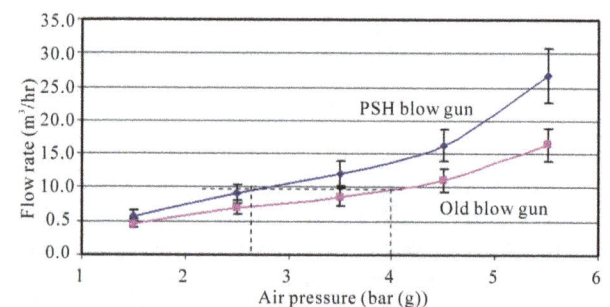

Figure 4. Comparison of old blow gun and PSH blow gun (new blow gun) performance.

and performance deteriorations.

4. Conclusions

Measurement of compressed air at usage point is important for maintenance and replacement of inefficient pneumatic end-use equipments. A test apparatus is developed based on measurement of reduction in compressed air pressure in a given time from a fixed volume cylinder. The test apparatus is easy to build and simple to operate in an industrial environment. The use of test apparatus is demonstrated in a foundry for performance characterization of a blow gun. The performance of an old and a new blow gun is measured with the test apparatus at different operating pressures. The experimental results indicates that the flow delivered by the old blow at 4 bar (g) can be delivered by the new blow gun at 2.7 bar (g). The coefficient of discharge estimated for the old gun is about 0.43 ± 0.1 and for the new gun is 0.75 ± 0.12. Based on measurement of flow it was possible to quantify the difference in the operating performance of the old and the new blow guns.

The test apparatus can be used for flow measurement of pneumatic end-use equipments for benchmarking. The test apparatus is also useful for measurement of overall compressed air consumption by different pneumatic end-use equipments together in a particular area or for a machine.

REFERENCES

[1] Y. Ming, "Air Compressor Efficiency in a Vietnamese Enterprise," *Energy Policy*, Vol. 37, No. 6, 2009, pp. 2327-2337.

[2] Bureau of Energy Efficiency, "Energy Efficiency in Electrical Utilities," Ministry of Power, New Delhi, 2003.

[3] Energy Efficiency and Renewable Energy, "Improving Compressed Air System Performance: A Sourcebook for Industry," United States Department of Energy, Washington DC, 2003.

[4] R. N. James and J. J. K. Peter, "Compressed Air System Best Practice Programmer: What Needs to Change to Secure Long-Term Energy Savings for New Zealand," *Energy Policy*, Vol. 37, No. 9, 2009, pp. 3400-3408.

[5] M. D'Antonio, G. Epstein, S. Moray and C. Schmidt, "Compressed Air Load Reduction Approaches and Innovations", *Proceedings of the Twenty-Seventh Industrial Energy Technology Conference*, New Orleans, 2005.

[6] M. A. Koski, "Compressed Air Energy Audit 'the Real Story'," *Energy Engineering*, Vol. 99, No. 3, 2002, pp. 59-70.

[7] S. K. Hajra Choudhary and A. K. Hajra Choudhary, "Workshop Technology: Volume I," Media Promoters and Publishers, Mumbai, 1989.

[8] GIPL, "Ex-Blow Gun Air Product Catalogs for Compressed Products," 2011.
http://www.giplindia.com/exblow.htm.

Energy Conservation in China's Road Transport: Policy Analysis

Xiaoyi He[1], Xunmin Ou[1,2*], Xiliang Zhang[1,2], Xu Zhang[1,2], Qian Zhang[1,2]

[1]Institute of Energy, Environment and Economy, Tsinghua University, Beijing, China
[2]China Automotive Energy Research Center, Tsinghua University, Beijing, China

ABSTRACT

Energy consumption for transport purposes has increased rapidly in China over the past decade. China's transport industry has undergone remarkable developments in energy conservation through structural, technological and managerial measures. The paper analyzes energy-conservation policies and measures related to road transport in China. The paper also identifies constraints for these policies and measures. The transport management authorities face a series of difficulties associated with methods, costs, public awareness, and management systems. Suggestions for improvement are also offered, including promotion of energy-efficient private vehicles, advances in business vehicle energy conservation, exploiting the energy potential of urban traffic and infrastructure development for energy-efficient clean vehicles.

Keywords: Energy Conservation; Road Transport; Policy Analysis; China

1. Introduction

Over the past decade, energy consumption for transport purposes has increased rapidly in China. This escalation is primarily accounted for by growing consumption of gasoline and diesel [1,2]. The freight transport sector has steadily improved its energy efficiency and reduced its energy intensity. The passenger transport sector, however, is likely to see a moderate increase in energy intensity in the near future because of a growing demand in service quality (i.e., speed, convenience, comfort) [3]. With the country's continuing industrialization and urbanization, energy consumption for transport purposes will continue to increase in the foreseeable future, and transport will gradually become a major energy user in the Chinese economy [4].

In recent years, China's transport industry has undergone remarkable developments in energy conservation through structural, technological and managerial measures. Nevertheless, a number of major problems remain to be solved. Through various policies, the local or national governments have actively encouraged public transport, inter-city rail transit construction, green travel, and energy conservation by means of structural and technological optimization. Despite positive initial results, the industry still faces enormous challenges, such as structural defects in the transport sector, inadequate public transport capability, and ineffec-

tive energy-management mechanisms and policies [5].

2. Main Measures

Energy conservation can be achieved primarily through two strategies (**Figure 1**): structural adjustment and technological advances. In practice, these strategies are refined into schemes (at a relatively macro-level) and measures (at a relatively micro-level). For these schemes and measures to take effect, governmental policies are required. With the continuing marketization of energy-related sectors, the government regulates the operation of such sectors primarily by three groups of measures: price policies, fiscal and taxation policies, and other managerial policies [6,7].

Government policies take the form of laws and bylaws, governmental plans, and guideline documents. For a new policy to be effectively enforced in China, the government has to consider carefully multiple aspects of the issues at hand. In particular, the following elements need to be clearly defined: 1) the objectives of the policy; 2) the target objects of the policy; 3) the governmental departments responsible for policy enforcement; 4) procedures required to enforce the policy; 5) mechanisms for monitoring policy enforcement; and 6) the relationship between the present policy and preceding and subsequent ones.

Energy conservation in road transport is the key task in

---

*Corresponding author.

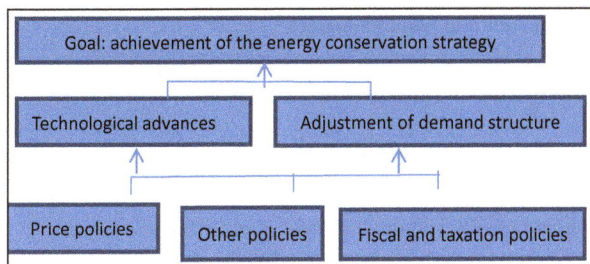

Figure 1. Relationship of the goal, strategies, and schemes (measures) in energy conservation.

transport energy conservation, and numerous policies, schemes, and measures are related to this task. The major ones are listed in **Table 1**.

3. Restraints Analysis

With respect to energy conservation, transport management authorities face a series of difficulties associated with methods, costs, public awareness, and management systems. They can be summarized as follows [8]: Finally, complete content and organizational editing before formatting. Please take note of the following items when proofreading spelling and grammar:

- Technological conditions. Some energy-conservation techniques are under development or at the demonstration stage and require funding support.
- Cost issues. The implementation of energy-conservation schemes involves increased costs and thus requires financial assistance.
- Limited public awareness. The importance of energy conservation and efficiency has not been widely recognized by corporate managers, appliances operators, and the general public.
- Weak management. Current systems for rating energy efficiency, energy conservation, and environmental protection are inadequate. Data are lacking. There are delays in the dissemination of energy conservation-related information. Management measures do not function well.

Table 2 lists the constraints faced by transport management authorities in enforcing energy-conservation policies.

In particular, the impact of these energy policies on non-business (*i.e.*, private-use) vehicles has been unsatisfactory. Achieving further improvement in the outcomes of these policies would appear to be an enormous challenge. The main factors that confine the outcomes of these policies are as follows. First, the ownership of private vehicles has increased rapidly. Correspondingly, the energy consumption by private vehicles has accelerated. Because of changing consumer habits, small vehicles are unlikely to become the first choice of vehicle for most people in the near future. Second, consumers are relatively insensitive to gasoline prices, which impede the transition to choosing more energy-efficient vehicles.

4. Policy Recommendations

4.1. Consistent Promotion of Energy-Efficient Private Vehicles

Motorized vehicles are the leading factor in rising fuel consumption. Experience in developed countries (Europe, Japan, United States) indicates that the most effective economic measures for improving the energy efficiency of vehicles are as follows: 1) issuing energy-efficiency standards and introducing fuel taxes; 2) requiring manufacturers to upgrade production techniques; and 3) encouraging consumers to purchase fuel-efficient vehicles. China should introduce energy-efficiency standards and promote the adoption of energy-conservation behavior and technologies as well as alternative-fuel vehicles.

4.2. Comprehensive Advances in Business Vehicle Energy Conservation

Energy conservation for business vehicles should proceed in terms of three aspects: vehicles, roads, and transport organization. First, it is necessary to reduce the energy consumption of vehicles. This can be achieved in two ways: energy-conservation techniques for new vehicles and the maintenance of existing vehicles. The former approach requires the regulation of manufacturers through relevant standards and policies. The latter approach is within the scope of transport management authorities and can be achieved in several ways: forced retirement of old energy-intensive vehicles; encouraging structural improvement for vehicles by means of economic incentives; rigorous monitoring of vehicle maintenance; and appropriate driver training.

Second, energy consumption can be attained by improving the road network structure and road conditions. It can also be achieved by optimizing transport organization and increasing transportation efficiency by improving the load efficiency.

Moreover, to ensure the long-term success of energy conservation in road transport, it is necessary to promote the inherent energy-conservation capacity in the transport sector and make energy conservation a customary procedure in this industry.

4.3. Exploiting the Energy Potential of Urban Traffic

Urban traffic involves many complex factors, such as management, policies and laws, planning, technology, operation management, and finance. Improvement in one or more of these factors can lead to advances in the traffic environment and efficiency, thereby contributing directly or indirectly to energy conservation. Thus, there is

Table 1. Measures relevant to energy conservation in road transport.

No	Energy conservation policies/measures	Objectives
1	Recommend vehicle types for freight transport	
2	Encourage the use of heavy-load vehicles and van-type vehicles.	Promote the use of energy-efficient vehicles
3	Promote diesel-fueled vehicles	
4	Eliminate old vehicles	Eliminate energy-intensive vehicles from the transport market
5	Upgrade vehicle maintenance and tests	Improve the energy efficiency of vehicles in operation
6	Recommend energy-efficient products	
7	Encourage corporatization of passenger transport organizations	Increase the transport efficiency
8	Promote the use of information technology in freight and passenger transport	
9	Quota management of energy use and reward/punishment	Increase the awareness of energy conservation in staff working in the transport sector
10	Driver training	
11	Construct national expressway networks	
12	Construct high-grade roads	Improve the road network structure, road conditions, and its traffic capacity
13	Improve the pavement of roads	
14	Control overweight and oversize in road transport	
15	Offer toll discounts for heavy-load vehicles	Encourage the use of energy efficient vehicles

Table 2. Energy-conservation measures and constraints for enforcement in transportation sector.

Scheme	Policy	Actual constraint	Type
Adjustment of transport structure	Ensure the proportion of non-motorized vehicles in traffic	Local bylaws discouraging electric bikes; inconvenience of traditional bikes	Public awareness
	Increase the proportion of train and water transport in total transport load	Unavailability of train passes; low speed of water transport	Cost
	Increase the adoption of public transit in traffic	Crowdedness and lack of comfort in buses; road congestion	Public awareness
	Increase the ratio of energy efficient vehicles in passenger vehicles	Energy benefit of small vehicles not necessarily predominant relative to their advantages (e.g., lack of impressive appearances)	Public awareness
	Increase the ratio of energy efficient vehicles in business vehicles	Energy benefit of large vehicles not necessarily predominant relative to their advantages (high costs)	Cost
Improvement of fuel economics of vehicles	Improve the energy standard of new vehicles	Differences in technological statuses of vehicle manufacturers	Technological factors
	Eliminate energy intensive in-use vehicles from the market	Vehicle owners unwilling to abandon existing vehicles	Cost
	Increase the proportion of diesel-fueled vehicles	Limited diesel availability; emission of black smoke and exhaust from diesel-fueled vehicles	Cost
	Increase the proportion of hybrid electric vehicles (HEVs)	High costs of HEVs	Cost
Promotion of alternative fuels	Increase the proportion of alternative-fuel vehicles such as electric vehicles (EVs)	Technological defects of alternative-fuel vehicles; inconvenience in recharging	Cost
Improvement of road networks	Improve the road network structure	Road congestion	Weak management
	Improve the road conditions	Road surface damage	Weak management
	Improve the traffic capability	Too many tolling points	Weak management
	Control energy use in subsectors	Lack of clear goals	Weak management
	Upgrade energy conservation-related monitoring and examination	Lack of a clear system	Weak management
Other managerial measures	Encourage the use of energy conserving products and techniques	Lack of motivation	Cost
	Perform comprehensive management of energy policies	Current laws need revisions	Weak management
	Improve the public concept of vehicle use	Lack of leaders	Public awareness

enormous potential for energy conservation in urban traffic.

Fuel use by urban vehicles accounts for over half of total vehicular energy consumption. Additionally, urban vehicles feature high spatial density, and there is a large proportion of small urban vehicles. As a result, urban vehicles represent a vast market potential for the application of alternative fuels.

Global experience has shown that the operation efficiency of urban traffic can be improved by such approaches as managing transport demand and improving the traffic supply. More specifically, improvements can be achieved by the following practices: limiting the use of private vehicles; introducing (or increasing) the fuel tax, exhaust emission tax, parking fees, and extra fees for traffic during peak hours and in those regions congested usually; creating bus-only lanes; using automatic traffic control systems; and developing rail transit networks. All these practices can contribute to improved energy efficiency.

4.4. Infrastructure Development for Energy-Efficient Clean Vehicles

The government needs to expedite the infrastructure development for clean vehicles. The infrastructure provides the essential conditions and support for growth of the alternative-fuel vehicle industry. Cleaner vehicles can reduce the emission of greenhouse gases and pollutants. Charging stations (or posts) and natural-gas fueling stations can be constructed in cities selected for the promotion of alternative-fuel vehicles, such as hybrid electric vehicles, electric vehicles, and vehicles fueled by compressed or liquefied natural gas. Moreover, intensive research should focus on the development of high-performance batteries and energy-storage devices. Efforts need to be made to develop technical capabilities and

standard systems for the manufacture, licensing, and quality control of energy-supplying equipment.

5. Acknowledgements

The project is co-supported by the China National Natural Science Foundation (Grant No.71103109, and 71073095) and the CAERC program (Tsinghua/ GM/SAIC-China).

REFERENCES

[1] Energy Research Institute of China (ERI), "China Energy Outlook," China Economic Publishing House, Beijing, 2012.

[2] China Automotive Energy Research Center, Tsinghua University (CAERC), "China Automotive Energy Outlook 2012," Scientific Press, Beijing, 2012.

[3] Ministry of Transportation of China (MOT), "2011 China Transportation Energy Saving, Emission Reduction and Low Carbon Development Annual Report," China Communications Press, Beijing, 2012.

[4] X.M. Ou, X.L. Zhang and S.Y. Chang, "Scenario Analysis on Alternative Fuel/Vehicle for China's Future Road Transport: Life-Cycle Energy Demand and GHG Emissions," *Energy Policy*, Vol. 38, No. 8, 2010, pp. 3943-3956.

[5] China Energy Research Association (CERS), "China Energy Development Report 2012," China Electric Power Press, Beijing, 2012.

[6] X. Y. Yan and R. J. Crookes, "Reduction Potentials of Energy Demand and GHG Emissions in China's Road Transport Sector," *Energy Policy*, Vol. 37, No. 2, 2009, pp. 658-668.

[7] Y. D. Dai and Q. Bai, "Overview of China's Energy Conservation Progress (2006-2010)," China Economic Publishing House, Beijing, 2012.

[8] International Energy Agency (IEA), "Energy Technology Perspective," IEA, Paris, 2008.

7

Quantifying Embodied Energy Using Green Building Technologies under Affordable Housing Construction

Nand Kishore Gupta[1], Anil Kumar Sharma[2], Anupama Sharma[2]
[1]M.P. Housing & infrastructure Development Board, Bhopal, India
[2]Maulana Azad National Institute of Technology, Bhopal, India

ABSTRACT

The building construction industry is a major contributor of environmental pollution, with high levels of energy consumption and greenhouse gas emissions, all of which contribute to climate change. Housing is the single largest subsector of the construction industry. It is also a basic need associated with social and economic benefits, and its demand in most emerging economies is substantial. Hence it is a sector with significant potential not to mitigate just the negative impact of climate change on buildings and people, but also to reduce the impact of the construction industry on the natural environment. Green buildings technology has advanced greatly in recent years, but most "high performance" green buildings are capital intensive, often with high-tech applications that are not in easy reach of the mass housing market. In the developing country context, where huge segments of the population lack access to essential services or housing, the green buildings approach to addressing climate change is perceived to be largely unaffordable. For green technology to be adopted in poorer nations and have scalable impact, it will have to be low-cost and affordable. According to a 2010 report, buildings in the commercial, office and hospitality sectors are poised to grow at 8% annually over the next 10 years in India. While the retail sector has been growing rapidly at 8% per annum, the residential sector has seen growth of 5% per annum during this period. It is estimated that over 70 million New Urban Housing Units will be required over the next 20 Years.

Keywords: Affordable Housing; Embodied Energy; Green Technology

1. Introduction

Work on green housing so far has been largely limited to standalone projects and projects catering to upper middle and high income groups, even mostly on commercial projects, with main stream developers only recently entering the green housing construction. In the affordable housing projects, the use of green building technology and systems has been limited, or even negligible.

For any significant environmental impact, green technologies and materials need to penetrate the mainstream housing industry. In other words, green housing appeals to a much wider audience, *i.e.* viewed as a socially responsible and commercially viable proposition for the common builder/developer, and an economically and socially viable proposition for the average end user. Once such transformation begins in the organized real estate market, the rest of the market is likely to pick up on it, with a catalytic effect on improving the environment.

The energy in buildings consumed in two different ways
- Energy that goes into the construction of the building using a variety of materials.
- Energy that is required to create a comfortable environment within the building.

Very few studies regarding the energy consumed during the maintenance of the building (heating, cooling and lighting) have been published. However the assessment of the embodied energy in buildings is still in its nascent stage in India and requires serious research.

The concept of green buildings is still at an emerging stage in India and concept of sustainable buildings and use of environmentally friendly construction materials like stones, timber, thatch, mud etc. have been practiced since ancient times. But the perception of people about strong and durable buildings have changed with the development and invention of modern materials like steel, cement, aluminium, glass etc. A large amount of fuel energy gets consumed in producing such materials.

These materials produced in industries further need to be transported to large distances before getting assemmbeled in the buildings, which further consumes energy. An estimate of the energy consumed in buildings using different permutations of materials and techniques will facilitate their appropriate selection and reduce the embodied energy consumption.

In this context, this study seeks to present the case for incorporation of green technologies, materials, and systems in the affordable housing sector in a tangible and quantifiable manner. To quantify the saving achieved in embodied energy using green technology, a case study for the construction of 600 EWS houses is considered under the climatic condition of Bhopal (M.P.). This study is focused on taking an integrated approach to address the issues of low cost sustainable technology system, their implication on large scale projects and their cultural acceptance for affordable housing construction. Study includes, identifying various available green technologies, which are technically viable, socially acceptable and quantifying embodied energy consumption including environmental impacts of conventional construction technology over green housing techniques.

2. Embodied Energy [1]

The embodied energy is the energy required to construct and maintain the campus, for example, in reinforced concrete construction, the energy required to quarry the coarse and fine aggregate, transport them to site, lay them, plaster them and (if necessary) paint and re-plaster over the life of the respective element. Best practice would also include energy calculations for demolition and recycling. A flowchart mentioning various activities involved from quarry of the material to the final finished product of the elements, required to estimate embodied energy is given in **Figure 1**. Debate continues about the boundaries that should be applied to calculate embodied energy. Commonly, the most influential components of embodied energy are those bounded by the cradle to gate approach, that is, all the energy required to deliver the product to the gate of the factory ready for transport to the construction site. Even within an embodied energy calculation bounded "cradle to gate" the complexity of embodied energy could be extreme.

For example, the energy used by the factory in the processing or manufacturing process may be easily identified, however what about the energy used by the employees:

- Transport Fuel (to and from work).
- Embodied energy of transport (to and from work).
- Energy of services (health, legal, accounting).
- Energy to produce food to feed the workforce (transport, agriculture, refrigeration).

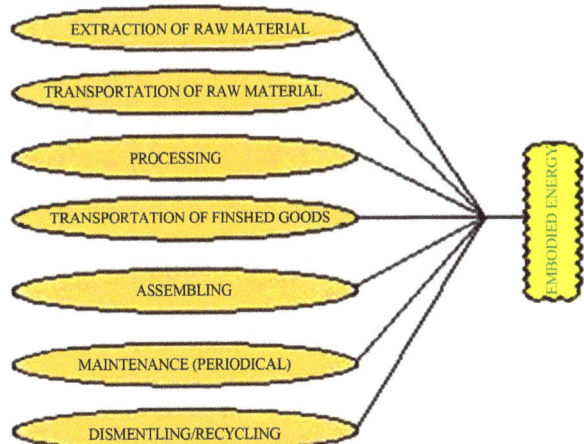

Figure 1. Components of embodied energy.

Any one point in the processing and manufacturing chain can be analyzed in detail chasing endless trail of energy calculations back to the Stone Age. With this in mind, it is important to remember the purpose of embodied energy calculations, it is to make informed decisions that lead to improvements in the way we use energy.

At present, order of magnitude accuracy would generally satisfy this purpose. The approach discussed in this paper is built on this principal.

3. Housing Demand in India [2]

Based on a study of residential housing demand in India, it is estimated that the additional demand for urban housing forecasted in India for 2012-13 is about 6.79 million. By 2015 the additional demand for housing is projected at 31 million, and that a large proportion of this will be required in the affordable sector. It is estimated that over 70 million new urban housing units will be needed over the next 20 years (see **Figure 2**).

The analysis is based on the 11th Year Plan report by Planning Commission and derives an annual growth rate of nearly 4% in housing. People migrating from peri-urban areas to urban areas will also add to this projection of housing demand.

Dr. P. S. Chani [3] in his study presented the embodied energy rates (EER) for a range of walling elements. These elements have been obtained by using combinations of alternative building blocks and mortar mixes. Through the tabulated data, a comparative analysis has been carried out to gauge the energy efficiency of the walling elements and to identify the most suitable option. The study also highlights the significant reasons, which lead to an energy efficient alternative.

Krishnakedar S. Gumaste [4] presented their work on computation of embodied energy in buildings and addressed the issues and problems with the materials and technologies used in building industry, in the study at-

Figure 2. Housing demand in India.

tempts have been made towards the assessment of embodied energy in various types of building with different number of stories of the buildings.

Talakonukula Ramesh *et al.* has [5] presents life cycle energy analysis of a multifamily residential house situated in Allahabad (U.P), India. The study covers energy for construction, operation, maintenance and demolition phases of the building. The selected building is a 4-storey concrete structured multifamily residential house comprising 44 apartments. The material used for the building structure is steel reinforced concrete and envelope is made up of burnt clay brick masonry. Embodied energy of the building is calculated based on the embodied energy coefficients of building materials applicable in Indian context. Operating energy of the building is estimated using e-Quest energy simulation software.

4. Case Study

A typical unit of EWS houses being constructed under affordable housing project by Madhya Pradesh Housing and Infrastructure Development Board in Bhopal (M.P.) India is considered as a prototype model. Details of the plan are as under

Building type:
- Ground floor row housing.
- 1 BHK Units ~30.5 sqm.
- Area of Openings ~6.5 sqm.
- Shading deviceschajja projection 400 mm.
Family profile
- Family of four:
- Husband, wife & 2 Children.
Appliances in use
- CFLs.
- 1 Refrigerator.
- 1 TV.
- Ceiling Fans 1 Evaporative Cooler.
Typical energy use
Average Electricity Bill Rs. 280/Month (~100 kWh/Month)
- A cooking gas cylinder of 14.5 kg lasts one month.
Financial profile

- Family Income ~Rs. 11,000/Month.
- Monthly Rental ~Rs. 1500/Month.
- Current Market Cost of the Apartment ~ Rs. 750,000.

Based on the architectural drawings **Figures 3** and **4** and specifications, the details of bill of quantities and embodied energy along with the CO_2 produced is calculated, represented in **Figures 5** and **6** respectively.

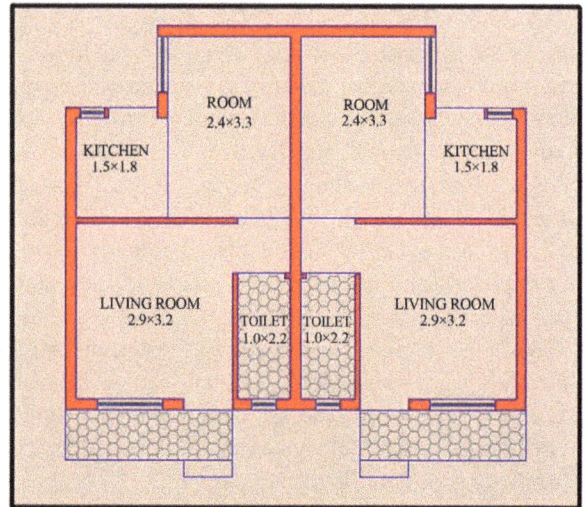

Figure 3. Case study building plan in Bhopal, India.

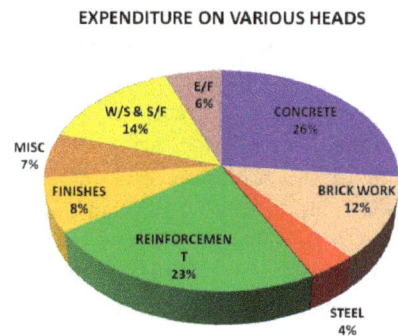

Figure 4. Percentage share of construction cost on various heads.

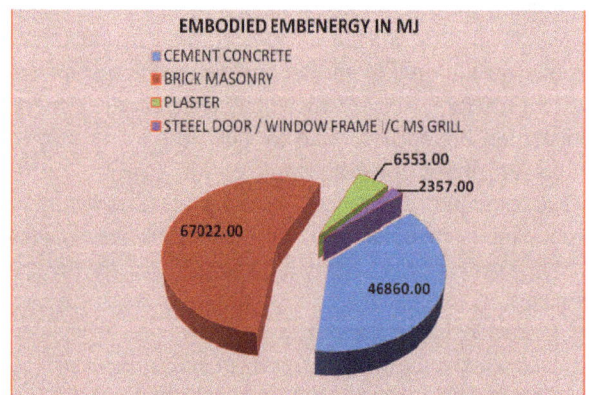

Figure 5. Amount of embodied energy for major construction activities.

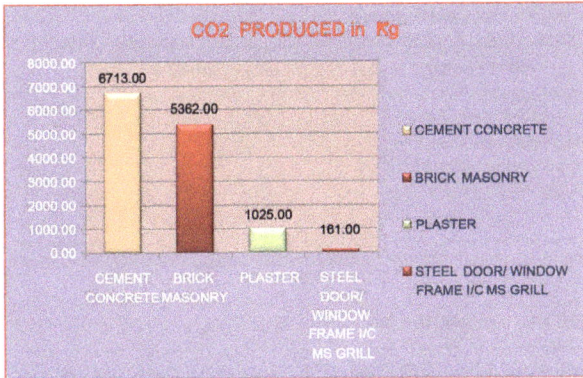

Figure 6. Amount of CO_2 produced for major construction activities.

5. Construction Cost

In building construction it is observed that wall systems, roofing and wood work for doors and windows accounts for almost 65% of the entire construction cost, since purpose of the study was to quickly arrive at the list of affordable green technologies, hence greater attenuation is provided towards green technologies available to these components, which contributes 65% (**Figure 3**) of the entire construction cost. Walls, Roofs, Floors, Fenestration, Framing, Partitions and Doors are the various building components which are essentially required for any type of building construction, a conventional technology is considered as a datum to compare green technology.

Quantity of embodied energy based on IIEC report calculated and represented in **Figure 5**, it is observed that embodied energy in bricks worked out to be the highest one and needs to be taken care of. Similarly another major contribution of embodied energy is from cement concrete, hence these two materials are required to be replaced with some other alternative materials, which can save significant amount of embodied energy. Similarly, amount of CO_2 emission for above two building materials is represented by **Figure 6**.

Based on various aspects with lowest environmental impacts in terms of embodied energy, some of the technologies are listed in the **Table 1**. For few technologies, marginally higher initial costs than the baseline technology are acceptable because of substantially lower environmental costs and short payback periods.

Above mentioned technologies are environmental friendly and consume least amount of embodied energy, comparative analysis for the wall and roofing material are mentioned in the succeeding tables.

Various green building technologies are listed in **Table 1**. In construction of low cost affordable housing units it can be observed that use of white reflective paint for the roof treatment can be one of the solution, which will not only cost effective, but also helps in reducing the

Table 1. Technologies for various elements of building [6].

Technology	Advantages	Challenges
White Paint	• Energy-efficient • Improved thermal comfort • Cost effective	Frequent application is required
Fly Ash Bricks	• Does not involve burning • Improved strength & durability • Reuse of Industrial wastes	Water requirement is more
Hollow Concrete Block	• Light Weight • Good Thermal Insulation • Low cost of mortor • Can be cast in situ • Low embodied energy	Equipment required
Reflective Roof Tiles	• Reduces operational energy • Improved thermal comfort • Low cost	Extra expenditure, as not a part of structure
Filler Slab	• Reduced quantity and weight of material • Cost effective • Enhanced thermal comfort • Low embodied energy	Required skilled labour for implementation
R C Frames	• Conserves precious natural resource–wood • Cost–effective • Can be pre-fabricated on site • Improved strength & durability	Availability is a problem in mass housing projects

indoor temperature and ultimately reduces operational cost, only its repetitive application will be a recurring expenditure. Similarly use of RC frames should also be encouraged, for which availability of the same should be ensured so that, it can be used in affordable dwelling units. Other building materials have also been mentioned along with its advantages and challenges to be faced for its use in building industry.

Tables 2-4 gives the quantity of embodied energy for various materials used in wall, roof and door/window frames respectively. In walls of the building use of tradional bricks should be replaced with hollow concrete blocks, which consumes only 31.45% energy as required to make traditional bricks, while, white reflective paint will require only 2.7% of energy. Further use of precast door frame will almost be produced with negligible amount of energy, which otherwise would be required to produce steel door frame.

6. Summary and Conclusions

Attempts in minimizing or replacing the conventional high energy materials like cement, steel, bricks with cheaper and local alternatives will lead to the reduction in the embodied energy in buildings.

Materials like Cement, Steel and Bricks and Glass are the major contributors to the total energy consumption in RC buildings.

1) The use of alternative building units like hollow

Table 2. Estimate of embodied energy for various wall material [7].

Material	Embodied Energy MJ/sqm of surface area	Embodied Energy MJ/sqm of floor area	Percentage with base case
Base Case: Bricks	615	1139.9	
Flyash Lime Gypsum Bricks	242	448.8	39.37%
Hollow Concrete Blocks 200 mm	193	358.39	31.45%
Compressed Stabilised Earth Bricks	195	361.64	31.75%

Table 3. Estimate of embodied energy for various roofing material [7].

Material	Embodied Energy MJ/sqm of surface area	Embodied Energy MJ/sqm of floor area	Percentage with base case
Base Case: RCC	847	847	
High Albedo Roof (Reflective Tiles)	34	34	4.0%
High Albedo Roof (White Coating)	23	23	2.7%
RCC Filler Slab	590	590	69.65%

Table 4. Estimate of embodied energy for various door window frame material [7].

Material	Embodied Energy MJ/sqm of surface area	Embodied Energy MJ/ sqm of floor area	percentage with base case
Base Case: Steel	8873	6651.49	
Pre Cast Door Window Frame	288	25.63	0.4%
Wood Plastic Composite	67	5.99	0.1%

concrete blocks for masonry construction reduces the energy consumption by 69% as compared to brick masonry.

2) The conventional RC roof is energy intensive with embodied energy values of 847 MJ/m^2. The RCC filler slab roof almost saves 31% energy as compared to the RC roof, where as White Paint and Reflective tiles consumes embodied as low as 4% and 2.7% of RCC roof.

3) Almost 99% of embodied energy can be saved using Pre Cast RCC Door Window frame or Wood Plastic Composite frame with respect to conventional steel door window frame.

REFERENCES

[1]   Embodied Energy.
      http://en.Wikipedia.org/wiki/Embodied Energy

[2]   A Bandyopaphyay, "A Study of Housing Demand in India," MPRA Paper Number 9339, NIBM-NHB, 2008.

[3]   P. S. Chani, et al., "Comparative Analysis of Embodied Energy Rates for Walling Elements in India," *IE(I) Journal—Architectural Engineering*, Vol. 84, 2003, pp. 47-50.

[4]   K. S. Gumaste, "Embodied Energy Computations in Buildings," Advances in Energy Research (AER-2006), pp. 404-409.

[5]   T. Ramesh, et al., "Life Cycle Energy Analysis of a Multifamily Residential House: A Case Study in Indian Context," *Open Journal of Energy Efficiency*, Vol. 2, No. 1, 2013, pp. 34-41.

[6]   "Identification of Low Cost Green Options and Their Macro-Environmental Impact," Final Report, 2011

[7]   Affordable Housing Energy Efficiency (AHEE), "Affordable Housing Energy Efficiency Handbook," 2007. http://www.h-m-g.com/multifamily/AHEEA/Handbook/default.htm

8

Life Cycle Energy Analysis of a Multifamily Residential House: A Case Study in Indian Context

Talakonukula Ramesh[1], Ravi Prakash[1*], Karunesh Kumar Shukla[2]

[1]Department of Mechanical Engineering, Motilal Nehru National Institute of Technology, Allahabad, India
[2]Department of Applied Mechanics, Motilal Nehru National Institute of Technology, Allahabad, India

ABSTRACT

The paper presents life cycle energy analysis of a multifamily residential house situated in Allahabad (U.P), India. The study covers energy for construction, operation, maintenance and demolition phases of the building. The selected building is a 4-storey concrete structured multifamily residential house comprising 44 apartments with usable floor area of 2960 m². The material used for the building structure is steel reinforced concrete and envelope is made up of burnt clay brick masonry. Embodied energy of the building is calculated based on the embodied energy coefficients of building materials applicable in Indian context. Operating energy of the building is estimated using e-Quest energy simulation software. Results show that operating energy (89%) of the building is the largest contributor to life cycle energy of the building, followed by embodied energy (11%). Steel, cement and bricks are most significant materials in terms of contribution to the initial embodied energy profile. The life cycle energy intensity of the building is found to be 75 GJ/m² and energy index 288 kWh/m² years (primary). Use of aerated concrete blocks in the construction of walls and for covering roof has been examined as energy saving strategy and it is found that total life cycle energy demand of the building reduces by 9.7%. In addition, building integrated photo voltaic (PV) panels are found most promising for reduction (37%) in life cycle energy (primary) use of the building.

Keywords: Residential House; Life Cycle Energy; Embodied Energy; Operating Energy; PV Panels

1. Introduction

A large number of buildings are built for residential, commercial and office purposes every year all over the world. Building construction sector experiences fast paced growth in developing countries, like India, due to growth economy and rapid urbanization. Worldwide buildings consume 30% - 40% amount of primary energy in their construction, operation and maintenance and held responsible for emitting 40% of global warming gases [1]. In India, 24% of primary energy and 30% of electrical energy is consumed in buildings [2]. Since, buildings are consuming large amount of energy; they need to be analyzed in the lifecycle perspective to develop strategies for reduction of their energy use and associated environmental impacts, and make them sustainable. Life cycle assessment (LCA) is the state of art tool in assessing the sustainability of buildings. In order to assess the environmental impact, it is necessary to perform an inventory analysis of building materials and the process of construction, and demolition. But, building materials production processes are much less standardized because of

the unique character of each building. There is limited quantitative information about the environmental impacts of the production and manufacturing of construction materials, and the actual process of construction and demolition particularly in developing countries like India.

Life cycle energy analysis (LCEA) of buildings can also give a useful indication of environmental impacts attributable to buildings, if energy use of the building is expressed in primary energy terms. The analysis also helps in identifying the phases of largest energy consumption and to develop strategies to make buildings sustainable. Life cycle energy analysis is an approach that accounts for all energy inputs to buildings in their life cycle. It includes direct energy inputs during construction, operation and demolition of building, and indirect energy inputs through the production of components, materials used in construction (embodied energy). LCEA has been applied to examine the relationship between embodied energy in construction materials and operational energy and also to analyze the influence of building characteristics like frame work, construction, thermal insulation, heat recovery etc., on life cycle energy use of residential houses in Sweden [3,4]. Winther and

*Corresponding author.

Hestnes [5] compared life cycle energy use of different versions of a residential building, containing varying active and passive energy saving measures, with that of conventional building in the Norwegian context. Citherlet and Defaux [6] analyzed and compared a family house by changing its insulation thickness and type, type of energy production system and use of renewable energy in Switzerland. Mitraratne and Vale [7] recommended provision of higher insulation to a timber framed house as energy saving strategy for low energy housing in New Zealand context. Treloar *et al.* [8] and Fay *et al.* [9] analyzed the life cycle energy of Australian residential buildings built in Melbourne. Utama and Gheewala [10] analyzed clay and cement based single landed houses in Indonesia. In Indian context Shukla *et al.* [11] evaluated embodied energy of an adobe house. Reddy and Jagadish [12] estimated the embodied energy of residential buildings using different construction techniques and materials. Debnath *et al.* [13] evaluated embodied energy of the load bearing single storey and multistory concrete structured buildings. It can be observed that studies limited to embodied energy analyses of buildings are reported in open literature in Indian context. However, in order to understand the total energy needs of the building, complete life cycle energy analysis covering all phases of its life cycle is required. Only recently, the authors have reported on life cycle energy analysis of single family residential houses [14,15].

In the present work an attempt is made to present life cycle energy profile of a multifamily residential building covering the embodied, operational and demolition energy aspects in the Indian context. Some energy saving measures have also been examined from life cycle energy perspective. A typical multifamily residential house (**Figure 1**) called as International House (I H), located in the campus of the Motilal Nehru National Institute of Technology (MNNIT), Allahabad, India is selected for the study to gain insight into the life cycle energy use of the residential house in Indian context. The study covered energy for construction, operation, maintenance and demolition phases of the building. India is a sub tropical

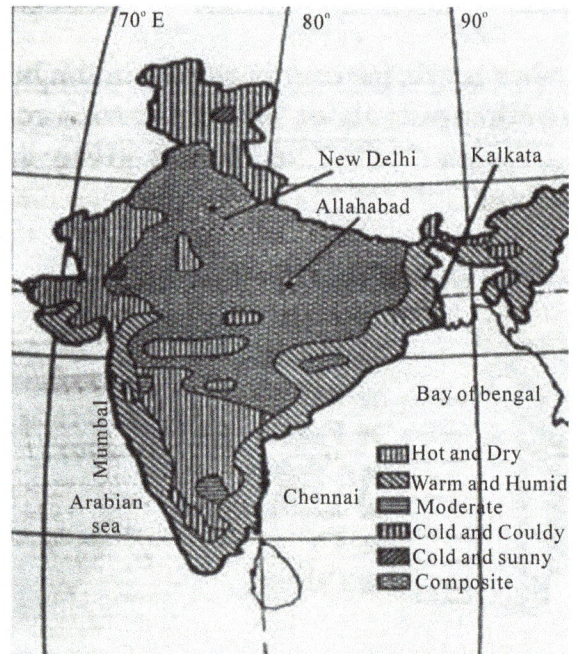

Figure 2. Climatic zones in India (Bansal, 2007).

country with 5 climatic zones viz: hot and dry, hot and humid, moderate, temperate and cold (includes cold and sunny and cold and cloudy climates) and composite (**Figure 2**). Allahabad falls in composite climate and is located at 25.45°N latitude and 81.84°E longitude. Allahabad experiences three seasons: hot dry summer, cool dry winter and warm humid monsoon. The summer season lasts from April to June with the maximum temperatures ranging from 40°C to 45°C.

Monsoon begins in early July and lasts till September. The winter season lasts from December to February. Temperatures rarely drop to the freezing point. Average maximum temperatures are around 22°C and minimum around 10°C.

2. Case Study

The selected building is a 4-storey concrete framed structured multifamily residential house comprising 44 apartments (**Table 1**). The material used for the building structure is reinforced cement concrete and envelope is made from brick masonry. Each flat consists of bed room, living room, kitchen and restroom in the floor area of 40 m^2. The calculated U-values (includes outside air film for exterior surfaces) using e-Quest simulation software [16] for various elements of the building are listed below:

- Roof: 5.08 W/m^2 K;
- Ceiling: 4.73 W/m^2 K;
- Window: 10.85 W/m^2 K;
- External wall: 2.15 W/m^2 K;
- Ground floor: 5 W/m^2 K.

The electricity from the national grid is the only oper-

Figure 1. Case study building in Allahabad, India.

Table 1. Basic parameters of the case study of residential building.

Building parameters	Specifications
Residential floors	4 floors above ground
Ceiling height	3.5 m
Service life	75 years
Usable floor area	2960 m^2
Structure	RCC
Envelope	Brick masonry
Foundation/basement	RCC
Walls (interior)	Brick masonry, plaster finish
Walls (exterior)	Brick masonry, external grit finish
Ground floor	Concrete, ceramic tiles
Roof	RCC flat roof

Table 2. Quantity and embodied energy of materials used.

Name of the material	Unit	Quantity	Embodied energy per Unit (GJ)	Embodied energy of material (GJ)
Cement	ton	553.3	6.7	5338.7
Steel	ton	256.3	28.212	7230.7
Bricks	m^3	2199	2.235	4914.8
Aggregate	m^3	1358.3	0.538	730.8
Aluminum	ton	0.16	236.8	39.6
Glass	ton	1.6	25.8	41.28
Copper	ton	0.29	110	32.5
Ceramic tiles	ton	237	3.333	789.9
PVC	ton	2.10	158	332.9
Marble/Granite	ton	137.2	1.08	148.1
Paint	ton	0.55	144	79.2
Flush door	m^2	940	0.482	453
Grey cast iron pipes	ton	11	38	418

ating energy used by the building systems. Bed room of the building is air conditioned using window air conditioner. The indoor operating set point temperature is around 25°C and all lighting controls of the building are manual. The life cycle energy of the selected building is evaluated based on an assumed service life of 75 years.

The system studied included the manufacture of building materials, construction, operation and maintenance, and demolition phases. The transportation for each life cycle stage is also included. All the materials are manufactured in India. Embodied energy coefficients of building materials are taken from Indian literature [11-13,17, 18], and are shown in **Table 2**. The main information on the types and quantities of materials as well as components of the building is obtained from the detailed estimates of the building, technical specifications and other relevant documents from the building consultant.

3. Life Cycle Energy (LCE)

Life cycle energy of the building is estimated by summing up the energy incurred for construction (initial embodied), operation, maintenance (recurring embodied) and finally demolition of the building at the end of its life.

3.1. Initial Embodied Energy

Embodied energy of the building materials are obtained by summing up the product of quantity of materials multiplied by their embodied energy coefficients (**Table 2**). Energy for construction included energy (electricity) used for lighting, water lifting and diesel fuel used by construction equipment at the site. These are subse-

quently aggregated with energy consumption for the transportation of building material to the construction site. The main building elements are building frames (beams, columns), slabs, floors, staircases, foundation, walls, windows, and finishes. Items such as fitments, sanitary fixtures, appliances, electrical and external items are excluded from the study due to the difficulty associated with obtaining their embodied energy data. All data relevant to construction machines and equipment used on site and transportation distances of construction materials to the construction site are obtained from the available records.

3.2. Building Operation

Operating energy of the building includes electrical energy used for cooling, heating, ventilation, lighting (6 W/m^2 for 100 lux), miscellaneous equipment operation and water supply. This is calculated by simulation through e-Quest energy simulation software [16].

Figure 3 shows 3D model of the building. The building is partially occupied during day time between 9.00 am to 5.00 pm and is fully occupied during night time and fully operated during weekend *i.e.*, Saturday, Sunday and other public holidays. Comfort indoor air temperature is set as 25°C for cooling and 18°C for heating. Coefficient of performance (COP) of window air conditioner is taken as 3 for cooling and 0.9 (taken as a conservative value) for electrical resistance heating. Thus, calculated annual electrical energy demand of the building for its operation (**Figure 4**) is then converted to primary energy using primary energy conversion factor. In

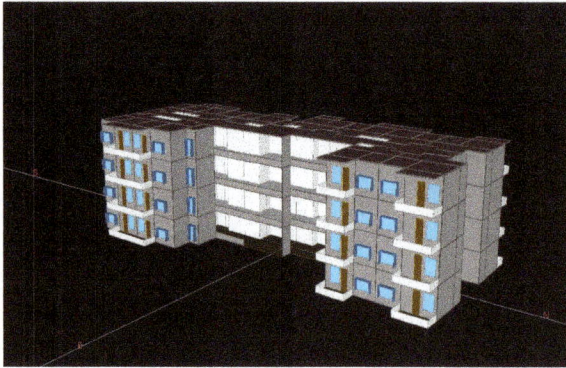

Figure 3. 3D model of the building studied.

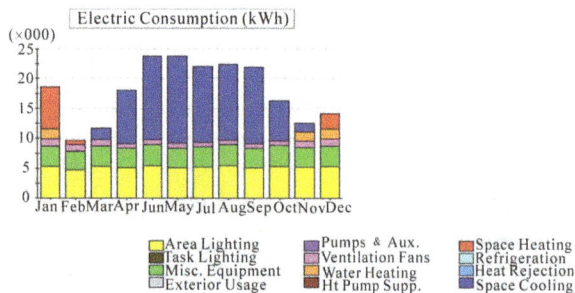

Figure 4. Summary of simulation output.

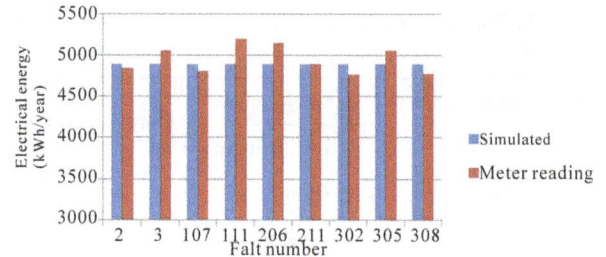

Figure 5. Comparisons of simulated and actual electricity readings.

Table 3. Life span of building components.

Component	Life span (years)
Substructure	100
Brick masonry	100
RCC structure	100
RCC Slab	100
Glass	50
G.I pipes	30
Marble/Granite	75
PVC	50
Ceramic tiles	75
Paint	15
White wash	5
Plaster	40

India 70% of electricity is derived from coal, oil and gas, 25% hydro and remaining from renewable energy resources [19]. A primary energy conversion factor of 3.4 [9] is adapted for electricity from national grid. Annual operating energy of the building is assumed to be constant throughout its life span. Due to changes in climatic conditions and occupants' behavior, operating energy of the building may change little in future but, this is not taken into consideration in the analysis. To verify the results obtained by energy simulation; annual electricity meter readings (actual electrical energy consumption) are compared with the simulated values (average of 44 apartments) for some of the apartments for which meter readings are available (**Figure 5**). As difference between the two values is small (1% - 6%); results of the simulation can be taken as reliable.

3.3. Building Maintenance

Energy consumption estimation for the future maintenance (recurring embodied energy) is computed based on the estimated life span of the building materials and components (**Table 3**) following same procedure as explained in evaluation of initial embodied energy of the building.

3.4. Building Demolition

The last phase of a building's life is demolition. The conventional demolition process often results in landfilldis-posal of majority of the materials. Energy consumption at this phase is mainly due to the operation of demolition machinery and transportation of waste materials to landfill site. Due to lack of data on the energy requirement of actual demolition process in India, it is assumed that 51.5 MJ/m^2 of energy as diesel fuel is required to demolish the building [20]. The major waste materials generated are masonry, concrete and mortar. Volume of materials generated during building demolition for m^3/m^2 [21] floor area is: Masonry: 0.3825 m^3/m^2, Concrete and mortar: 0.5253 m^3/m^2. Volume of waste materials generated is converted to mass in ton by material density (Masonry: 1600 kg/m^3, Concrete and mortar 1920 kg/m^3). To estimate the energy required for transportation of waste material, it is considered that waste material is transported from the site via a diesel powered dump truck. Truck requires 2.85 MJ energy per ton of material for one km travel [21]. The transportation distance for this study is assumed to be the distance from the MNNIT campus where the building under study is located to the landfill site at the city outskirts at about 15 km. Finally, the energy requirements for demolition equipment and transportation of demolition waste to landfill site are summed up to yield the total energy requirement for the demolition phase.

3.5. Life Cycle Energy

The building's life cycle energy (LCE) is calculated using following relations [4,10]:

$$LCE = EBE_i + EBE_r + (OPE \times building\ lifetime) + DE \tag{3.1}$$

where EBE_i = Initial Embodied Energy;

EBE_r = Recurring Embodied Energy;

OPE = Operating Energy per year;

DE = Demolition Energy.

4. Results and Discussion

The results of the life cycle energy analysis are presented in this section. The initial embodied energy of the building is calculated to be 7.35 GJ/m^2. Also, a comparison of the embodied energy and operating energy of the building over its life span indicated that embodied energy is about 11% of the operating energy. This is considerable and is equivalent to about 9 years of the building's operating energy requirement. Further analysis of the initial embodied energy profile reveals that steel, cement, and bricks are the most dominant materials (**Figure 6**).

Steel accounted for 34% of the initial embodied energy. This was followed by cement (25%) and brick (24%). A summary of the quantities of the main materials used for constructing the building (**Figure 7**) showed that bricks accounted for about 38% of the materials by weight, followed closely by aggregate with 29%. Though steel contributed 2.4% and cement 7.6% by weight, due to their large embodied energy coefficients, as given in **Table 2**, its contribution to initial embodied energy is large. Operating energy of the building is evaluated as 66.86 GJ/m^2 over its life span. The LCE attributable to the studied building based on 2960 m^2 gross floor area for 75 years is 75.07 GJ/m^2 (**Table 4**). The analysis of the LCE profile of the building revealed that the operation phase dominated other life cycle phases. It accounts for 89% of the LCE consumption. The LCE share of energy consumption of the manufacturing phase is also not negligible (≈11%). All other phases do not contribute significantly to the building LCE profile.

Further, activity wise energy split (**Figure 8**) for operating phase reveals that energy for space cooling is the largest contributor (45%) to operating energy followed by area lighting with 29%. Due to climatic conditions of the place, cooling energy part is much more than heating energy. Electricity derived mostly from fossil fuels is used for running space cooling equipment (window air conditioners) and due to which operating energy (primary) of the building is quite high. As operating phase

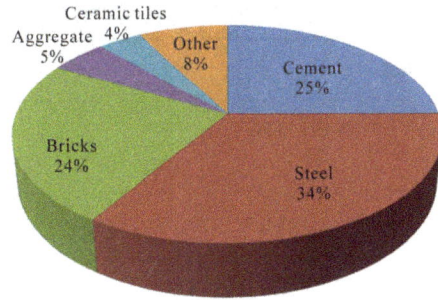

Figure 6. Material percentage share of initial embodied energy.

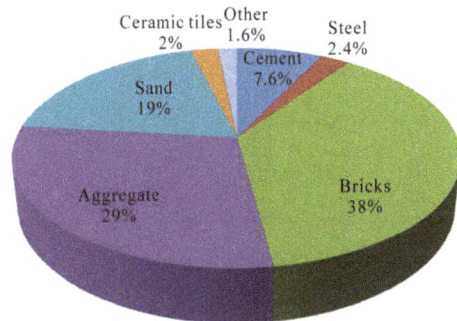

Figure 7. Percentage share of materials (by weight) used in the construction.

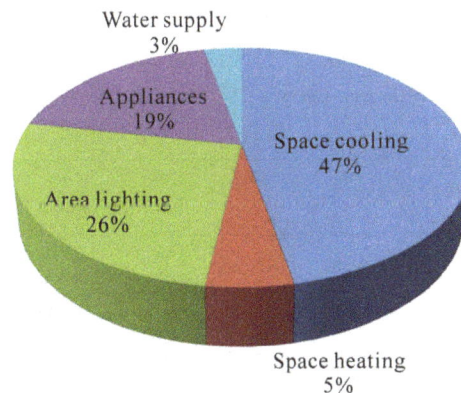

Figure 8. Operating energy distribution by activity.

(89%) is the largest contributor to life cycle energy of the building, strategies aiming at reducing it would result in significant energy savings and will make building sustainable. Hence, energy for space cooling needs to be targeted first to achieve reduction in operating energy of the building.

As energy for space cooling is directly proportional to the cooling load of the space, attention must be focused on reducing it by changing the building design and/or construction materials. **Table 5** shows that building walls (27%) and roof (23%) contribute large to cooling load of the building. Hence, thermal conductance of walls and roof need to be reduced by applying some insulation to them and/or using materials having low thermal conduc-

Table 4. Life cycle energy of the studied building (Usable floor area ≈ 2960 m^2).

Phase			Initial (I)	Recurring (R)	I + R	Energy in GJ	%	kWh/m^2·year (primary)
1		Embodied Energy						
	(a)	Materials	20,551	2094	22,645		10.19	
	(b)	Transportation	791	26.2	817.2		0.37	
	(c)	Construction	438	-	438		0.2	
		(a + b + c)	21,780	2120.2	23900.2		10.76	30
2		Operating Energy						
	(d)	HVAC				98,419	44.29	
	(e)	Lighting				57,750	25.99	
	(f)	Water supply				5138	2.31	
	(g)	Miscellaneous equipment				36,591	16.47	
		(d + e + f + g)				197,898	89.06	248
3		Demolition Energy						
	(h)	Destruction				213	0.1	
	(i)	Transportation				205	0.1	
		(h + i)				418	0.2	1
4		Life cycle Energy	(1 + 2 + 3)			222216.2		279
						≈75.07 GJ/m^2		

Table 5. Thermal loads on the building in GJ per year.

Load	Walls	Roofs	Under surface	Window conduction	Window solar radiation	Lights	Equipment	Infiltration	Occupancy	Total
Heating	−8.49	−13.43	0.16	−3.91	1.69	2.04	1.32	−1.59	8.39	−13.83
Sensible cooling	231.47	195.32	63.94	37.67	111.29	38.37	25.13	23.28	129.1	855.58
Latent cooling	0	0	0	0	0	0	0	52.92	100.63	153.54
Sensible cooling load %	27.1	22.8	7.5	4.4	13	4.5	2.9	2.7	15.1	100

tivity at parity of other properties, in their construction. Also, as India is a tropical country with large potential for solar energy; part or total of electrical energy demand of the building can be met from renewable sources like photovoltaic cells, wind mills etc.

4.1. Energy Saving Measures

To explore the potential for reduction in the life cycle energy consumption, the selected building is studied with two modifications: Case A: Burnt clay bricks of the wall replaced with aerated concrete (A.C) blocks and roof covered with 25 mm A.C tiles; Case B: Photovoltaic

(PV) panels integrated to case A to meet part of electrical energy demand of the building during its operation phase. Case A: Thermal conductivity, density and embodied energy of A.C blocks are lower than burnt clay bricks (**Table 6**). Compressive strength of the A.C blocks is also comparable with bricks; hence, they are considered as suitable materials to replace burnt clay bricks so as to reduce embodied energy and cooling load of the building. Case B: Photo voltaic units (44), each unit consisting of 16 modules are integrated with building to meet part of the electrical energy demand of the building. **Table 7** shows specifications of PV modules used in the building. Sizing of the photo voltaic panels is done as explained in

Table 6. Thermo physical properties of wall materials.

Name of the material	Thermal conductivity (W/mK)	Density (kg/m^3)	Specific heat (J/kgK)	Compressive strength (N/mm^2)
Burnt clay Brick	0.77	1600	840	4.9
Aerated concrete block	0.16	500	1000	4.5

Table 7. PV Module specifications.

Parameter	Value
Wattage (Wp)	75 Watt
Short circuit current Isc	4.8 A
Open circuit voltage Voc	21V
Maximum current Imax	4.5A
Maximum voltage Vmax	16.5V
Number of PV modules	$16 \times 44 = 704$
Area of single module	0.6 m^2
Type of cell	Single crystalline silicon
Number of cells in a module	36
Life span	30 years
Embodied energy	1710 kWh/m^2

reference [22]. **Table 8** shows that building life cycle energy demand decreases 9.7% when burnt clay bricks are replaced with A.C blocks and roof is covered with A.C tiles (case A). Embodied energy of the building increases 20% with integration of PV panels to the building (case B) due to high energy intensity of materials used in PV cells of the panels. But, there is 37% decrease in the overall life cycle energy (primary) of the building as PV panels develop 75 MWh electricity per annum which is used to meet part of the electrical energy demand of building during its operation phase.

4.2. Sensitivity Analysis

Life cycle energy of the building is also evaluated for different life spans of the building to assess its impact on LCE demand of the building. LCE demand of the building is decreasing with increase in life span of the building (**Figure 9**). Actually, embodied energy of the building which decreases with increase in life span of the building, causes LCE of the building to come down.

However, decrease in embodied energy of the building slows down over the life span of 50 years and so is the LCE of the building. Generally, buildings in India are designed for 60 to 100 years. Hence, the results presented herein, holds good for range of life span of buildings from 50 to 100 years (**Table 9**).

Table 8. Life cycle energy comparison (GJ/m^2).

Case	Features	Embodied	Operating	Life cycle	Life cycle energy savings (%)
Base case (As built)	Envelope: Burnt clay brick masonry	8.07	66.85	75.07	
	Roof: RCC				
Case A	Envelope: Aerated concrete block masonry	7.04	60.58	67.76	9.7
Roof: RCC covered with aerated concrete tiles (25 mm thick)					
Case B	Envelope: Aerated concrete block masonry	9.68	37.3	47	37.4
Roof: RCC covered with aerated concrete tiles (25 mm thick) PV panels					

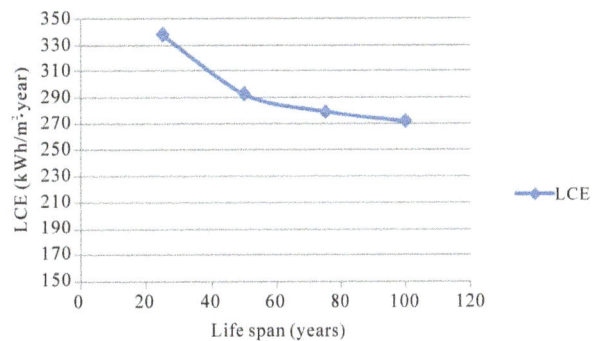

Figure 9. Variation of LCE of the building with change in life span.

5. Conclusion

The paper presents the life cycle energy analysis of a multifamily residential house in Allahabad, India. It shows that while the operating phase of building is very significant (89%), the manufacturing phase is also not negligible (11%). Energy for on-site construction and demolition of the building is minute (<1%) and can be ignored in evaluation of LCE of the building. Steel, cement, and bricks are most significant materials in terms of contribution to the initial embodied energy profile of the building. The life cycle energy intensity of the building is found to be 75 GJ/m^2 and energy index as 279 kWh/m^2 per year. Multy storey houses can be preferred over single storey houses as LCE of single storey houses (300 - 330 kWh/m^2 year) [15] is higher than multy storey houses (270 - 280 kWh/m^2 year). Electricity consumption during operation phase of the building is to be reduced to lower its life cycle energy demand and make it sustainable. Use of aerated concrete blocks in the construction of walls and for covering roof reduces building's life cycle energy demand by 9.7%. Building integrated photo voltaic panels are found most promising for reduction in life cycle energy use of the building as it decreases 37% when part of electrical energy demand (75 MWh per annum) of the building is met through PV panels. Though embodied energy of the buildings accounts only 11% of the LCE of the building, opportunity for its reduction through low embodied energy materials should also be considered.

REFERENCES

[1] M. Asif, T. Muneer and R. Kelley, "Life Cycle Assessment: A Case Study of a Dwelling Home in Scotland," *Building and Environment*, Vol. 42, No. 3, 2007, pp. 1391-1394.

[2] N. K. Bansal, "Energy Security, Climate Change and Sustainable Development," In J. Mathur, H. J. Wagner and N. K. Bansal Ed., *Science, Technology and Society: Energy Security for India.* Anamaya Publishers, Inc., New Delhi, 2007, pp. 15-23.

[3] K. Adalberth, "Energy Use during the Life Cycle of Single-Unit Dwellings: Examples," *Building and Environment*, Vol. 32, No. 4, 1997, pp. 321-329.

[4] K. Adalberth, "Energy Use in Four Multi-Family Houses During Their Life Cycle," *International Journal of Low Energy and Sustainable Buildings*, Vol. 1, 1999, pp. 1-20.

[5] B. N. Winther and A. G. Hestnes, "Solar Versus Green: The Analysis of a Norwegian Row House," *Solar Energy*, Vol. 66, No. 6, 1999, pp. 387-393.

[6] S. Citherlet and T. Defaux, "Energy and Environmental Comparison of Three Variants of a Family House during Its Whole Life Span," *Building and Environment*, Vol. 42, No. 2, 2007, pp. 591-598.

[7] N. Mithraratne and B. Vale, "Life Cycle Analysis Model for New Zealand Houses," *Building and Environment*, Vol. 39, No. 4, 2004, pp. 483-492.

[8] G. Treloar, R. Fay, P. E. D. Love and U. Iyer-Raniga, "Analysing the Life—Cycle Energy of an Australian Residential Building and its Householders," *Building Research & Information*, Vol. 28, No. 3, 2000, pp. 184-195.

[9] R. Fay, G. Treloar and U. Iyer-Raniga, "Life-Cycle Energy Analysis of Buildings: A Case Study," *Building Research & Information*, Vol. 28, No. 1, 2000, pp. 31-41.

[10] A. Utama and S. H. Gheewala, "Life Cycle Energy of Single Landed Houses in Indonesia," *Energy and Buildings*, Vol. 40, No. 10, 2008, pp. 1911-1916.

[11] A. Shukla, G. N. Tiwari and M. S. Sodha, "Embodied Energy Analysis of Adobe House," *Renewable Energy*, Vol. 34, No. 3, 2009, pp. 755-761.

[12] B. V. V. Reddy and K. S. Jagadish, "Embodied Energy of Common and Alternative Building Materials and Technologies," *Energy and Buildings*, Vol. 35, No. 2, 2003, pp. 129-137.

[13] A. Debnath, S. V. Singh and Y. P. Singh, "Comparative Assessment of Energy Requirements for Different Types of Residential Buildings in India," *Energy and Buildings,* Vol. 23, No. 2, 1995, pp. 141-146.

[14] T. Ramesh, R. Prakash and K. K. Shukla, "Life Cycle Energy Analysis of a Single Family Residential House with Different Envelopes and Climates in Indian Context," *Applied Energy*, Vol. 89, No. 1, 2012, pp. 193-202.

[15] T. Ramesh, R. Prakash and K. K. Shukla, "Life Cycle Approach in Evaluating Energy Performance of Residential Buildings in Indian Context," *Energy and Buildings*, Vol. 54, 2012, pp. 259-265.

[16] e-Quest, "The Quick Energy Simulation Tool," 2009.

[17] Development Alternatives (DA), "Energy Directory of Building Materials," Building Materials & Technology Promotion Council, New Delhi, 1995.

[18] T. N. Gupta, "Building Materials in India," Building Materials & Technology Promotion Council, WordSmithy, New Delhi, 1998.

[19] Ministry of Power, Government of India, "Power Sector at a Glance All India," 2012. http://www.powermin.nic.in/JSP_SERVLETS/internal.jsp

[20] B. Thomas, A. Jonsson and A. M. Tillman, "LCA of Building Frame Structures," *Technical Environment Planning Report*, Goteborg, 1996.

[21] The Energy and Resources Institute (TERI), "Sustainable Building Design Manual," Vol. 2, New Delhi, 2004.

[22] C. Arvind, G. N. Tiwari and A. Chandra, "Simplified Method of Sizing and Life Cycle Cost Assessment of Building Integrated Photovoltaic System," *Energy and Buildings*, Vol. 41, No. 11, 2009, pp. 1172-1180.

9

Energy-Saving and Economical Evaluations of a Ceramic Gas Turbine Cogeneration Plant

Satoru Okamoto

Department of Mathematics and Computer Science, Shimane University, Matsue, Japan

ABSTRACT

A ceramic gas turbine can save energy because of its high thermal efficiency at high turbine inlet temperatures. This paper deals with the thermodynamic and economic aspects of a ceramic gas turbine cogeneration system. Here cogeneration means the simultaneous production of electrical energy and useful thermal energy from the same facility. The thermodynamic performance of a ceramic gas turbine cycle is assessed using a computer model. This model is used in parametric studies of performance under partial loads and at various inlet air temperatures. The computed performance is compared to the measured performance of a conventional gas turbine cycle. Then, an economic evaluation of a ceramic gas turbine cogeneration system is investigated. Energy savings provided by this system are estimated on the basis of the distributions of heat/power ratios. The computed economic evaluation is compared to the actual economic performance of a conventional system in which boilers produce the required thermal energy and electricity is purchased from a utility.

Keywords: Energy; Exergy; Energy-Saving; Economical Evaluation; Ceramic Gas Turbine; Cogeneration Plant

1. Introduction

Adaptations of aircraft engines for industrial, utility, and marine-propulsion applications have long been accepted as means for generating power with high efficiency and for ease of maintenance. Because of their heritage, aero-derivative gas turbines typically require less space and supporting structure than other industrial gas turbines of equivalent output power. These features also equate to reduced plant construction time and adaptability to meet unique requirements dictated by the site or application.

To improve the performance of gas turbines in general and the overall thermal efficiency in particular, it is necessary to increase the turbine inlet temperature. Consequently, thermal loads on turbine blades and the combustion chamber become extremely high. In such cases, modern ceramics are the best-suited materials due to their excellent high-temperature strength and other attractive properties.

In Japan, the development of ceramic components for gas turbines in cogeneration applications was initiated in 1988 by New Energy and Industrial Technology Development Organization (NEDO) sponsored by the Ministry of International Trade and Industry (MITI) [1,2]. Two types of ceramic gas turbine engines for cogeneration

were built and tested. CGT301 is a restored single-shaft CGT that has characteristics for continuous full-load applications. CGT302 is a restored two-shaft CGT suitable for partial-load applications in facilities such as hotels, hospitals, and office buildings (**Table 1**).

Cogeneration is frequently defined as the sequential production of useful thermal energy and shaft power from a single energy source. The shaft power can be used to drive mechanical loads such as compressors, pumps, and electric generators. For applications that generate electricity, the power can either be used internally or supplied to the utility grid.

The thermal benefits of cogeneration are discussed as follows. A gas turbine cogeneration cycle is arranged to reject a portion of its exhaust energy at the temperature required in the process. The resultant system achieves approximately 75% utilization of input thermal energy compared to approximately 35% for a fossil-fuel-fired steam plant designed to provide only power. This significant energy savings is a primary factor contributing to favorable economics for many gas-turbine-based cogeneration systems.

In evaluating a power cycle, thermodynamics cannot be the only consideration. There are five general areas of

Table 1. Performance of a ceramic gas turbine (CGT302) [1,2].

Characteristics	Units	Data
Maximum power output	kW	300
Gas turbine type		Heat-exchange twin-shaft
Thermal efficiency	%	42
Pressure ratio		8
Air flow rate	kg/s	0.89
Turbine inlet temperature	°C	1350
Shaft rotational speed	rpm	3000/3600
Compressor type		Single-stage centrifugal
Gas generator turbine type		Single-stage axial
Gas generator turbine rotational speed	rpm	76,000
Power turbine type		Single-stage axial
Power turbine rotational speed		57,000
Combustor type		Cannular type
Heat exchanger type		Recuperator type

concern in evaluating ceramic gas turbine cogeneration: 1) first-law efficiency, 2) second-law efficiency, 3) system performance, 4) energy savings, and 5) economic evaluation. This paper briefly describes the system under consideration, and then summarizes computational results from parametric studies.

2. Energy-Saving Evaluation

This section deals with the thermodynamic aspects of a ceramic gas turbine cogeneration system. It presents expressions involving relevant variables for fuel-utilization efficiency (first-law efficiency), electrical to thermal energy ratio (power-to-heat ratio), and second-law efficiency (exergetic efficiency). The study included the impact of atmospheric temperature on ceramic gas turbine cogeneration performance. A computer program was especially designed to calculate overall thermal efficiency and the net specific work from a simple-cycle gas turbine cogeneration system. These calculations were conducted for various combustor discharge temperatures (such as *TIT*) and pressure ratios. During these calculations, both partial and full loads were studied.

2.1. System Description

Currently, the simple-cycle gas turbine is the most widely used topping-cycle cogeneration systems due to its simple design. **Figure 1** shows the flow diagram for the

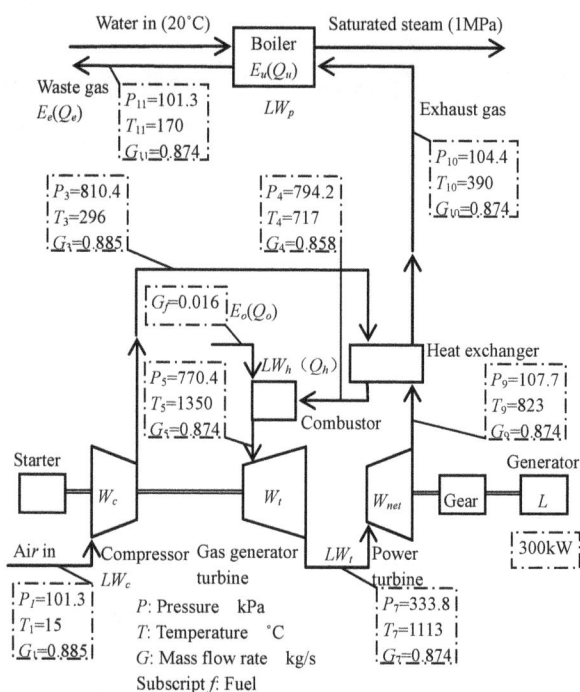

Figure 1. Schematic diagram of a ceramic gas turbine (CGT302) cogeneration system.

cycle under consideration; its corresponding thermodynamic state points are described in the figure.

The operating principle of the ceramic gas turbine can be simplified as follows. Ambient air is drawn into a single-stage centrifugal compressor where it is compressed to approximately 8 atm. The compressed air then passes to the combustion chamber where fuel is injected and burned. The products of combustion enter the turbine and expand to approximately atmospheric pressure. Part of the work developed by the gas generator turbine is used to drive the compressor, while the remainder is delivered to a power turbine. The power turbine exhaust entering the heat-recovery steam generator is the waste-heat source for process heat production.

The quantity and quality of the process steam produced depends on the temperature of air entering and the temperature of steam produced in the heat-recovery steam generator. Therefore, the performance of a gas turbine cogeneration system varies significantly with compressor inlet air conditions, mainly atmospheric temperature [3]. Gas turbine design ratings are usually based on standard conditions. A popular standard is that of the International Standards Organization (ISO). The site conditions for this standard are sea-level altitude, 101.325 kPa, and 15°C.

2.2. Performance Parameters of a Cogeneration System

The useful products from a cogeneration system are elecical energy W_{net} and thermal energy or process heat Q_u. A parameter used to assess the thermodynamic performance of such a system is the fuel-utilization efficiency, which is simply the ratio of overall energy in the useful products, W_{net} and Q_u, to the energy of input fuel Q_o,

$$\eta_q = \frac{W_{net} + Q_u}{Q_o}. \tag{1}$$

Another parameter commonly used to assess the therdynamic performance of a cogeneration system is the power-to-heat ratio R_{ph},

$$R_{ph} = \frac{W_{net}}{Q_u}. \tag{2}$$

In these parameters, power and process heat are treated equally. This reflects the first law of thermodynamics, which is concerned with the quantity and not the quality of energy. Thus, the fuel-utilization efficiency is also known as the first-law efficiency. However, according to the second law of thermodynamics, electric power is significantly more valuable than process heat. Exergy, the central concept in a second-law analysis, is always consumed or destroyed in any real process. A process is better thermodynamically if less exergy is consumed. Consequently, the ratio of the amount of exergy in the products to that supplied is a more accurate measure of the thermodynamic performance of a system [4]. By definition,

$$\eta_e = \frac{W + E_u}{E_o}, \tag{3}$$

where W is the overall exergy, E_u is the exergy content of process heat produced, and E_o is the exergy content of input fuel. The quantity η_e is a second-law efficiency.

The exergy factor of process heat λ_u and the exergy factor of fuel input λ_o can be defined by the following expressions:

$$\lambda_u = \frac{E_u}{Q_u}, \tag{4}$$

$$\lambda_o = \frac{E_o}{Q_o}. \tag{5}$$

Then, the second-law efficiency may be written as

$$\eta_e = \frac{\eta_q}{\lambda_o} \times \frac{R_{ph} + \lambda_u}{R_{ph} + 1}. \tag{6}$$

The exergy factor of process heat is always less than unity. In the case of saturated steam, it increases with the pressure of steam produced. This is consistent with the second law of thermodynamics because the quality of the energy content in high-pressure saturated steam is greater

than the quality of the energy content in low-pressure saturated steam. The exergy factor of fuel input is close to unity for most fuels because the chemical energy in fuel is essentially overall exergy [5]. Thus, the second-law efficiency is not very sensitive to the exergy factor of the fuel used in cogeneration systems.

For a typical cogeneration system with process heat in the form of saturated steam, λ_u is in the range 0.25 - 0.4, and R_{ph} is usually less than unity. Thus, η_e is significantly less than η_q, and an evaluation of thermodynamic performance of a cogeneration system based on the first-law efficiency alone could be misleading [5].

2.3. Results and Discussion

The thermodynamic performance of a ceramic gas turbine cogeneration system was studied. Pertinent data are shown in **Table 1** [1,2]. With this information, only a procedure for calculating the quantity and quality of process heat produced is required. Then, the fuel-utilization efficiency, power-to-heat ratio, and second-law efficiency can be calculated.

Overall thermal efficiency and net specific work were calculated for various values of *TITs* and pressure ratio. The values of maximum *TIT* were taken to be 900°C, 1000°C, 1100°C, 1200°C, 1300°C, 1350°C, and 1400 C, while the values of pressure ratios used in the calculations were 2, 4, 6, 8, and 10.

Figure 1 shows the simple-cycle ceramic gas turbine arrangement considered in this study. Fuel gas was used in this analysis; however, the properties of any type of fuel can be fed to the computer program. Efficiencies of the compressor, turbine, and combustion chamber were assumed to be 82%, 84%, and 99%, respectively. These assumptions were taken from catalogs provided by manufacturers and are expected to lead to realistic computational results.

Figure 2 shows thermal efficiencies at different maximum *TITs*. *TIT* increases with the thermal efficiency.

The effect of atmospheric air temperature on thermal efficiency is shown in **Figure 3**. These values of thermal

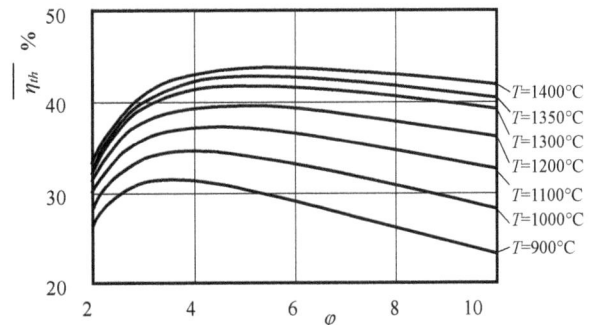

Figure 2. Overall thermal efficiency with various turbine inlet temperatures.

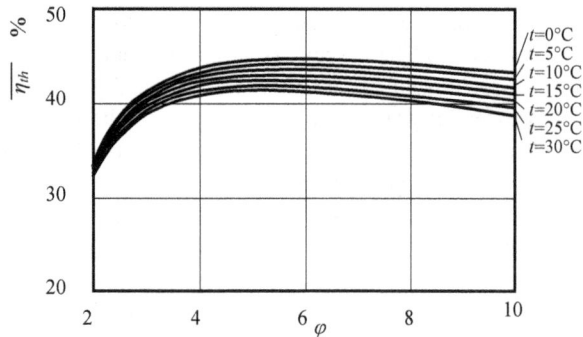

Figure 3. Overall thermal efficiency with various atmospheric temperatures (*TIT* = 1350˚C).

efficiency decrease as the atmospheric air temperature increases for the same *TIT* (1350˚C).

Increases in the pressure ratio increase the thermal efficiency in some cases considered here. For pressure ratios less than 8, the increase in the pressure ratio causes an increase in the thermal efficiency for a constant maximum inlet temperature of 1350˚C, as shown in **Figure 3**. Above a pressure ratio of 8, increases in the pressure ratio decrease the thermal efficiency at all values of the assumed atmospheric temperature. However, at higher values of the atmospheric temperature, the rate of increase in thermal energy with pressure ratio becomes smaller.

The calculated exergy flow and energy flow are represented graphically in **Figures 4** and **5**, respectively. The lower heating value of the fuel represents 100% of the exergy input into the process. This input to the fuel is converted into two exergy flows and a loss. The exergy fed to the ceramic gas turbine is converted into electrical

energy, waste heat in the gas, and a loss because of combustion. The heat-recovery system produces saturated steam from the heat of the waste gas. For the ceramic gas turbine cogeneration process, it is 49.4 J per 100-Joule input, which means an overall efficiency of 49.4%.

The exergy flow of a conventional gas turbine cogeneration process differs from a ceramic gas turbine cogeneration process because the maximum *TIT* (900˚C) for the conventional process is lower than that in the ceramic turbine process (1350˚C). Only the amounts of the exergy flows differ. The conversion efficiency from fuel to electricity is 25.8%. Apart from this, the exergy flow is similar to that in the ceramic gas turbine cogeneration process.

To compare the two processes, characteristic data are condensed in **Table 2**. The boundary conditions for the two plants are the same. Large exergy losses become visible in all combustion process, e.g., for the combustion in gas turbines. The other conversion losses are comparatively small.

According to the second law of thermodynamics, the useful heat and power delivered by a cogeneration plant do not have the same unitary value, although they are quantified in the same physical unit, Joule.

Electricity is a form of "pure exergy", while the heat contained in process steam has an exergy content (or economic value) that depends on the temperature at which it is available. **Figure 4** presents exergetic diagrams relative to the production of heat and power by typical process units. These diagrams clearly show that some of the heat cannot be converted into useful work.

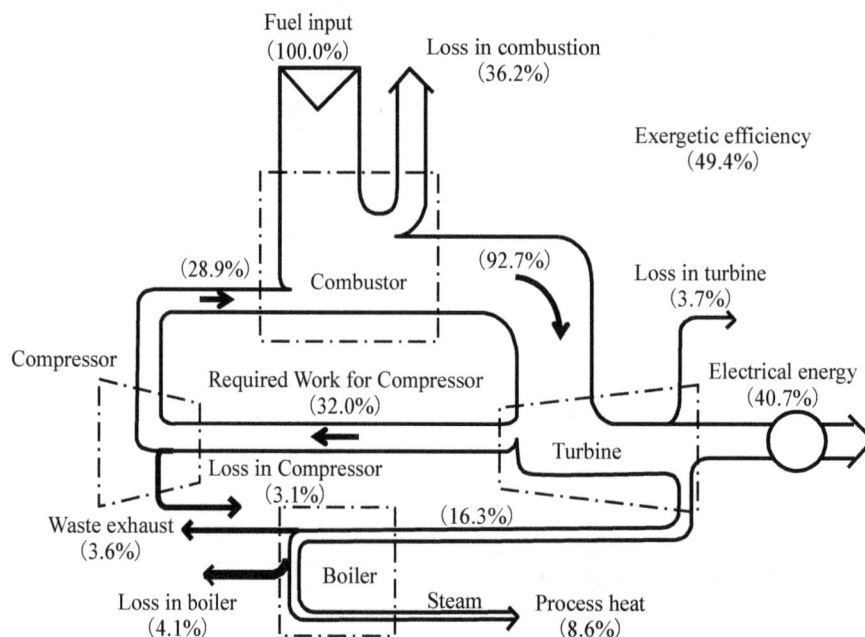

Figure 4. Exergy flow diagram of a ceramic gas turbine cogeneration system.

This results from internal irreversibilities caused by imperfections in the conversion process; it measures the degradation of energy entering the process (fuel) whose exergy content is close to unity.

The situation is different in electricity production where only the exergy from the process is taken as the useful output. This explains why thermal efficiencies based on the first law of thermodynamics easily reach 67% in **Figure 5**, while this sophisticated power plant hardly achieves 50% efficiency.

3. Economic Evaluation

3.1. Definition of Energy Demand

In most cogeneration plants, both thermal and electric power demands experience wide variations over time. Since the selected time step is 1 h and the plant simulation is to be performed over an entire year, the most general format of input data necessary to define each case would comprise an array of 8760 values. Besides making load specification very unpractical, such a format would also require exhaustive computing times. On the basis of the author's experience, the following simplified assumptions yield a sufficiently accurate load description for most practical situations.

Monthly load variations can be described by specifying the minimum and maximum electric and thermal demands. Hourly fluctuations between the minimum and maximum demands are described by means of a daily load profile that is made dimensionless with respect to the total demand. Notice that the magnitude of the total

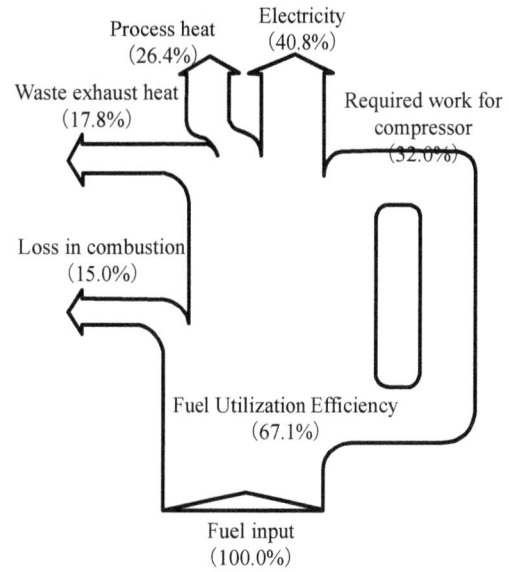

Figure 5. Energy flow diagram of a ceramic gas turbine cogeneration system.

Table 2. Exergy and energy balances.

	Items	Conventional gas turbine cogeneration		Ceramic gas turbine cogeneration	
		Q, E, W kJ/(kg/s)	%	Q, E, W kJ/(kg/s)	%
E X E R G Y	Exergy content of fuel input: E_o	684.7	100.0	891.2	100.0
	Electrical energy: W_{net}	176.4	25.8	363.0	40.7
	Exergy content of process heat: E_u	80.6	11.8	77.0	8.6
	Exergetic loss in combustion: LW_h	303.7	44.4	322.1	36.2
	Exergetic loss in turbine: LW_t	28.5	4.2	33.4	3.7
	Exergetic loss in compressor: LW_c	24.5	3.6	27.3	3.1
	Exergetic loss in boiler: LW_p	39.2	5.7	36.6	4.1
	Waste exhaust exergy: E_e	31.8	4.6	31.8	3.6
	Exergetic efficiency: $\overline{\eta_e}$ %	37.5		49.4	
E N E R G Y	Energy content of fuel input: Q_o	684.7	100.0	891.2	100.0
	Required work for compressor: W_e	235.2	34.4	285.4	32.0
	Energy loss in fuel combustion: Q_h	102.7	15.0	133.7	15.0
	Turbine output work: W_t	411.6	60.1	648.4	72.8
	Energy content of process heat: Q_u	246.3	36.0	235.2	26.4
	Waste exhaust heat: Q_e	159.3	23.3	159.3	17.8
	Fuel utilization efficiency: $\overline{\eta_q}$ %	61.7		67.1	

demand used for scaling daily profiles can change from month to month. It is assumed that there are only two types of daily dimensionless profiles, one for weekdays and the other for weekends. A further distinction is introduced between summer and winter. Thus, the entire year can be described by four daily profiles.

Under these assumptions, an entire year can be simulated by optimizing only 24 days (two days for each month); this results in substantial savings in computing time. In most practical cases, the accuracy of the results is much better than that of the load description.

The maximum and annual demands at a hotel, a hospital, and an office building in Tokyo are shown in **Table 3** [6]. Monthly energy usage was allocated to both peak and off-peak periods. Hourly data on electric demand were provided by the electric utility. The seasonality of demand variations is the average weekday demand shown for each of the two types of area of buildings of months. Monthly fuel usage was taken from daily boiler logs. While all boilers were capable of firing oil, only natural gas was used.

3.2. Estimation of Energy Saving

To estimate the energy savings available from ceramic gas turbine cogeneration, the "energy saving ratio" is defined as the ratio of energy saved by cogeneration to fuel consumption in a "conventional" system. In the conventional system, electric power is purchased from a utility (with 35% efficiency), and heat is produced by a boiler (with 90% efficiency). From the anticipated energy savings by cogeneration, a relationship was obtained between energy saving ratio and electric power supply per maximum demand in buildings. Using this relationship and the demand distribution in buildings, energy savings were estimated.

In the simulation model of the ceramic gas turbine cogeneration plant shown in **Figure 6**, the source energy consumption of the system was estimated hourly. These estimates were based on shaft power and recovered heat, which were calculated from the partial-load efficiency of a ceramic gas turbine, waste-heat recovery boiler, and economizer.

Using parameters of the generating efficiency of electric power, the overall efficiency of energy utilization under partial loads, and the energy saving ratio, a plot of potential energy savings can be created, as shown in **Figure 7**. **Figure 7** indicates the values for energy saving index of the hotel (20,000 m^2), the hospital (20,000 m^2) and the office building (20,000 m^2). This figure relates the energy saving ratio to the energy demand distribution. It is assumed that there are only two types of ce-

Table 3. Energy demand in buildings [6].

Buildings	Area of buildings m^2	Peak demand of electricity kW	Annual demand of electricity MWh/y	Peak demand of heat Gcal/h			Annual demand of heat Gcal/y		
				Cooling	Heating	Hot water	Cooling	Heating	Hot water
Hotel	5000	350	1015	0.34	0.22	0.19	203	244	607
	20,000	1200	4060	1.40	0.90	0.80	812	976	2428
Hospital	5000	400	640	0.36	0.29	0.10	549	234	328
	20,000	1500	2560	1.45	1.14	0.40	2,197	936	1313
Office building	5000	350	810	0.45	0.30	0.004	209	112	4
	20,000	1200	3240	1.81	1.20	0.015	835	446	16

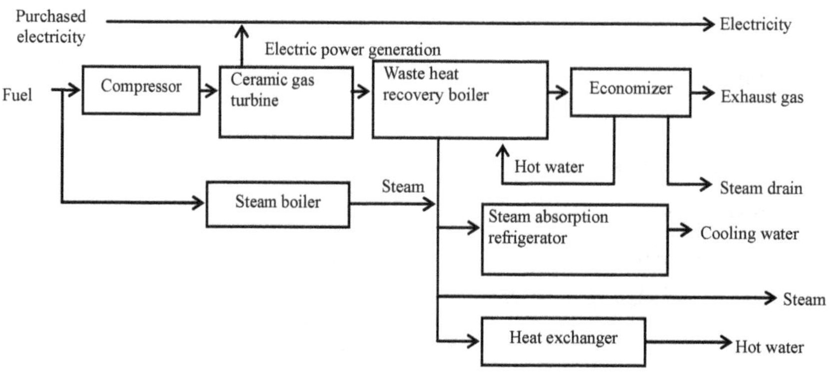

Figure 6. System diagram of a ceramic gas turbine cogeneration plant.

ramic gas turbines, as shown in **Figure 7**: fully loaded and partially load. The fully loaded engine is CGT301, whereas the partially loaded engine is CGT302. The maximum saving energies in the partially loaded type for the hotel and hospital are higher than those in the fully loaded type. In the range of electric power supplied per maximum demand, especially at lower values of supplied electric power, a large energy saving is not expected.

3.3. Analysis of Energy Cost

When comparing the ceramic gas turbine cogeneration

system with a conventional system, care must be taken to properly evaluate the unit energy costs in each case. With cogeneration, electric utilization and contractual power are generally lower, implying higher electricity costs. Moreover, the "value" for heat depends on the user's characteristics. Considering an example of business and commercial use, a "value" for heat may be simply obtained as the product of the heat demand and the unit cost of fuel used in the conventional system; but such a fuel cost can be the same as, lower than, or higher than that for the cogeneration plant. Considering the user's characteristics, the computer program requires a "value" for the cogenerated heat that must consider all differences in fuel costs and efficiencies between the cogeneration and conventional cases [7].

Supplemental power costs, which are defined as the cost of power purchased from the utility on a regular

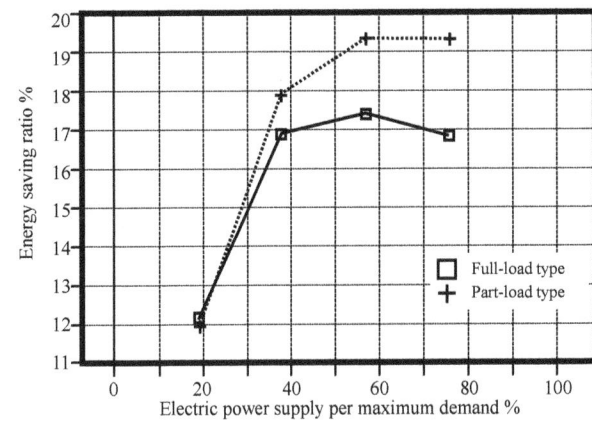

(a)

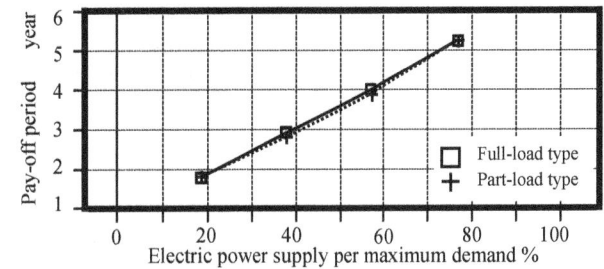

(a)

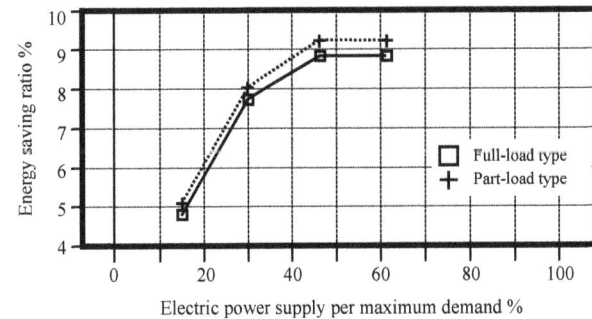

(b)

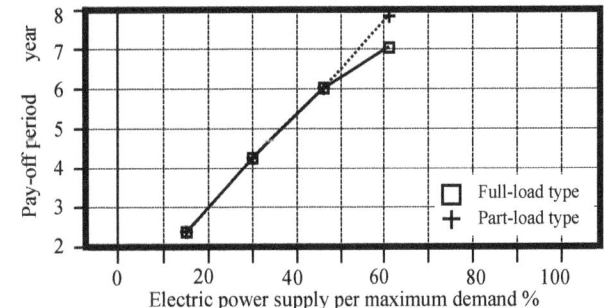

(b)

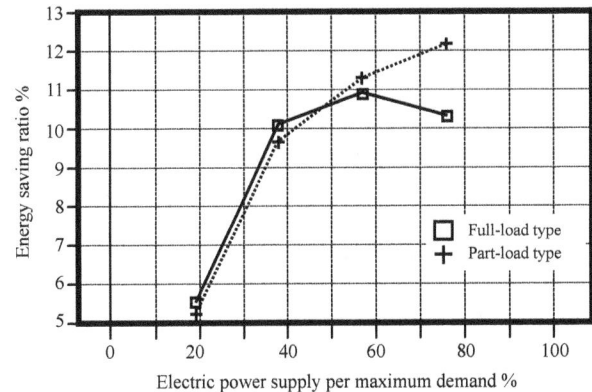

(c)

Figure 7. Values for energy saving index (20,000 m^2) in a (a) hotel, (b) hospital, and (c) office building.

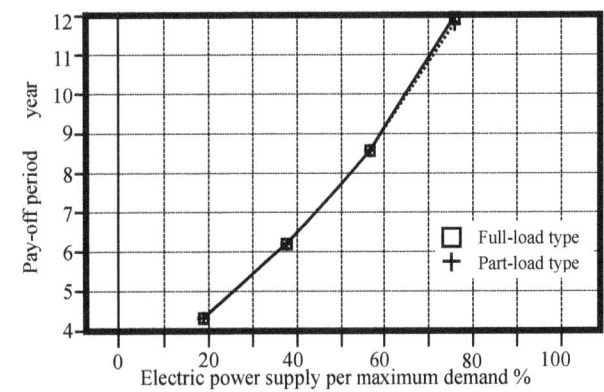

(c)

Figure 8. Values for economical index (20,000 m^2) in a (a) hotel, (b) hospital, and (c) office building.

basis, and supplemental fuel costs, which are the costs of fuel required for the conventional boiler and the duct burner, were computed. The payback periods through cost savings derived from the efficiency of the ceramic gas turbine cogeneration system for each building are shown in **Figure 8**. **Figure 8** indicates the values for economic index of the hotel (20,000 m^2), the hospital (20,000 m^2) and office building (20,000 m^2). The incremental costs of operating this system are also presented. In the low range of electric power supply per maximum demand, the capacity of the ceramic gas turbine cogeneration system is designed for payback periods of less than five years in the hotel and hospital.

4. Conclusions

Many useful expressions have been developed for the study of a ceramic gas turbine cogeneration system. Some of the importance conclusions are as follows:

1) Specific output power, specific process heat production, fuel-utilization efficiency, and second-law efficiency improve with increases in the maximum inlet temperature.

2) Second-law efficiency and power-to-heat ratio are better indicators of thermodynamic performance than fuel-utilization efficiency.

3) In the high range of electric power supply per maximum demand, maximum energy savings are realized.

4) The capacity of the system is designed for payback periods of less than five years through cost savings derived from improved system efficiencies.

REFERENCES

[1] K. Honjo, R. Hashimoto and H. Ogiyama, "Current Status of 300 kW Industrial Ceramic Gas Turbine R & D in Japan," *Transactions of ASME*, Vol. 115, No. 1, 1993, pp. 51-57.

[2] Gas Turbine Society of Japan, "Small Scale Ceramic Gas Turbine: Research and Development of Advanced Gas Turbine," NTS Inc., Tokyo, 2003. (in Japanese)

[3] A. A. El Hadik, "The Impact of Atmospheric Conditions on Gas Turbine Performance," *Transactions of ASME*, Vol. 112, No. 4, 1990, pp. 590-596.

[4] F. F. Huang, "Performance Evaluation of Selected Combustion Gas Turbine Cogeneration Systems Based on First and Second-Law Analysis," *Transactions of ASME*, Vol. 112, No. 1, 1990, pp. 117-121.

[5] M. A. El-Masri, "On Thermodynamics of Gas Turbine Cycles: Part 1-Second Law Analysis of Combined Cycles," *Transactions of ASME*, Vol. 107, No. 4, 1985, pp. 880-889.

[6] New Energy and Industrial Technology Development Organization, "A Study of the Effect of Ceramic Gas Turbine Development," NEDO-P-8723, NEDO, 1988, p. 65. (in Japanese)

[7] S. Consonni, Lozza and E. Macchi, "Optimization of Cogeneration Systems Operation Part B: Solution Algorithm and Examples of Optimum Operating Strategies," *Proceedings of the* 1989 *ASME COGEN-TURBO III*, 1989, pp. 323-331.

Nomenclature

E_e : waste exhaust exergy kJ/(kg/s);

E_o : exergy content of fuel input kJ/(kg/s);

E_u : exergy content of process heat kJ/(kg/s);

G_i : mass flow rate in gas turbine cycle, $i = 1 - 11$ kg/s;

G_f : mass flow rate of fuel in gas turbine cycle kg/s;

LW_c : exergetic loss in compressor kJ/(kg/s);

LW_h : exergetic loss in combustion kJ/(kg/s);

LW_P : exergetic loss in boiler kJ/(kg/s);

LW_t : exergetic loss in turbine kJ/(kg/s);

P_i : pressure in gas turbine cycle, $i = 1 - 11$ kPa;

Q_e : waste exhaust heat kJ/(kg/s);

Q_h : energy loss in fuel combustion kJ/(kg/s);

Q_o : energy content of fuel input kJ/(kg/s);

Q_u : energy content of process heat kJ/(kg/s);

R_{ph} : power-to-heat ratio;

T or t : temperature °C;

T_i : temperature in gas turbine cycle, $i =$ 1°C - 11°C;

TIT : turbine inlet temperature °C;

W_c : required work for compressor kJ/(kg/s);

W_{net} : electrical energy kJ/(kg/s);

$\underline{W}_t$: turbine output work kJ/(kg/s);

$\underline{\eta}_e$: exergetic efficiency (second-law efficiency) %;

$\underline{\eta}_q$: fuel-utilization efficiency %;

η_{th} : overall thermal efficiency %;

λ_o : exergy factor of fuel input ($\equiv E_o/Q_o$);

λ_u : exergy factor of process heat ($\equiv E_u/Q_u$);

φ : pressure ratio.

Energy Efficiency Analysis of White Button Mushroom Producers in Alburz Province of Iran: A Data Envelopment Analysis Approach

Habib Reyhani Farashah, Seyed Ahmad Tabatabaeifar*, Ali Rajabipour, Paria Sefeedpari

Department of Agricultural Machinery Engineering, Faculty of Agricultural Engineering and Technology,
University of Tehran, Karaj, Iran

ABSTRACT

The aim of this study was to determine energy consumption pattern and specifically to measure and benchmark the efficiency for white button mushroom production in Alburz province of Iran. The data used in this study were collected by interviewing mushroom producers in the region. In the surveyed farms, average yield and total energy consumption were calculated as around 208.46 kg·ton^{-1} compost and 133.25 MJ·ton^{-1}, respectively. The results revealed that fossil fuel (40.43%), compost (30.45%) and electricity (27.42%) consumed the bulk of energy. The results of DEA approach also showed that 12 and 14 farmers had efficiency score of unity. Electricity and fossil fuel were found to be used in excess in target mushroom production farms. Moreover, we came to the conclusion that the total energy consumption can be reduced to 120.15 MJ·ton^{-1} for mushroom production in which diesel fuel energy (50.89 MJ·ton^{-1}), FYM (37.32 MJ·ton^{-1}) and electricity (30.34 MJ·ton^{-1}) energies were considerably significant.

Keywords: Energy; Technical Efficiency; Data Envelopment Analysis; White Button Mushroom Production

1. Introduction

Button mushroom (*Agaricus bisporus*) is regarded as a high protein, low-calorie food with medicinal properties [1]. It is the most widely cultivated, harvested, and distributed mushroom in the world. Mushrooms contain substances of high nutritional value such as minerals and a variety of vitamins. White button mushroom is the most prevalent mushroom in the world [2] and needs to be fresh or dried for various uses in the food industry. Mushrooms also have very low energy levels; five medium-sized button mushrooms added together only have twenty calories (80 kilojoules).

Agricultural sector between other parts has a significant place in producing food requirements of the growing population of the world. On the other hand, agriculture itself uses a large amount of energy in form of inputs in order to produce larger amounts of energy output. Based on these facts, a harmony between resource scarcity and food safety is needed in practice. Namely, sustainability should be introduced to today's agricultural systems. A sustainable agriculture, like all other sustainable devel-

opment, must meet the needs of the present without diminishing opportunities for the future.

Energy use in agriculture has developed in response to increasing populations, a limited supply of arable land, and a desire for an increasing standard of living. As the population continues to rise, the agricultural sector may face the risk of breaking down in the near future from a lack of energy. However, reducing the dependence on fossil fuels will gain in importance as efforts continue to reduce GHG emissions and as fuel supplies eventually do dwindle. Fortunately, the development of alternative farming methods and advances in biotechnology and mechanization of production systems will likely achieve this end. An example of fossil fuel use in agriculture is to manufacture inputs such as fertilizers, pesticides, and herbicides and also to power industrial farm machinery.

Apart from the above mentioned factors in regard of energy importance, increasing input costs consisting fuel cost, irrigation cost, application of chemicals and transportation cost has caused high a raise in energy costs.

Although considerable data exist in the literature regarding the energy consumption for energy consumption in button mushroom production system such as kiwi [3],

*Corresponding author.

apple [4,5], grapes [6], sunflower [7,8], oilseed [9], greenhouse vegetable production [10,11], little information is available on the analysis of energy consumption in white button mushroom production.

Efficiency was defined by Sherman [12] as the ability to produce the outputs using a required minimum resource level. In production, efficiency is defined as the ratio of weighted sum of outputs to inputs or as the actual output to the optimal output ratio. The weights for inputs and outputs are estimated to the best advantage for each unit so as to maximize relative efficiency. In order to measure the optimal input or output amounts, it is necessary to first specify the production frontier [13].

Iranian agricultural sector needs taking a serious look at the ways of energy expenditure and improving efficiency on farms to reduce their ongoing costs and accordingly improve their bottom line. A wrong belief among farmers causing the inefficient use of energy sources is the excess use of resources to get higher productivity, particularly when they are priced low, free or accessible in plenty [14].

One of the proposed and well-established ways to evaluate the relative technical efficiency (TE) of entities by some mathematical programming models is Data Envelopment Analysis (DEA) [15]. By applying this approach firstly the efficient and inefficient decision making units (DMUs) would be detected and then the optimized amount of inputs use can be decided [16].

In recent years, DEA has become renowned in agricultural researches. Recently, Sefeedpari [16] used DEA to assess the technical efficiency in industrial dairy farms of Tehran province in Iran. In this study the efficiency of farmers based on the constant and variable returns to scale models were found to be 0.88 and 0.93, respectively. It was also concluded that DEA was a useful tool to improve the productive efficiency of farms. In another study [17], DEA was applied to investigate the efficiency of individual farmers and identify the efficient units in citrus production in Spain. Mousavi-Avval et al. [18] optimized energy use and energy costs for apple production using data envelopment analysis. They reported that 54% of farmers are technically efficient. The technical efficiency score was calculated as 0.78. Kiwi production was investigated by Mohammadi et al. [19] from the view of energy efficiency. The technical, pure technical and scale efficiencies of farmers were calculated as 0.942, 0.993 and 0.948, respectively. In spite of the careful literature review on DEA application in agricultural production systems, no study was found with this criterion in edible mushroom production.

Considering the importance of efficient use of energy in Iran, the main objective of the present study was to investigate the energy efficient farms during operations of white button mushroom production in Alburz province of Iran. To achieve this, as the first step, energy use pattern for button mushroom was specified. Also, it ranks efficient and inefficient farmers and sketches the optimum footprints of input use and its potential to energy saving.

2. Material and Methods

2.1. Case Study Region Selection and Data Collection

This study was conducted in Alburz province of Iran within 35°31' and 36°32' north latitude and 50°18' and 51°18' east longtitude with total area of 17,953 km^2 (1.09% of total country area) [20]. This province is situated 1300 meter above sea levels. Karaj is the central city of this province. Constitution of the farms of agricultural products in this city has caused the upgradability of this sector in recent years. The adjacency to markets of Tehran province and its geographical location can be enumerated as the main reasons of high demand for agricultural and livestock products in Alburz province [21].

Required data used in this study were obtained from all button mushroom producers in Alburz province, Iran. The survey to collect quantitative information was conducted on energy inputs used for the production of white button mushroom in the production period of October 2011-December 2011. In fact, this period is a representative of a production period of button mushroom which takes about two months and the collected data were then converted and so they can be attributed to the average input use in a whole year. To achieve this, a comprehensive and definite questionnaire was provided and completed by farmers. In order to use the needed software programs (DEA Solver), we were supposed to collect sufficient data in accordance to number of input and output parameters (at least 20 DMUs). Hence, questionnaires were sent to the whole white button mushroom production units (26 units).

2.2. Energy Balance Analysis Method

Data collection included the quantity of various inputs use in the form of chemicals, chemical fertilizers, farmyard manure (FYM), diesel fuel, natural gas, electricity, water for irrigation, human labor or machinery and equipment. A standard procedure was used to convert each agricultural input and output into energy equivalents. The energy equivalent may thus be defined as the energy input taking into account all forms of energy in agricultural production. The energy equivalents were computed for all inputs and outputs using the conversion factors (obtained from previous studies) indicated in **Table 1**. Accordingly, the energy equivalents were calculated by multiplying the quantity of the inputs used and output

Energy Efficiency Analysis of White Button Mushroom Producers in Alburz Province of Iran: A Data Envelopment Analysis Approach

59

Table 1. Energy equivalent coefficients of inputs and output in white button mushroom production in Alburz, Iran.

Item	Unit	Energy equivalent (MJ·unit^{-1})	References
Inputs			[22]
1. Human labour	h	2.2	
2. Machinery	kg		[23]
Tractor		93.61	[23]
Other machinery		62.7	[24]
3. Diesel fuel	L	56.3	[25]
4. Natural gas	m^3	49.5	
5. Chemicals	kg		[26]
Herbicides		238	[26]
Insecticides		101.2	[26]
Fungicides		216	
6. Fertilizers	kg		[26]
Nitrogen		66.14	[26]
Farm yard manure		0.3	[26]
Compost (kg)	kg	8.034	Calculated
7. Water for irrigation	m^3	1.02	[26]
8. Electricity	kWh	12	[25]
9. Straw	kg	12.5	[24]
Outputs			
White button mushroom	kg	27	[27]

harvested per hectare with their conversion factors.

It should be noted here that energy equivalent of human labor varies considerably, depending on the approach chosen; it must be adapted to the actual living conditions in the target region [28]. In this study the energy coefficient of 1.96 MJ·h^{-1} was applied. It means only the muscle power used in different field operations of crop production. Also, in order to make an analysis of the embodied energy in the farm machinery, it was assumed that the embodied energy of tractors and agricultural machinery be depreciated during their economical life time [29]; Also, the embodied energy in machinery was calculated by multiplying the depreciated weights of machinery (kg·ha^{-1}) with their energy equivalents (MJ·kg^{-1}) using the following Equation (1) [30]:

$$ME = \frac{G \times M_p \times t}{T} \qquad (1)$$

where ME is the machinery energy per unit area (MJ·ha^{-1}); G is the machine mass (kg); M_p the production energy of machine (MJ·kg^{-1}), t is the time that machine used per unit area (h·ha^{-1}) and T is the economic life time of machine (h).

For compost energy, we did not find any energy equivalent coefficient indicating energy content coefficient of compost fertilizer as a significant input in white button mushroom production. For this reason, we in study were supposed to measure this coefficient. For this purpose, the compost production process was thoroughly investigated and its energy use was determined. The energy inputs to compost production were human labour, diesel fuel, electricity, stationary equipment, wheat straw, animal manure, urea fertilizer and water. Here, also, each input use amount has been multiplied by their energy equivalents and finally the energy consumption for each kilogram of compost was found.

2.3. Data Envelopment Analysis Technique

In this study, in order to evaluate the technical, pure technical and scale efficiencies of individual farmers, a nonparametric method of DEA was employed. It is believed that farmers use similar inputs and produce the same product (white button mushroom) and operate in a relatively homogeneous region (e.g., topography, soil type, climatic conditions, etc.). In the above methodology, the energy consumed from different energy sources including: human labor, machinery, diesel fuel, fertilizer, chemicals, water for irrigation, electricity and natural gas were defined as input variables; while, the white button mushroom yield was the single output parameter; also each farmer called a DMU. Data analysis was done by DEA Solver software.

In DEA, two models would be applied in this study including CCR and BCC models. Charnes *et al.* [31] developed the CCR DEA model by and assumes constant returns to scale. On the other hand, the BCC model developed by Banker *et al.* [32] and assumes variable returns to scale conditions. Also applying each model would lead to a specific efficiency value. Efficiency score obtained from CCR model is called technical efficiency and the efficiency value calculated by BCC model is pure technical efficiency. Scale efficiency gives quantitative information of scale characteristics; it is the potential productivity gain from achieving optimal size of a DMU. In Equation (2), scale efficiency is calculated by the relation between technical and pure technical efficiencies as below [33]:

$$\text{Pure Technical Efficiency} = \frac{\text{Technical Efficiency}}{\text{Pure Technical Efficiency}} \qquad (2)$$

In DEA, there are two different analyses for converting an inefficient unit to efficient one. An inefficient DMU can be made efficient either by minimizing the input levels while maintaining the same level of outputs (input oriented), or, symmetrically, by increasing the output levels while holding the inputs constant (output oriented). In fact, the input-oriented is commonly utilized in DEA applications because efficiency profitability depends on the efficiency of operations. Besides that, a farmer is able to take the control of inputs use more easily than output level [15,34].

Let the DMU_j to be evaluated on any trial be designated as DMU_o ($o = 1, 2, \cdots, n$). To measure the relative efficiency of a DMU based on a series of n DMU_s, the model is structured as a fractional programming problem as follows [35]:

$$\text{Maximize} \, \theta = \frac{\sum_{r=1}^{s} u_r y_{r0}}{\sum_{i=1}^{m} v_i x_{i0}}$$

$$\begin{aligned} s.t. & \\ & \frac{\sum_{r=1}^{s} u_r y_{rj}}{\sum_{i=1}^{m} v_i x_{ij}} \leq 1, \ j = 1, 2, \cdots, n \\ & u_r \geq 0, \ v_i \geq 0 \end{aligned} \quad \text{Model (1)}$$

Using a linear programming (LP) problem, Model (1) can be equivalently written as follows [35]:

$$\text{Maximize} \, \theta = \sum_{r=1}^{n} u_r y_{ri}$$

$$\begin{aligned} s.t. & \\ & \sum_{s=1}^{m} v_s x_{si} = 1, i = 1, 2, \cdots, k \\ & \sum_{i=1}^{n} u_r y_{ri} - \sum_{s=1}^{m} v_s x_{si} = 0 \\ & u_r \geq 0, \ r = 1, 2, \cdots, n \\ & v_s \geq 0, \ s = 1, 2, \cdots, m \end{aligned} \quad \text{Model (2)}$$

where θ is the technical efficiency. Dual linear programming (DLP) problem is simpler to solve than Model (2) due to fewer constraints. Mathematically, the DLP can be written in vector matrix notation [35]:

$$\text{Maximize} \, \theta$$

$$\begin{aligned} s.t. & \\ & Y\lambda \geq y_0 \\ & X\lambda - \theta x_0 \leq 0 \\ & \lambda \geq 0 \end{aligned} \quad \text{Model (3)}$$

where y_o is the $s \times 1$ vector of the value of original outputs produced and x_o is the $m \times 1$ vector of the value of original inputs used by the oth DMU. Y is the $s \times n$ matrix of outputs and X is the $m \times n$ matrix of inputs of all n units included in the sample. λ is a $n \times 1$ vector of weights and θ is a scalar with boundaries of one and zero

which determines the technical efficiency score of each DMU. Model (3) is known as the input-oriented CCR model. It assumes constant returns to scale (CRS), inferring that an increase in inputs would lead in a proportionate increase in outputs.

The results of DEA models divide the DMUs into two sets of efficient and inefficient units; the inefficient units can be ranked according to their efficiency scores. Since DEA lacks the capacity to discriminate among efficient units; a number of methods are in use to enhance the discriminating capacity of DEA [36]. In this study, the benchmarking method was applied to overcome this problem. In this method, an efficient unit which is chosen as a useful target for many inefficient DMUs, and so appears frequently in the referent set, is highly ranked.

For the purpose of estimating the optimized energy input level of each input for inefficient units, λ coefficient was applied. The optimal input use calculation can be defined as follows (Equation (3)):

$$\theta^* x_0 - s^- = X_n \lambda_n \qquad (3)$$

where θ^* is technical efficiency score of inefficient units, x_0 energy input of nth referent DMU and s^- is input slack of inefficient units compared to efficient DMUs [37].

In this study for data analysis, the Microsoft Excel spreadsheet and the DEA Solver programs were employed.

3. Results and Discussions

3.1. Energy Balance Analysis for White Button Mushroom Production

Prior to energy use pattern analysis of white button mushroom production, energy use amount in compost production (among the most important inputs of white button mushroom production) was estimated. **Table 2** shows the inputs use and their energy use equivalent values.

As the result of energy use pattern analysis, the amounts of energy inputs use, mushroom yield and their energy equivalents were determined. These results are presented in **Table 3**.

As it is obvious from **Table 3**, total energy consumption in white button mushroom production is about 133 MJ/kg of product. Based on the results, fossil fuels, compost fertilizer and electricity are mostly contributed inputs to energy consumption in which fossil fuels (diesel fuel and natural gas) had the first rank (40.43%). Fossil fuel was normally used for heating. It should be noted here that the fossil fuel energy input to compost production process has been focused in our formerly calculations. Moreover, energy input of mushroom seed was not considerable for energy analysis. As a matter of fact, ten kilograms of seed is normally required for one ton of compost. Due to the short planting period of mushroom

Table 2. Mean input use and energy equivalent values in compost production process in Alburz, Iran.

Inputs	Mean input use (unit·ton⁻¹)	Energy equivalent (MJ·ton⁻¹)	Share (%)
Human labour (h)	10.6	23.32	0.29
Diesel fuel (L)	16.6	934.75	11.64
Stationery equipment (kg)	10.41	72.78	0.91
Electricity (kWh)	833.3	2999.88	37.35
Straw and stubble (kg)	310	3875	48.25
Poultry manure (kg)	320	96	1.19
Urea fertilizer	0.003	0.22	0.002
Water (L)	5	29.4	0.37
Total input energy (MJ)		8034.44	

Table 3. Mean input use and energy equivalent values in white button mushroom production process in Alburz, Iran.

Items	Mean input use (unit·ton⁻¹)	Energy equivalent (MJ·ton⁻¹)	Share (%)
Inputs			
Compost (kg)	4.86	40.57	30.45
Fuel		53.87	40.43
Diesel fuel (L)	0.95	45.5	
Natural gas (m³)	1.1	8.37	
Chemicals	0.0085	1.05	0.78
Human labour (h)	0.21	0.47	0.47
Electricity (kWh)	3.05	36.55	27.42
Equipment (kg)	0.049	0.45	0.37
Water for irrigation (L)	49.23	0.29	0.22
Total input energy		133.25	
Output			
White button mushroom (kg)	1	27	

and the need for keeping the temperature at 25℃, too, there is a high need to fossil fuel use in mushroom production farms for heating purposes. Moreover, since mushrooms do not require oxygen in their first month, air ventilation is not essential unless carbon dioxide content rise from its usual and standard rate. Hence, isolation of planting places would eventually lead in energy saving. Compost was reported as the second high contributing

input to energy consumption amount in this study. The amount of compost use was 4.86 kg per ton of yield (white button mushroom) with energy input of 40.57 MJ·ton⁻¹ (30.45%).

In the study conducted by Ghojebeig [38] in greenhouse products of Tehran province, total energy input use was calculated to be 1.57 MJ for each kilogram of cucumber. They also reported fuel (for heating the greenhouses) and animal manure as the high energy consuming inputs in cucumber production. Moreover, Banaeian *et al.* [39] investigated energy use pattern in 25 greenhouses of strawberry production in Tehran province, Iran. In our study, results indicated that a total energy of 121891.33 MJ·ha⁻¹ (about 2 MJ·ton⁻¹) was consumed in greenhouse units. Diesel fuel (78%), chemical fertilizers (10%) and electricity (4.5%) were the highest energy consuming inputs. On the contrary, water for irrigation contribution was relatively low (0.47%). This is mainly due to lower water need of mushroom in contrast with other products.

3.2. Data Envelopment Analysis Findings

3.2.1. Technical Efficiency Analysis

Data obtained from questionnaire approach performed in white button mushroom farms were analyzed for the purpose of measuring farmers' technical efficiency in light of energy use by applying data envelopment analysis approach. In the following sections, the efficient and inefficient farmers, their efficiency scores and the required changes for the better resources use management strategies are given. It should also be noted here that both explained DEA models (BCC and CCR models) have been utilized in the present study.

Figure 1 illustrates technical efficiency scores of applied data using CCR and BCC models. The results revealed that, from the total of 26 farmers considered for the analysis, 14 farmers (53.8%) had the pure technical efficiency score of 1. Moreover, from the technically efficient farmers 12 farmers (27.66%) had the technical efficiency score of 1; showing that they were globally efficient and were operating at the most productive scale size of production [40]. In addition, 9 and 7 farmers had their technical efficiency and pure technical efficiency score in the 0.9-1 range, respectively. Both DEA models, reported 5 farmers in the efficiency range of 0.8 - 0.9. It can be derived from **Figure 1** that the BCC model reduces the number of feasible units compared to the CCR model and yields a comparatively higher number of efficient DMUs.

Table 4 specifies the mean value of the three efficiency types (technical efficiency (TE), pure technical efficiency (PTE) and scale efficiency (SE)) for inefficient farmers. The results showed that these average values

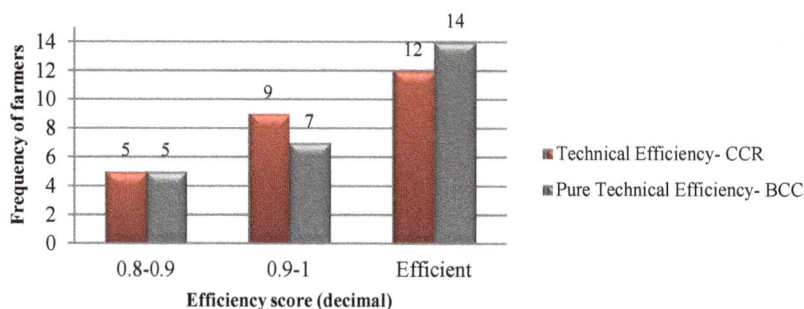

Figure 1. Frequency distribution of farmers from CCR and BCC models in white button mushroom production.

Table 4. Mean value of efficiency types for white button mushroom farmers.

Efficiency type	Mean	SD*
Technical efficiency	0.955	0.0571
Pure technical efficiency	0.956	0.0573
Scale efficiency	0.999	0.0017

*SD: Standard Deviation.

were 0.955, 0.956 and 0.999 respectively. These inefficient DMUs were in the efficiency range of 0.8 - 0.99. PTE had the highest variation indicating the farmers were not fully aware of the right optimal quantity and production techniques or did not applied them at the proper time.

Omid *et al.* [37] applied the nonparametric approach of DEA to determine the technical, pure technical and scale efficiencies of farmers in some selected greenhouses production in Iran. They considered 18 greenhouse units and 6 of them had the efficiency score of unity. They reported that the TE of the inefficient DMUs, on average, was calculated as 91.5%. In another study [19], the technical, pure technical and scale efficiency of kiwi fruit farmers were calculated as 0.942, 0.993 and 0.948, respectively.

3.2.2. Prioritization of Efficient Units
In this study for ranking the efficient farmers, the number of times they appear in a referent set based on CCR model was counted [36]. The efficient DMUs can be selected by inefficient DMUs as best practice DMUs, making them a composite DMU instead of using a single DMU as a benchmark. The results revealed that farmers No. 13, 9 and 22 with the average cross efficiency scores of 1 had the highest appearance times in referent set (**Table 5**). Therefore, after this step, these farms can be used as terms of benchmarking and establishing the best practice management.

3.2.3. Input Slacks Analysis
We know that when the pure TE score of a producer is

less than one, at present, he is using more energy than required from the different sources. Therefore, it is desired to suggest realistic levels of energy to be used from each source for every inefficient grower in order to avoid energy wastes without reducing the yield level (input-oriented). This can be done by using the value of slacks. **Table 6** demonstrates the results of applying BCC model (input-oriented) for estimating the overuses and the slacks for the inputs amount. The slack values indicate that apart from reducing inputs by the amount of $(1-\theta)$, the inefficient DMUs have to reduce their inputs by the amounts indicated by the respective slacks in order to become allocatively efficient. The sources of allocative inefficiency at the white button mushroom production units were identified as the overuse of electricity and diesel fuel. Equipment and water for irrigation sources had the right proportions of input use by all the DMUs (zero slacks). This implies the fact that white button mushroom producers have put emphasis on optimal use of water. Optimized water use is generally due to the specific water requirement amount of mushroom; hence, farmers do their best for optimal consumption of water.

Table 5. Ranking 10 superior efficient farmers in white button mushroom production.

Rank	Farmer No.	Frequency in referent set
1	13	10
2	9	6
3	22	6
4	21	5
5	3	4
6	15	4
7	17	3
8	23	3
9	14	2
10	19	2

Energy Efficiency Analysis of White Button Mushroom Producers in Alburz Province of Iran: A Data Envelopment Analysis Approach

63

Table 6. Input slacks in BCC model.

Units	Efficiency score	Compost	Fuel	Chemicals	Electricity	Equipment and water for irrigation	Human labour	Output
1	1	0	0	0	0	0	0	0
2	0.94	0	0	0.37	0	0	0.05	0
3	1	0	0	0	0	0	0	0
4	0.9	0	0	0	0	0.36	0.34	0
5	0.89	0.50	0	0	0	0	0	0
6	0.91	0	0	0.50	3.08	0	0.08	0
7	0.82	0	0	0.80	4.20	0	0	0
8	1	0	0	0	0	0	0	0
9	1	0	0	0	0	0	0	0
10	0.91	0	1.12	0	3.10	0.15	0.01	0
11	1	0	0	0	0	0	0	0
12	0.93	0	34.55	1.98	29.85	0.04	0	0
13	1	0	0	0	0	0	0	0
14	1	0	0	0	0	0	0	0
15	1	0	0	0	0	0	0	0
16	1	0	0	0	0	0	0	0
17	1	0	0	0	0	0	0	0
18	0.83	0	0	1.29	3.29	0.30	0	0
19	1	0	0	0	0	0	0	0
20	0.98	0	8.24	0	11.84	0.17	0	0
21	1	0	0	0	0	0	0	0
22	1	0	0	0	0	0	0	0
23	1	0	0	0	0	0	0	0
24	0.89	0	0	0.89	0.47	0.08	0	0
25	0.98	0	0	0.08	6.64	0	0	0
26	0.87	0	4.97	0	3.46	0.17	0.05	0

On the basis of given results in **Table 6**, a single unit found with overuse of compost fertilizer. The high buildings and planting room with more stories was the potential reason for inefficient performance of compost resource use in such units. It is worth mentioning that the input orientation of analysis caused the output slacks of all units to be zero in the last column of **Table 6**.

3.2.4. Setting Realistic Input Levels for Inefficient Producers

In order to calculate the total energy input saving, the average pure technical efficiency of inefficient farmers (0.955) was multiplied by the present use of energy inputs (133 MJ/kg) (Equation (3)). Namely, the total energy input should be reduced about 16%. **Table 7** shows the optimum energy requirement for white button mushroom production, based on the results of BCC model. Also the quantity and percentage of energy saving with respect to present use of energy is illustrated. As can be seen, optimum energy requirement for diesel fuel energy (50.89 MJ·ton^{-1}) was the highest, followed by FYM (37.32 MJ·ton^{-1}), electricity (30.34 MJ·ton^{-1}) energies.

Table 7. Energy optimization from different sources of energy provided following the results of this study.

Input	Present use (MJ·ton^{-1})	Target use (MJ·ton^{-1})	Energy saving (MJ·ton^{-1})	Contribution to energy saving (%)
1. Compost	41.41	37.32	4.08	9.68
2. Fossil fuel	60.85	50.89	9.95	14.5
3. Chemicals	1.18	0.57	0.62	41.92
4. Electricity	39.71	30.34	9.37	21.8
5. Water and equipment	0.78	0.60	0.18	21.21
6. Human labour	0.51	0.41	0.09	15.58
Total	144.47	120.13	24.3	15.63

The total optimum energy requirement for mushroom production was calculated to be 120.15 MJ·ton^{-1}. In the last column of **Table 7** the contribution of each energy input in energy saving is presented. It is evident that 41.92% of chemicals energy, 21.8% of electricity energy and 21.21% of water for irrigation and equipment energy, which had the highest inefficiencies, could be saved. The high percentage of chemicals energy saving resulted from the low efficiency of chemical fertilizer while it is more than plant needs. There was a common belief between the farmers that increased input use especially chemical fertilizers and chemicals will increase the yield. Moreover, energy usage from human labour, machinery, fossil fuels and compost inputs could be saved by 15.58%, 14.5%, 9.68%, respectively. The results also showed that total energy input for mushroom production could be saved by 15.63% (24.3 MJ·ton^{-1}), if the recommendations of this study are followed.

Figure 2 shows the shares of the various sources from total input energy saving. It is evident that the highest contribution to the total energy saving was 42% for fossil fuel energy. This shows that in the case of fossil fuel energy, there is a greater scope to increase the efficiency of energy consumption. It was due to the relatively high contribution of fossil fuel energy from total energy consumption in present condition. Apart from fossil fuel energy, it is evident that, the contributions of compost and electricity energy inputs from total energy saving were 31% and 25%, respectively. Also, energy saving from human labour had the lowest share from total energy saving.

4. Conclusions

To sum it up, the results of DEA application suggested that there was a great potential for improving energy and economical efficiencies of edible mushroom producers in Alburz, Iran. Based on the performed analysis in this study the following results were drawn:

- Fossil fuels (53.87 MJ·ton^{-1}), compost fertilizer

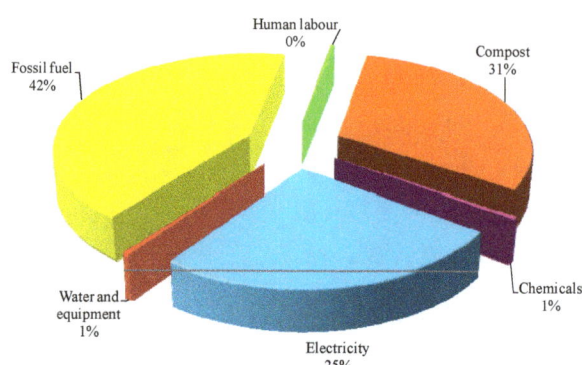

Figure 2. Distribution of energy inputs in energy saving for white button mushroom production.

(40.57 MJ·ton^{-1}) and electricity (40.57 MJ·ton^{-1}) are mostly contributed inputs to energy consumption in which fossil fuels (diesel fuel and natural gas) had the first rank (40.43%).

- From the total of 26 farmers, 14 farmers had the pure technical efficiency score (BCC model) of 1. Moreover, from the technically efficient farmers (CCR model) 12 farmers had the technical efficiency score of unity.
- The results showed that the average values of TE, PTE and SE were 0.955, 0.956 and 0.999, respectively.
- Efficient farmers No. 13, 9 and 22 with had the highest appearance times in referent set.
- The sources of allocative inefficiency (with overuse) at the white button mushroom production units were identified as electricity and diesel fuel.
- The total optimum energy requirement for mushroom production was calculated to be 120.15 MJ·ton^{-1} in which diesel fuel energy (50.89 MJ·ton^{-1}), FYM (37.32 MJ·ton^{-1}) and electricity (30.34 MJ·ton^{-1}) energies were in the highest degree.

As a result of this study, effectiveness of use of all inputs can be achieved in an informed and efficient production system. In this regard, extension programs toward the development of such systems should be taken into consideration. Here, the extension officers of agricultural institutions would be striking for managing and establishing the energy and economically efficient and environmentally sensitive edible mushroom production systems in this and other similar socio-economic regions. Moreover, providing farmers with suitable compost fertilizer can lead to highly efficient producers.

REFERENCES

[1] S. G. Walde, V. Velu, T. Jyothirmayi and R. G. Math, "Effects of Pretreatments and Drying Methods on Dehydration of Mushroom," *Journal of Food Engineering*, Vol. 74, No. 1, 2006, pp. 108-115.

[2] M. Goltape and I. Pourjam, "Principles of Button Mushrooms," Tarbiat Modares Univesity Publication, Tehran, 2001. (in Persian)

[3] A. Mohammadi, S. Rafiee and S. Mohtasebi and H. Rafiee, "Energy Inputs—Yield Relationship and Cost Analysis of Kiwifruit Production in Iran," *Renewable Energy*, Vol. 35 No. 5, 2010, pp. 1071-1075.

[4] R. Fadavi, A. Keyhani and S. S. Mohtasebi, "Estimation of a Mechanization Index in Apple Orchard in Iran," *Journal of Agricultural Science*, Vol. 2, 2010, pp. 180-185.

[5] S. Rafiee, S. H. Mousavi Avval and A. Mohammadi, "Modeling and Sensitivity Analysis of Energy Inputs for Apple Production in Iran," *Energy*, Vol. 35, No. 8, 2010, pp. 3301-3306.

[6] B. Ozkan, C. Fert and C. F. Karadeniz, "Energy and Cost Analysis for Greenhouse and Open-Field Grape Production Grape Production," Vol. 32, 2007, pp. 1500-1504.

[7] M. Uzunoz, Y. Akcay and K. Esengun, "Energy Input-Output Analysis of Sunflower Seed (*Helianthus annuus* L.) Oil in Turkey," *Energy Sources Part B-Economics Planning and Policy*, Vol. 3, No. 3, 2008, pp. 215-223.

[8] L. Kallivroussis, A. Natsis and G. Papadakis, "The Energy Balance of Sunflower Production for Biodiesel in Greece," *Biosytems Engineering*, Vol. 81, No. 3, 2002, pp. 347-354.

[9] S. H. Mousavi Avval, S. Rafiee, A. Jafari and A. Mohammadi, "Investigating the Energy Consumption in Different Operations of Oilseed Productions in Iran," *Journal of Agricultural Technology*, Vol. 7, No. 3, 2011, pp. 557-565.

[10] M. Canakci and I. Akinci, "Energy Use Pattern Analyses of Greenhouse Vegetable Production," *Energy*, Vol. 31, No. 8-9, 2006, pp. 1243-1256.

[11] S. A. Hatirli, B. Ozkan and C. Fert, "Energy Inputs and Crop Yield Relationship in Greenhouse Tomato Production," *Renewable Energy*, Vol. 31, No. 4, 2006, pp. 427-438.

[12] H. D. Sherman, "Service Organization Productivity Management," The Society of Management Accountants of Canada, Hamilton, 1988.

[13] K. Mukherjee, "Energy Use Efficiency in the Indian Manufacturing Sector: An Interstate Analysis," *Energy Policy*, Vol. 36, No. 2, 2008, pp. 662-672.

[14] G. Singh, S. Singh and J. Singh, "Optimization of Energy Inputs for Wheat Crop in Punjab," *Energy Conversion and Management*, Vol. 45, No. 3, 2004, pp. 453-465.

[15] P. Sefeedpari, "Assessment and Optimization of Energy Consumption in Dairy Farm: Energy Efficiency," *Iranica Journal of Energy & Environment*, Vol. 3, No. 3, 2012, pp. 213-224.

[16] P. Sefeedpari, S. Rafiee and A. Akram, "Selecting Energy Efficient Poultry Egg Producers: A Fuzzy Data Envelopment Analysis Approach," *International Journal of Applied Operational Research*, Vol. 2, No. 2, 2012, pp. 77-88.

[17] E. Reig-Martínez and A. J. Picazo-Tadeo, "Analysing Farming Systems with Data Envelopment Analysis: Citrus Farming in Spain," *Agricultural Systems*, Vol. 82, No. 1, 2004, pp. 17-30.

[18] S. H. Mousavi-Avval, S. Rafiee and A. Mohammadi, "Optimization of Energy Consumption and Input Costs for Apple Production In Iran Using Data Envelopment Analysis," *Energy*, Vol. 36, No. 2, 2011, pp. 909-916.

[19] A. Mohammadi, S. Rafiee, S. S. Mohtasebi, S. H. Mousavi Avval and H. Rafiee, "Energy Efficiency Improvement and Input Cost Saving in Kiwifruit Production Using Data Envelopment Analysis Approach," *Renewable Energy*, Vol. 36, No. 9, 2011, pp. 2573-2579.

[20] H. Reyhani Farashah, "Study on Energy Efficiency and Assessment of Economical Indices of the Button Mushroom Production Units in The Provinces of Alborz Using Data Envelopment Analysis (DEA) Approach," M.S. Thesis, University of Tehran, Tehran, 2012. (in Persian)

[21] Jamshidi, "Sustainability Analysis of Greenhouse Systems in Alburz Province of Iran," M.S. Thesis, University of Tehran, Tehran, 2011. (in Persian)

[22] D. Pimentel and M. Pimentel, "Food, Energy and Society," Resource and Environmental Science Series, Edward Arnold Publ., London, 1979.

[23] M. Canakci, M. Topakci, I. Akinci and A. Ozmerzi, "Energy Use Pattern of Some Field Crops and Vegetable Production: Case Study for Antalya Region, Turkey," *Energy Conversion and Management*, Vol. 46, No. 4, 2005, pp. 655-666.

[24] S. Singh and J. P. Mital, "Energy in Production Agriculture," Mittal Pub, New Delhi, 1992.

[25] Kitani, "CIGR Handbook of Agricultural Engineering," Vol. V, Energy and Biomass Engineering, ASAE Publication, ST Joseph, 1999.

[26] S. Rafiee, S. H. Mousavi Avval and A. Mohammadi, "Modeling and Sensitivity Analysis of Energy Inputs for

Apple Production in Iran," *Energy*, Vol. 35, No. 8, 2010, pp. 3301-3306.

[27] M. Ghafarpour, A. Houshyarrad, H. Kiyanfar and B. Bani Eqbal, "Nutrients Album," Nutrition World Pub., Institute of Nutritional and Food Science, 2007.

[28] S. R. Moore, "Energy Efficiency in Small-Scale Biointensive Organic Onion Production in Pennsylvania, USA," *Renewable Agriculture and Food Systems*, Vol. 25, No. 3, 2010, pp. 181-188.

[29] B. Tabar, A. Keyhani and S. Rafiee, "Energy Balance in Iran's Agronomy (1990-2006)," *Renewable and Sustainable Energy Reviews*, Vol. 14, No. 2, 2010, pp. 849-855.

[30] I. Gezer, M. Acaroglu and H. Haciseferogullari, "Use of Energy and Labor in Apricot Agriculture in Turkey," *Biomass & Bioenergy*, Vol. 24, No. 3, 2003, pp. 215-219.

[31] A. Charnes, W. W. Cooper and E. Rhodes, "Measuring the Efficiency of Decision Making Units," *European Journal of Operational Research*, Vol. 2, No. 6, 1978, pp. 429-444.

[32] R. D. Banker, A. Charnes and W. W. Cooper, "Some Models for Estimating Technical Scale Inefficiencies in Data Envelopment Analysis," *Management Science*, Vol. 30, No. 9, 1984, pp. 107-192.

[33] S. M. Nassiri and S. Singh, "Study on Energy Use Efficiency for Paddy Crop Using Data Envelopment Analysis (DEA) Technique," *Applied Energy*, Vol. 86, No. 7-8, 2009, pp. 1320-1325.

[34] S. H. Mousavi-Avval, S. Rafiee, A. Jafari and A. Mohammadi, "Improving Energy Use Efficiency of Canola Production Using Data Envelopment Analysis (DEA) Approach," *Energy*, Vol. 36, No. 5, 2011, pp. 2765-2772.

[35] L. M. Cooper, L. M. Seiford and K. Tone, "Introduction to Data Envelopment Analysis and Its Uses," Springer, New York, 2006.

[36] N. Adler, L. Friedman and Z. Sinuany-Stern, "Review of Ranking Methods in the Data Envelopment Analysis Context," *European Journal of Operational Research*, Vol. 140, No. 2, 2002, pp. 249-265.

[37] M. Omid, F. Ghojabeige, M. Delshad and H. Ahmadi, "Energy Use Pattern and Benchmarking of Selected Greenhouses in Iran Using Data Envelopment Analysis," *Energy Conversion and Management*, Vol. 52, No. 1, 2011, pp. 153-162.

[38] F. Ghojebeig, "A Decision Support System for Optimizing Energy Consumption in Vegetable Production Greenhouses," M.S. Thesis, University of Tehran, Karaj, 2010.

[39] N. Banaeian, M. Omid and H. Ahmadi, "Energy and Economic Analysis of Greenhouse Strawberry Production in Tehran Province of Iran," *Energy Conversion and Management*, Vol. 52, No. 2, 2011, pp. 1020-1025.

[40] S. H. Mousavi-Avval, S. Rafiee, A. Jafari and A. Mohammadi, "Optimization of Energy Consumption for Soybean Production Using Data Envelopment Analysis (DEA) Approach," *Applied Energy*, Vol. 88, No. 11, 2011, pp. 3765-3772.

11

Energy Analyses of Thermoelectric Renewable Energy Sources

Jarman T. Jarman[1], Essam E. Khalil[2], Elsayed Khalaf[2]
[1]Ministry of Interior, Riyadh, KSA
[2]Faculty of Engineering, Cairo University, Giza, Egypt

ABSTRACT

The recent energy crisis and environmental burden are becoming increasingly urgent and drawing enormous attention to solar-energy utilization. Direct solar thermal power generation technologies, such as, thermoelectric, thermionic, magneto hydrodynamic, and alkali-metal thermoelectric methods, are among the most attractive ways to provide electric energy from solar heat. Direct solar thermal power generation has been an attractive electricity generation technology using a concentrator to gather solar radiation on a heat collector and then directly converting heat to electricity through a thermal electric conversion element. Compared with the traditional indirect solar thermal power technology utilizing a steam-turbine generator, the direct conversion technology can realize the thermal to electricity conversion without the conventional intermediate mechanical conversion process. The power system is, thus, easy to extend, stable to operate, reliable, and silent, making the method especially suitable for some small-scale distributed energy supply areas. Also, at some occasions that have high requirements on system stability, long service life, and noiselessness demand, such as military and deep-space exploration areas, direct solar thermal power generation has very attractive merit in practice. At present, the realistic conversion efficiency of direct solar thermal power technology is still not very high, mainly due to material restriction and inconvenient design. However, from the energy conversion aspect, there is no conventional intermediate mechanical conversion process in direct thermal power conversion, which therefore guarantees the enormous potential of thermal power efficiency when compared with traditional indirect solar thermal power technology [1].

Keywords: Solar Energy; Two Stage Concentrator; Mathematical Modeling; Thermal Analysis; Thermoelectric Power Generation

1. Introduction

Thermoelectric power generation is one of the current interests in clean energy research in view of direct solar power generation. Thermoelectric power generation becomes an attractive application. Recent research analyses were proposed in the open literature to cover the various aspects of energy generation [2-23]. Efforts have been devoted to investigate the thermoelectric power conversion theory and practical applications. These efforts have been under way from the following four aspects [24]:

1) Maximizing the temperature difference between two sides of thermoelectric devices by increasing the heat flow through thermoelectric devices with methods such as raising the solar concentration ratio.

2) Enhancing material thermoelectric characteristic. Seeking more suitable thermoelectric materials, such as nanometer materials, is the most useful and effective way to improve thermoelectric conversion efficiency at present.

3) Implementing effective heat dissipation on the cold side, so that thermoelectric materials can work in the most suitable temperature range.

4) Carrying out design optimization and computer simulation to optimize the structure of thermoelectric elements and improve the packaging technology of devices.

This review is intended to present an account of the recent advances in developing the thermoelectric technologies specially, for direct solar thermal power generation. Both the fundamental issues and latest application research are illustrated and critical issues are discussed.

2. Thermoelectric Applications

Thermoelectric applications are broad. Thermoelectric

materials had their first decisive long-term test with the start of intensive deep-space research. During the Apollo mission, thermoelectric materials were responsible for the power supply, and currently, radioisotope thermoelectric generators (RTEGs) are the power supplies (350 W) used in deep-space missions beyond Mars. Recently, the Cassini satellite was launched with three RETGs using 238Pu as the thermal energy source and SiGe as the thermoelectric conversion material. Smaller self-powered systems such as thermoelectric-powered radios were first mentioned in Russia around 1920; a thermoelectric climate-control system in a 1954 Chrysler automobile shows the scope of this technology. Currently, millions of thermoelectric climate-controlled seats that serve as both seat coolers and seat warmers are being installed in luxury cars. In addition, millions of thermoelectric coolers are used to provide cold beverages. Even wristwatches marketed by Seiko and Citizen and biothermoelectric pacemakers are being powered by the very small temperature differences within the body or between a body and its surroundings.

Thermoelectric materials were previously used primarily in niche applications, but with the advent of broader automotive applications and the effort to utilize waste-heat-recovery technologies; thermoelectric devices are becoming more prominent. The rising costs of fossil fuels have helped spawn a program between the Energy Efficiency and Renewable Energy office of the US Department of Energy and several automotive manufacturers to incorporate thermoelectric waste-heat-recovery technology in the design of heavy trucks. Indeed, without such systems, more than 60% of the primary energy of fossil fuels is lost worldwide as unusable waste energy; the loss is as high as 70% in some automobiles [25].

This field of thermoelectric also covers forthcoming applications and markets for remote "self-powered" systems for wireless data communications in the microwatt power range, as well as automotive systems and deep-space probes in the intermediate range of hundreds of watts. Researchers hope to produce systems of several kilowatts using waste heat energy from stand-alone woodstoves and also transform the huge amounts of waste energy from industrial furnaces and power plants.

Thermoelectric power generation is one of the current interests in clean energy research. Thermoelectric power generation technology has been widely used for many years power generation, heating and cooling applications. Although recent developments in nanotechnology have helped to improve the efficiency of the thermoelectric generators, they are not yet competitive with other electrical energy generation technologies from the efficiency perspective. Thermoelectric efficiency of these generators has generally been limited to about 5% - 6% [2,21, 24]. However, they are easy to operate, compact, longer-lived, and require low maintenance cost. Also, the thermoelectric devices can utilize the waste heat to generate electricity; thus, the efficiency of the solar thermal power plant can be improved. Thermoelectric devices are friendly to the environment, so they have attracted increasing attention as a green and flexible source of electricity. Research in the area of thermoelectric power generation assists in identifying the best fields for implementation of this technology, and helps in reducing the time between the development of advanced materials and cost-effective thermoelectric power generation. The field of thermoelectricity began in 1822 with the discovery of the thermoelectric effect by Thomas Seebeck [3]. Seebeck found that, **Figure 1(a)**, when the junctions of two dissimilar materials are held at different temperatures (ΔT), a voltage (V) is generated that is proportional to ΔT. The proportionality constant is the Seebeck coefficient or thermo power: $\alpha = -\Delta V/\Delta T$. When the circuit is closed, this couple allows for direct conversion of thermal energy (heat) to electrical energy. The conversion efficiency, η_{TE}, is related to a quantity called the figure of merit, ZT that is determined by three main material parameters: the thermo power α, the electrical resistivity ρ, and the thermal conductivity κ. Heat is carried by both electrons (k_e) and phonons (κ_{ph}), and $k = k_e + k_{ph}$. The quantity ZT itself is defined as

$$ZT = \frac{\alpha^2 \sigma T}{k} = \frac{\alpha^2 \sigma T}{\left(k_e + k_{ph}\right)} \tag{2.1}$$

where σ is the electrical conductivity.

A related effect was discovered a little bit later in 1834 by a French scientist Jean Peltier [4,5]. Peltier found, **Figure 1(b)**, that when he applied a current in two pieces of coupled materials, a temperature gradient can be developed and the rate of heat absorbed or released at the junctions followed a relationship of $Q/dt = \Pi \cdot I$, where Π is Peltier coefficient and I is the applied current. Two decades later, Lord Kelvin connected these two effects

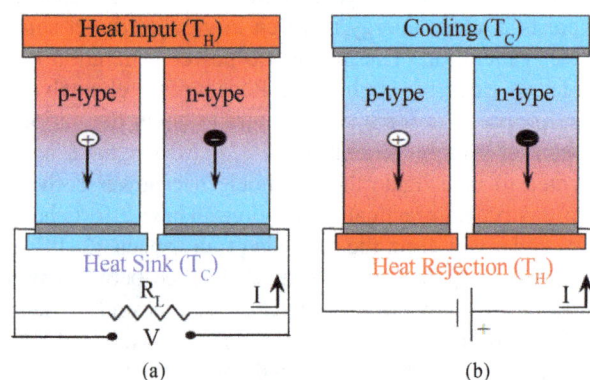

Figure 1. Arrangement of thermoelectric modules for (a) power generation by theSeebeck effect and (b) refrigeration by the Peltier effect.

together by Kelvin relationship: $\Pi = \alpha T$, which was theoretically proved almost 80 years later by Lars Onsager in 1931. In the first 30 years of thermoelectric history, scientists discovered the basic effects and started to understand it macroscopically.

It was not until in 1911, that Altenkirch [3] derived thermoelectric efficiency and provided a microscopic understanding of thermoelectric phenomena. The knowledge of thermoelectricity was mainly developed in the 20 years after his derivation and led to the progress in the discovery of good bulk thermoelectric materials by 1970. The efficiency of thermoelectricity (η_{TE}) is determined by the dimensionless figure of merit (ZT):

$$\eta_{TE} = \eta_C \left(\frac{\sqrt{1+ZT}-1}{\sqrt{1+ZT}+\dfrac{T_C}{T_H}} \right) \qquad (2.2)$$

where η_C is the Carnot efficiency, $\eta_C = (T_H - T_C)/T_H$ and T_H and T_C are the hot and cold temperatures, respectively. Thus, a significant difference in temperature (large ΔT) is also needed to generate sufficient electrical energy. Improvement of thermoelectric generator performance has been the objective of many of investigations. Kassas [6] carried out thermodynamic analysis of a thermoelectric device. He considered the thermoelectric diode and calculated the electronic as well as the Carnot efficiency of the device. He found that increasing emitter temperature increases the Carnot efficiency of the thermoelectric device. He also found that the second law efficiency increases with emitter to collector temperature ratio and reduces with increasing collector temperature as a result of the increase in collector current flow.

Yamashito [7] derived some thermal rate equations by considering the temperature dependencies of the electrical resistivity and thermal conductivity of the thermoelectric materials into the thermal rate equations. He derived the energy conversion efficiencies from both the new and conventional thermal rate equations and discussed the effect of the temperature dependency of the properties. He concluded that the temperature dependence of the electrical resistivity and thermal conductivity significantly influences the efficiency of the thermoelectric generator. Performance of thermoelectric devices was also studied by Yilbas and Sahin [8]. They formulated the optimum values of the slenderness ratio and external load parameter for maximizing the device efficiency. They found that for a fixed thermal conductivity ratio, the external load parameter increases with increasing the slenderness ratio while the electrical conductivity ratio of the p and n pins in the device reduces. Hsiao et al. [9] studied the development of thermoelectric generators for implementation in the internal combustion engine.

They found that installing the thermoelectric generators on the exhaust pipe would yield better results that installing them on the radiator. Champier et al. [10] studied the feasibility of using thermoelectric module in a biomass cook stove for generating electricity to power the fan and give light. They discussed the feasibility of adding commercial thermoelectric modules to the biomass cook stove to come up with the best position of the modules.

Design optimization for the thermoelectric generators has been subject of some previous investigations. Kubo et al. [11] carried out experimental and numerical investigation on the performance of thermoelectric device and studied the relationship between electrical power, conversion efficiency, and incision size and the cold side temperature. They found that the electrical power generated, the conversion efficiency, and the incision size depend on the cold side temperature. Guo et al. [12] studied the design performance of a low-temperature waste heat thermoelectric generator both analytically and experimentally. They found that expanding heat sink surface area in a proper range and enhancing cold-side heat transfer capacity can enhance the performance. Omer and Infield [26], investigated the geometrical optimization of the thermoelectric elements. The model considers the effect of the parameters that contribute to the heat transfer process associated with the thermoelectric devices in power generation mode. Their optimization of the element length was based on maximum power output from the device, rather than efficiency.

3. Solar Thermoelectric Power Generation (STEG)

The growing demand for energy throughout the world has caused great importance to be attached to the exploration of new sources of energy. Among the unconventional sources, solar energy is one of the most promising energy resources on earth and in space, because it is clean and inexhaustible. Applications of solar thermoelectric generator are attractive. The use of the solar thermoelectric generator usually combines a solar thermal collector with a thermoelectric generator, which delivers the electric energy. Tirtt et al. [13] reported that the infrared (IR) region of the solar spectrum can supply the needed hot temperature, T_H. With regard to solar energy conversion, thermoelectric devices will likely utilize the IR spectrum of solar radiation as shown in **Figure 2**.

Telkes [15] gave a brief summary of the work in STEGs before 1954, and reported STEGs constructed of ZnSb and Bi-Sb alloys. The maximum efficiency achieved of a flat-panel STEG was 0.63%. Fewer than 50 times optical concentration, the efficiency reached 3.35%. However, these efficiencies may be inaccurate; Telkes

	Wavelength	Spectrum	%
Photovoltaic	~200-800nm	UV & visible light	58
Thermoelectric	~800-3000nm	IR	42

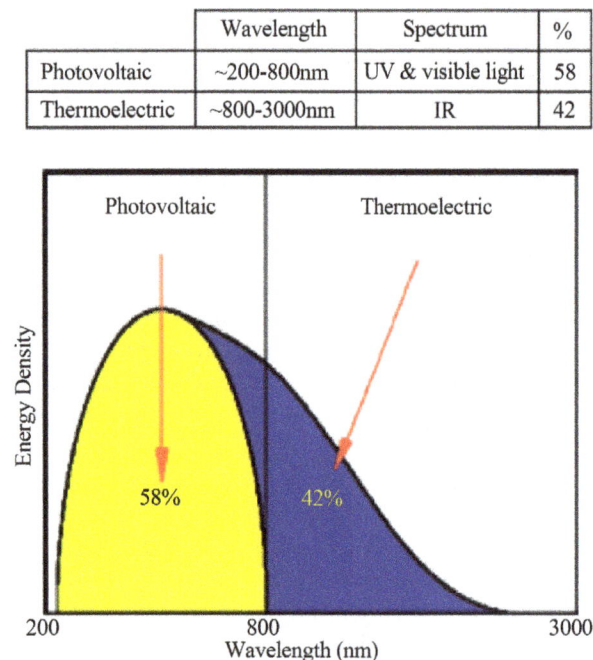

Figure 2. Sun radiates energy as a 6000 K blackbody radiator with part of the energyin the ultraviolet (UV) spectrum and part in the infrared (IR) spectrum [3].

used the average regional solar flux (800 W/m^2) to calculate efficiency, and not the actual incident flux. Telkes included air conduction and convection heat losses in analyzing the STEG cell performance. She treated radiation loss by an effective convection heat transfer coefficient. These analyses are valid only when the temperature rise on the absorber is small. Few papers have been published on STEGs since Telkes' report. Rush [16] investigated a flat-panel STEG configuration for space applications with an estimated maximum efficiency of 0.7%. Goldsmid et al. [17] carried out experiments on STEGs using both flat-panel and optically concentrated configurations. For the flat-panel configuration, the reported generator efficiency is about 1.4% but the system efficiency is only about 0.6% due to radiation and conduction losses, comparable to that of Telke's report. Under 4 times optical concentration, their reported system efficiency decreased to 0.5% due to optical losses. Dent and Cobble [18] used a sun-tracking heliostat directing the solar radiation on a parabolic dish, which focused the sunlight onto thermoelectric generators using PbTe thermoelectric elements. A thermodynamic efficiency of 4% was estimated when the hot side was heated to 510°C, based on the calculated heat input to the thermoelectric generators. Considering the optical concentration losses, the system efficiency of Dent and Cobble's generator at best would be similar to that of Telkes.

A thermodynamic analysis is presented by Amatya et al. [14] for predicting the thermal-to-electrical conver-

sion efficiency for the generator. With solar concentration of 669 suns, a system efficiency of 3% was measured for a commercial Bi_2Te_3 module with output power of 1.8 W. Using novel thermoelectric materials such as n-type ErAs: $(InGaAs)_{1-x}(InAlAs)_x$ and p-type $(AgSbTe)_x(PbSnTe)_{1-x}$, a conversion efficiency of 5.6% can be achieved for a STG at 120× suns. Recently, Li et al. [19] proposed an experimental prototype concentration solar thermoelectric generator with improved total conversion efficiency. They developed a theoretical model of the concentration solar thermoelectric generator system to predict system performance based on the best available properties of different bulk thermoelectric materials found in the literature, including Bi_2Te_3, skutterudite, and silver antimony lead telluride alloys. They showed that the highest possible efficiency of the concentration solar thermoelectric generator can attain 9.8%, 13.5%, and 14.1% for Bi_2Te_3, skutterudite, and silver antimony lead telluride alloys, respectively.

System-level theoretical analysis of solar thermoelectric generation is of great use and can give important theoretical support for developing practical applications. Chen [20] established a thermodynamic model to investigate the optimal performance of a STEG. In the model, four irreversibilities including finite-rate heat transfer between the thermoelectric devices and the external heat reservoirs, heat leak via the thermoelectric devices, Ohmic heat production inside the thermoelectric devices, and heat loss in the solar collector were taken into consideration. Some important theoretical results, such as the efficiency upper bound of solar TEGs, optimal operation temperature of solar collectors, maximum power output, and optimal load matching, were determined. The results are meaningful and could offer theoretical support for experimental investigation of solar-driven TEGs.

Chen [22] investigated the theoretical efficiency of solar thermoelectric generators (STEGs). A model was established including thermal concentration in addition to optical concentration. Based on the model, the maximum efficiency of STEGs is a product of the opto-thermal efficiency and the device efficiency. The device efficiency increases but the opto-thermal efficiency decreases with increasing hot side temperature, leading to an optimal hot-side temperature that maximizes the STEG efficiency. For a given optical concentration ratio, this optimal hot-side temperature depends on the thermoelectric materials' nondimensional figure-or-merit, the optical properties of wavelength-selective surface and the efficiency of the optical system. Operating in an evacuated environment, STEGs can have attractive efficiency with little or no optical concentration working in the low temperature range (150°C - 250°C) for which Bi_2Te_3-based materials are suitable.

Solar thermoelectric generation is a significant option

for space power supply. In the work of Lenoir et al., [23] the skutterudites thermoelectric materials were proposed for space solar TEG. Additionally, two space solar thermoelectric generator (STEG) concepts, namely, the flat plate configuration with identical collector and radiator areas which had a relatively small area and low mass and the STEG configuration with concentrators which had high concentration coefficient, were considered suitable for the space requirement. The theoretical comparison and discussion about the electrical performance as a function of sun-spacecraft distance of these two concepts were performed. The results showed that the output power higher than 400 W could be achieved when the distance was smaller than 0.45 AU. However, if the spacecraft moved to 0.30 AU, the collector temperature would increase up to 1000 K, which might exceed the allowable temperature range of skutterudites thermoelectric materials. Therefore, the sun-spacecraft distance greatly affects power generation and reasonable protection is needed to maintain the temperature within the allowable range.

Omer et al. [25] presented a design procedure and thermal performance analysis of a two stage solar energy concentrator suited to combined heat and thermoelectric power generation. The concentrator is comprised of a primary one axis parabolic trough concentrator and a second stage compound parabolic concentrator mounted at the focus of the primary. The thermoelectric device is attached to the absorber plate at the focus of the secondary. A cooling tube is fitted to the cold side of the thermoelectric device to extract the waste heat and maintain a high temperature gradient across the device to improve conversion efficiency. The key requirements of the concentrator design are to be tolerant of tracking misalignment, maintain temperature gradients to suit thermoelectric generation and minimize heat losses. A design methodology is presented which allows interception of rays within an angular region ($\pm\delta$). The results in a wider receiver for the parabolic trough concentrator would usually be used for a similar concentration ratio. The role of the second stage concentrator in limiting heat losses from the absorber plate is evaluated. Results indicate that in addition to improving the concentration efficiency, the second stage compound parabolic concentrator of the proposed design also inhibits convective air movement and, consequently, improves the overall performance of the solar concentrator.

Solar thermoelectric technology can also be used to improve the indoor environment and save building energy consumption. Maneewan et al. [27] performed a numerical and laboratory-scale investigation on attic heat gain reduction by means of thermoelectric roof solar collector (TERSC). Due to the incident solar radiation, the thermoelectric modules had a temperature difference between its hot and cold sides and generated direct current to drive a ventilating fan for cooling the TE-RSC and achieving better indoor ventilation so as to reduce ceiling heat gain. The analytical result showed that the decrease in the roof heat gain of 25% - 35% and the corresponding induced air change rate of about 20 - 45 ACH (air changes per hour) could be achieved. According to the economical analysis, the annual electrical energy saving was about 362 kWh and the payback period was about 4.36 years. When compared to commercial insulations, although the ceiling heat transfer reduction in commercial insulations was relatively higher, the initial cost of TE-RSC was lower than microfiber or radiant barrier insulations by about 50% and 27%, respectively. Therefore, the TE-RSC system is an attractive option by reason of its relatively low initial cost and simplicity.

A model of a two-stage semiconductor thermoelectric-generator with external heat-transfer is built by Chen et al. [28]. Performance of the generator, assuming Newton's heat-transfer law applies, is analyzed using a combination of finite-time thermodynamics and non-equilibrium thermodynamics. The analytical equations about the power output versus the working electrical current, and the thermal efficiency versus working electrical-current are derived. For a fixed total heat-transfer surface-area for two heat-exchangers, the ratio of heat-transfer surface area of the high-temperature side heat-exchanger to the total heat-transfer surface-area of the heat-exchangers is optimized for maximizing the power output and the thermal efficiency of the thermoelectric-generator. For a fixed total number of thermoelectric elements, the ratio of number of thermoelectric elements of the top stage to the total number of thermoelectric elements is also optimized for maximizing both the power output and the thermal efficiency of the thermoelectric-generator. The effects of design factors on the performance are analyzed.

Active building envelope (ABE) systems are a new enclosure technology which integrate photovoltaic (PV) and thermoelectric (TE) technologies. In ABE systems, a PV system supplies electrical power to a TE heat-pump system, which can transfer heat in one direction or another depending on the direction of the current. Both the TE and PV systems are integrated by Xu et al. [29] into one enclosure surface. Hence, ABE systems have the ability to actively control the flow of heat across their surface when exposed to solar radiation. Applications for this technology include all types of enclosures that require cooling or heating, such as building enclosures. They developed various ABE system prototypes by using commercially available PV and TE technologies. In this study, two types of commercial available TE modules are studied for their potential application in an ABE prototype window system. They performed various experiments to determine the coefficient of performance for TE modules when operating under different voltage regimes, and have tested different electrical connection diagrams. Based upon the measured data, and results based on the computational models of a TE system, the most suitable type of TE modules, the voltage and current, and the preferable connection diagrams are discussed.

Smith [30] presented a model describing the performance of a thermoelectric system is developed and designed to operate over a large range of system configurations. The theoretical model is compared to the experimental results obtained from a Thermoelectric Power Generation System testing box tested under several configurations and conditions. Discrepancies between model and experiments are described with several model improvements developed and implemented. Finally, the model is incorporated with a heat transfer model and a pricing model to develop a preliminary optimization tool. The optimization tool is then used to analyze the viability of thermoelectric power generation in a hypothetical automotive application when compared with the operating costs of an alternator to develop viability curves based off the price of fuel.

Lertsatitthanakorn *et al.* [31] investigated the thermoelectric (TE) solar air collector, sometimes known as the hybrid solar collector, generates both thermal and electrical energies simultaneously. A double-pass TE solar air collector has been developed and tested. The TE solar collector was composed of transparent glass, air gap, an absorber plate, thermoelectric modules and rectangular fin heat sink. The incident solar radiation heats up the absorber plate so that a temperature difference is created between the thermoelectric modules that generate a direct current. Only a small part of the absorbed solar radiation is converted to electricity, while the rest increases the temperature of the absorber plate. The ambient air flows through the heat sink located in the lower channel to gain heat. The heated air then flows to the upper channel where it receives additional heating from the absorber plate. Improvements to the thermal and overall efficiencies of the system can be achieved by the use of the double-pass collector system and TE technology. Results showed that the thermal efficiency increases as the air flow rate increases. Meanwhile, the electrical power output and the conversion efficiency depend on the temperature difference between the hot and cold side of the TE modules. At a temperature difference of 22.8°C, the unit achieved a power output of 2.13 W and the conversion efficiency of 6.17%. Therefore, the proposed TE solar collector concept is anticipated to contribute to wider applications of the TE hybrid systems due to the increased overall efficiency.

Fan *et al.* [32] presented design details, theoretical analysis, and outcomes of a preliminary experimental investigation on a concentrator thermoelectric generator (CTEG) utilizing solar thermal energy. The designed CTEG system consisted of a parabolic dish collector with an aperture diameter of 1.8 m used to concentrate sunlight onto a copper receiver plate with 260 mm diameter. Four BiTe-based thermoelectric cells (TEC) installed on the receiver plate were used to convert the concentrated solar thermal energy directly into electric energy. A micro channel heat sink was used to remove waste heat from the TEC cold side, and a two-axis tracking system was used to track the sun continuously. Experimental tests were conducted on individual cells and on the overall CTEG system under different heating rates. Under maximum heat flux, a single TEC generator was able to produce 4.9 W for a temperature difference of 109°C, corresponding to 2.9% electrical efficiency. The overall CTEG system was able to produce electric power of up to 5.9 W for a 35°C temperature difference with a hot-side temperature of 68°C. The results of the investigation help to estimate the potential of the CTEG system and show concentrated thermoelectric generation to be one of the potential options for production of electric power from renewable energy sources.

Sark *et al.* [33] proposed to use the thermal waste by attaching thermoelectric (*TE*) converters to the back of PV modules, to form a PV–TE hybrid module. Due to the temperature difference over the TE converter additional electricity can be generated. Employing present day thermoelectric materials with typical figure of merits (*Z*) of 0.004 K^{-1} at 300 K may lead to efficiency enhancements of up to 23% for roof integrated PV-TE modules, as is calculated by means of an idealized model. The annual energy yield would increase by 14.7% - 11%, for two annual irradiance and temperature profiles studied, *i.e.*, for Malaga, Spain, and Utrecht, the Netherlands, respectively. As new TE materials are being developed, efficiency enhancements of up to 50% and annual energy yield increases of up to 24.9% may be achievable. The developed idealized model, however, is judged to overestimate the results by about 10% for practical PV-TE hybrids. Sahin *et al.* [34] studied thermal efficiency of the topping cycle is analyzed and compared with its counterpart without the presence of the thermoelectric elements. Thermodynamic analysis for the efficiency of both the systems with and without thermoelectric generator was presented. The fluid flow and heat transfer in a tube with presence of thermoelectric elements resembling the solar heating system incorporated in the topping cycle are simulated numerically. It is found that, for a certain combination of operating and thermoelectric device parameters, thermal efficiency of the topping cycle becomes slightly higher than that of the same system without the presence of the thermoelectric generators.

A three-dimensional thermoelectric generator model is

proposed and implemented (Chen et al. [35]) in a computational fluid dynamics (CFD) simulation environment (FLUENT). This model of the thermoelectric power source accounts for all temperature dependent characteristics of the materials, and includes nonlinear fluid-thermal-electric multi-physics coupled effects. In solid regions, the heat conduction equation is solved with ohmic heating and thermoelectric source terms, and user defined scalars are used to determine the electric field produced by the Seebeck potential and electric current throughout the thermo elements. The current is solved in terms of the load value using user defined functions but not a prescribed parameter, and thus the field-circuit coupled effect is included. The model is validated by simulation data from other models and experimental data from real thermoelectric devices. Within the common CFD simulator FLUENT, the thermoelectric model can be connected to various CFD models of heat sources as a continuum domain to predict and optimize the system performance.

He et al. [36] presented an experimental and analytical study on incorporation of thermoelectric modules with glass evacuated-tube heat-pipe solar collectors. The integrated solar heat-pipe/thermoelectric module (SHP-TE) can be used for combined water heating and electricity generation. The experimental prototype unit comprises a glass evacuated-tube, a heat-pipe and a thermoelectric module with its one side attached to the condensation section of the heat-pipe and other side attached to a water channel. The heat-pipe transfers the solar heat absorbed within the glass evacuated-tube to the thermoelectric module. Under the condition of given solar irradiation and water temperature, the current, voltage and power outputs of the thermoelectric module are given for variable external electrical resistance. An analytical model of the prototype unit is presented to relate its thermal and electrical efficiencies with the solar irradiation, ambient temperature, water temperature, areas of the glass evacuated-tube and the thermoelectric module, and the length, cross-section area and number of thermo elements in the thermoelectric module. The analytic model is validated against the experimental data before it is used to optimize the design and operating parameters of the prototype for combined water heating and additional electricity generation.

Wotan et al. [37] reported on the design, fabrication and proof of concept of a multilayer fluidic packaging system enabling an increase in the output power performance of micro thermoelectric generators (μTEGs). The complete integration of the micro fluidic heat transfer system (μHTS) with a μTEG is successfully demonstrated. The fabricated prototype is characterized with respect to its thermal and hydrodynamic performance as well as the generated output power. At a very low

pumping power of 0.073 mW/cm^2, a heat transfer resistance of 0.74 cm^2 K/W is reached. The assembled device generated up to 1.47 mW/cm^2 at an applied temperature difference of 50 K and a fluid flow rate of 0.1 l/min. Further system improvements and the potential of the proposed packaging approach are discussed.

It is well known that photovoltaic (PV) cells can only convert a portion of solar energy into electric power, and a large amount of remaining solar radiation mainly produces heat energy. Therefore, a lot of effort has been expended to combine the PV and thermoelectric technology in an efficient and powerful way. Most of these inventions are focused on the structure design [24]. Micallef [38] presented a Seebeck solar cell device, in which the materials used to form conductors in the n-type and p-type regions of the cells were chosen for their different thermoelectric characteristics. Therefore, electric power could be produced not only from the PV cells but also from the temperature gradient in the conductors resulting from solar radiation and waste heat generated in the PV cell. Multiple devices could be connected in series or parallel so as to enhance the output power. Hunt [39] presented a simple hierarchical structure, which had at least one thermoelectric module thermally attached to the PV module and could produce electricity both from the PV cell and thermoelectric module. This simple structure combines both PV and thermoelectric conversion.

Hecht [40] proposed another solar-energy conversion package which combined PV cell, thermoelectric conversion unit, and thermal heating system together. One of its embodiments unlike the system described previously, the thermoelectric cell and PV cell in this system were separated, and the irradiated surface of the PV cell was treated with selective spectrum reflective coating to allow high conversion PV wavelength energy to be absorbed by the PV cells, while the rest of less effective wavelength radiation would be reflected to the thermoelectric cell surface to produce electricity from heat; therefore higher efficiency could be achieved. Apart from that, making full use of solar radiation can also be achieved by performing a wavelength band division of solar light. The introduced solar radiation could be divided by wavelength band dividers. At first, the ultraviolet light would be separated by the wavelength band divider and reach the wavelength converter to be converted into visible light. Then the rest of solar radiation would pass through the wavelength band divider to be divided into visible light and infrared light. After these wavelengths band division processes, the visible light would reach the photoelectric converter where it could be converted into electric energy, and the infrared light would be converted by a thermoelectric converter into electric energy through thermal energy. Therefore, more efficient solar-energy utilization can be achieved using this compound system.

4. Thermoelectric Materials

Thermoelectric material which greatly affects the efficiency is of huge importance for solar thermoelectric power generation. Apart from the large Seebeck coefficient, good electrical conductivity, and small thermal conductivity, the thermoelectric materials must present excellent thermal and chemical stability at high temperature when used under the concentrated solar radiation. A great deal of research on thermoelectric material has been conducted over the past 50 years, and the literature is rich. The three factors α, σ and k are interrelated and make it quite challenging to optimize ZT. Equation (1.2) emphasizes that high Seebeck coefficients are important for a good thermoelectric material. Nevertheless, an increase in α is almost always accompanied with a decrease in σ. Typically semiconductors and semimetals have higher α but lower σ than metals because of their rather lower carrier concentrations At room temperature, $T = 300$ K, desired values for the thermoelectric parameters are $\alpha = 225$ μV/K, $\sigma = 10^5$ $\Omega^{-1} \cdot m^{-1}$, and $k = 1.5$ W/m·K, which results in a $ZT \approx 1$. These values are typical for the best TE materials such as Bi_2Te_3 and Sb_2Te_3 alloys, which are presently used by industry in devices that operate near room temperature and are well investigated. Current TE devices operate at an efficiency of about 5% - 6%. By increasing ZT by a factor of 4 predicted efficiencies can increase to 30% [5].

In order to achieve a sufficient conversion efficiency η at the given temperature, values of at least $ZT \sim 1$ are required. The maximum conversion efficiency is thermodynamically limited by the Carnot efficiency. As was shown by (Yang and Caillat [41]), a figure of Merit in the range of $2 < ZT < 3$ results in conversion efficiencies of ~50% of the Carnot efficiency. The real conversion efficiency depends not solely on the materials properties, but also on the construction and geometry of the TE device, as well as on the macroscopic heat and electronic transport. Commercial thermoelectric devices are based on Bi_2Te_3 because this material exhibits a relatively high figure of Merit [41]. Disadvantages of Bi_2Te_3 compounds are their limited chemical stability at high temperatures in air and their toxicity. Therefore, complex metal oxide ceramics as alternative materials are promising candidates for high temperature applications as they are inert at high temperatures in air, non-toxic, and low cost materials [41]. Among these oxides, $Na_xCo_2O_4$ is especially interesting as it shows a high figure of Merit, $ZT \sim 0.8$ at $T = 800$ K. The production of single crystals with defined and stable stoichiometry is difficult, though. In contrast, perovskite-type materials based on manganate and cuprate can be easily synthesized with controllable composition and TE properties.

Current thermoelectric materials, as shown in **Figure 3** [13], have $ZT = 1$, and new materials with ZT values of 2 - 3 are sought to provide the desired conversion efficiencies. The current materials exhibit conversion efficiencies of 7% - 8% depending on the specific materials and the temperature differences involved. With regard to solar energy conversion, thermoelectric devices will likely utilize the IR spectrum of solar radiation. For example, a thermoelectric power conversion device with $ZT = 3$ operating between 500°C and 30°C (room temperature) would yield about 50% of the Carnot efficiency.

Tirtt et $al.$ [42] proved that a value of $ZT > 4$ does not significantly increase the conversion efficiency over that of a material with $ZT = 2$ - 3.5 Therefore, they believed that the "Holy Grail" of thermoelectric materials research is to find bulk materials (both n-type and p-type) with a ZT value on the order of 2 - 3 (efficiency = 15% - 20%) with low parasitic losses (e.g., contact resistance, radiation effects, and interdiffusion of the metals) and low manufacturing costs. With respect to solar energy, these materials would need to operate at about 1000 K (≈700°C). The solar energy conversion process could be envisioned where a high-efficiency solar collector turns the sunlight (from the IR spectrum) into heat that is then transformed by the thermoelectric devices into usable electricity. In addition, the solar energy could be stored in a thermal bath and transformed into electricity through thermoelectrics when the sun was not shining.

Sano et $al.$ [43] presented an attempt to improve the performance of a thermoelectric element/module for power generation by taking the following two approaches:

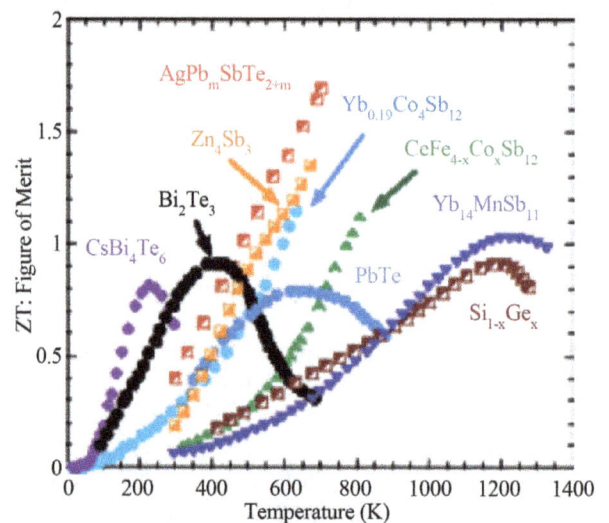

Figure 3. Figure of merit (ZT) as a function of temperature forseveral high-efficiency bulk thermoelectric materials [29].

- Improving the basic characteristic (Z) of thermoelectric element (Improve the characteristics of the element at room temperature and thereby improve them at higher temperatures).
- Shifting the peak of Z toward the high temperature side (Even if the value of Z in the low temperature region decreases, the overall efficiency of the element improves as the value of Z in the high temperature region increases).

With the aim of improving the figure of merit, Z, of a thermoelectric element, they applied plasma treatment to the raw material powder as described in [43]. Concerning the shift of the peak of Z toward the high temperature side, their experimental results introduced that the n-type maintains an improved characteristic in the high-temperature region without causing the low-temperature characteristic to decrease significantly. The figure of merit of the p type at low temperatures decreased markedly. However, the average performance index in the entire working temperature range improved.

5. The Future of Thermoelectric Materials

The future expansion of thermoelectric energy conversion technologies is tied primarily to enhanced materials performance along with better thermal management design. The best thermoelectric material should behave as a so-called phonon-glass-electron-crystal; that is, it should minimally scatter electrons, as in a crystalline material, whereas it should highly scatter phonons, as in an amorphous material. Materials researchers are now investigating several systems of materials including typical narrow-band gap semiconductors (half-Heusler alloys), oxides, and cage-structure materials (skutterudites and clathrates). More exotic structures that exhibit reduced dimensionality and nanostructures have been the focus of much recent research, including super lattices, quantum dots, and nanodot bulk materials. Also, recent progress in nanocomposites, mixtures of nano materials in a bulk matrix, has generated much interest and hope for these materials. The emerging field of these thermoelectric nanocomposites appears to be one of the most promising recent research directions. Such nanocomposites could allow for higher ZT values by reducing thermal conductivity while maintaining favorable electronic properties. With new higher efficiency materials, the field of harvesting waste energy through thermoelectric devices will become more prevalent [44-47].

The most stable, long-term, and readily available worldwide energy source is that of solar energy. The issue has always been low-cost transformation and storage. Other alternative energy technologies such as fuel cells, wind energy, and thermoelectric will provide some assistance in meeting our future energy needs. Many hybrid systems will be needed, and thermoelectric is able to work in tandem with many of these other technologies, especially solar as it can use the heat source provided by solar radiation. Over the past decade, thermoelectric materials have been developed with ZT values that are a factor of 2 larger than those of previous materials. Another 50% increase in ZT (to $ZT \approx 3$) with the appropriate material characteristics and costs will position thermoelectric to be a significant contributor to our energy needs, especially in waste heat or solar energy conversion. The likelihood of achieving these goals appears to be within reach in the next several years. Furthermore, some contribution from many of these alternative energy technologies such as thermoelectric will be needed in order to fulfill the world's future energy needs.

6. Concluding Remarks

It can be seen that although much effort has been made to develop direct solar thermal power generation technologies, the conversion potential and practical applications are still not widely used. In order to make full use of its advantages and develop practical civil devices, more effort should be devoted to material research, structure optimization, and practical application development.

REFERENCES

[1] S. B. Riffat and X. L. Ma, "Thermoelectrics: A Review of resent and Potential Applications," *Applied Thermal Engineering*, Vol. 23, No. 8, 2003, pp. 913-935.

[2] K. Soteris, "Solar Energy Engineering: Processes and Systems," Elsevier Inc., Berlin, 2009.

[3] X. F. Qiu, "Nano-Structured Materials for Energy Conversion Case," Ph.D. Thesis, Western Reverse University, Cleveland, 2008.

[4] D. E. Demirocak, "Thermodynamic and Economic Analysis of a Solar Thermal Powered Adsorption Cooling System," MSc. Thesis in Mechanical Engineering Department, Middle East Technical University, Ankara, 2008.

[5] R. H. Hyde, "Growth and Characterization of Thermoelectric Ba8Ga16Ge30 Type-I Clathrate Thin-Films Deposited by Pulsed Dual-Laser Ablation," Ph.D. Thesis, College of Arts and Sciences, University of South Florida, Tampa, 2011.

[6] M. Kassas, "Thermodynamic Analysis of a Thermoelectric Device," *International Journal of Exergy*, Vol. 4, No. 2, 2007, pp. 168-179.

[7] O. Yamashita, "Effect of Linear Temperature Dependence of Thermoelectric Properties on Energy Conversion Efficiency," *Energy Conversion and Management*, Vol. 49, No. 11, 2008, pp. 3163-3169.

[8] B. S. Yilbas and A. Z. Sahin, "Thermoelectric Device and

Optimum External Load Parameter and Slenderness Ratio," *Energy*, Vol. 35, No. 12, 2010, pp. 5380-5384.

[9] Y. Y. Hsiao, W. C. Chang and S. L. Chen, "A Mathematic Model of Thermoelectric Module with Applications on Waste Heat Recovery from Automobile Engine," *Energy*, Vol. 35, No. 3, 2010, pp. 1447-1454.

[10] D. Champier, J. P. Bedecarrats, M. Rivaletto and F. Strub, "Thermoelectric Power Generation from Biomass Cook Stoves," *Energy*, Vol. 35, No. 2, 2010, pp. 935-942.

[11] M. Kubo, M. Shinoda, T. Furuhata and K. Kitagawa, "Optimization of the Incision Size and Cold-End Temperature of a Thermoelectric Device," *Energy*, Vol. 30, No. 11-12, 2005, pp. 2156-2170.

[12] X. Gou, H. Xiao and S. Yang, "Modeling, Experimental Study and Optimization on Low-Temperature Waste Heat Thermoelectric Generator System," *Applied Energy*, Vol. 87, No. 10, 2010, pp. 3131-3136.

[13] T. M. Tritt, H. Böttner and L. Chen, "Thermoelectrics: Direct Solar Thermal Energy Conversion," *MRS Bulletin*, Vol. 33, No. 4, 2008, pp. 366-368.

[14] R. Amatya and R. J. Ram, "Solar Thermoelectric Generator for Micro powerapplications," *Journal of Electronic Materials*, Vol. 39, No. 9, 2010, pp. 1735-1740.

[15] M. Telkes, "Solar Thermoelectric Energy Generators," *Journal of Applied Physics*, Vol. 23, No. 6, 1954, pp. 765-777.

[16] R. Rush, "Solar Flat Plate Thermoelectric Generator Research," Tech. Doc. Rep. Air Force, AD 605931, 1964.

[17] H. J. Goldsmid, J. E. Giutronich, and M. M. Kaila, "Solar Thermoelectric Generation Using Bismuth Telluride Alloys," *Solar Energy*, Vol. 24, 5, 1980, pp. 435-440.

[18] C. L. Dent and M. H. Cobble, Proceedings of the 4*th* *International Conference on Thermoelectric Energy Conversion*, New York, 1982, pp. 75-78.

[19] P. Li, L. Cai, P. Zhai, X. Tang, Q. Zhang and M. Niino, "Design of Concentration Solar Thermoelectric Generator," *Journal of Electronic Materials*, Vol. 39, No. 9, 2010, pp. 1522-1530.

[20] J. Chen, "Thermodynamic Analysis of a Solar-Driven Thermoelectric Generator," *Journal of Applied Physics*, Vol. 79, No. 5, 1996, pp. 2717-2721.

[21] D. M. Rowe, "Thermoelectrics, an Environmentally-Friendly Source of Electrical Power," *Renewable Energy*, Vol. 16, No. 1-4, 1999, pp. 1251-1256.

[22] G. Chen, "Theoretical Efficiency of Solar Thermoelectric Energy Generators," *Journal of Applied Physics*, Vol. 109, No. 10, 2011, pp. 104908-104908-8.

[23] B. Lenoir, A. Dauscher, P. Poinas, H. Scherrer, and L.

Vikhor, "Electrical Performance of Skutterudites Solar Thermoelectric Generators," *Applied Thermal Engineering*, 23, No. 11, 2003, pp. 1407-1415.

[24] Y. Deng and J. Liu, "Recent Advances in Direct Solar Thermal Power Generation," *Journal of Renewable and Sustainable Energy*, Vol. 1, No. 5, 2009, Article ID: 052701.

[25] S. A. Omer and D. G. Infield, "Design and Thermal Analysis of a Two Stage Solar Concentrator for Combined Heat and Thermoelectric Power Generation," *Energy Conversion & Management*, Vol. 41, No. 7, 2000, pp. 737-756.

[26] S. A. Omer and D. G. Infield, "Design Optimization of Thermoelectric Devices for Solar Power Generation," *Solar Energy Materials and Solar Cells*, Vol. 53, No. 1-2, 1998, pp. 67-82.

[27] S. Maneewan, J. Hirunlabh, J. Khedari, B. Zeghmati and S. Teekasap, "Heat Gain Reduction by Means of Thermoelectric Roof Solar Collector," *Solar Energy*, Vol. 78, No. 4, 2005, pp. 495-503.

[28] L. Chen, J. Li, F. Sun and C. Wu, "Performance Optimization of a Two-Stage Semiconductor Thermoelectric-Generator," *Applied Energy*, Vol. 82, No. 4, 2005, pp. 300-312.

[29] X. Xu, S.V. Dessel and A. Messacb, "Study of the Performance of Thermoelectric Modules for Use in Active Building Envelopes," *Building and Environment*, Vol. 42, No. 3, 2007, pp. 1489-1502.

[30] K. D. Smith, "An Investigation into the Viability of Heat Sources for Thermoelectric Power Generation Systems," MSc. Thesis, Department of Mechanical Engineering, Rochester Institute of Technology, Rochester, 2009.

[31] C. Lertsatitthanakorn, N. Khasee, S. Atthajariyakul, S. Soponronnarit, A. Therdyothin and R. O. Suzuki, "Performance Analysis of a Double-Pass Thermoelectric Solar Air Collector," *Solar Energy Materials & Solar Cells*, Vol. 92, No. 9, 2008, pp. 1105-1109.

[32] H. Fan, R. Singh and A. Akbarzadeh, "Electric Power Generation from Thermoelectric Cells Using a Solar Dish Concentrator," *Journal of Electronic Materials*, Vol. 40, No. 5, 2011, pp. 1311-1320.

[33] W. G. J. H. M. Van Sark, "Feasibility of Photovoltaic—Thermoelectric Hybrid Modules," *Applied Energy*, Vol. 88, No. 8, 2011, pp. 2785-2790.

[34] A. Z. Sahin, B. S. Yilbas, S. Z. Shuja and O. Momin, "Investigation into Topping Cycle: Thermal Efficiency with and without Presenceof Thermoelectric Generator," *Energy*, Vol. 36, No. 7, 2011, pp. 4048-4054.

[35] M. Chen, L. A. Rosendahl and T. Condra, "A Three-

Dimensional Numerical Model of Thermoelectric Generators in FluidPower Systems," *International Journal of Heat and Mass Transfer*, Vol. 54, No. 1-3, 2011, pp. 345-355.

[36] W. He, Y. Su , Y. Q. Wang, S. B. Riffat and J. Ji, "A Study on Incorporation of Thermoelectric Modules with Evacuated-Tubeheat-Pipe Solar Collectors," *Renewable Energy*, Vol. 37, No. 1, 2012, pp. 142-149.

[37] N. Wojtas, E. Schwytera, W. Glatzb, S. Kühnea, W. Escherc and C. Hierolda, "Power Enhancement of Micro Thermoelectric Generators by Micro Fluidic Heat Transfer Packaging," *Sensors and Actuators A*: *Physical*, Vol. 188, 2012, pp. 389-395.

[38] J. A. Micallef, US Patent No. US2008053514-A1, 2008.

[39] R. D. Hunt, Patent No. WO2004004016-A1, 2004.

[40] D. H. Hecht, US Patent No. US2007289622-A1, 2007.

[41] P. Tomeš, M. Trottmann, C. Suter, M. Aguirre, A. Steinfeld, P. Haueter and A. Weidenkaff, "Thermoelectric Oxide Modules (TOMs) for the Direct Conversion of Simulated Solar Radiation into Electrical Energy," *Materials*, Vol. 3, No. 4, 2010, pp. 2801-2814.

[42] T. M. Tritt and M. A. Subramanian, "Thermoelectric Materials, Phenomena, and Applications: A Bird's Eye View,"

MRS Bulletin, Vol. 31, No. 3, 2006, pp. 188-198.

[43] S. Sano, H. Mizukami and H. Kaibe, "Development of High-Efficiency Thermoelectric Power Generation System," *Komatsu Technical Report*, Vol. 49, No. 152, 2003, pp. 1-7.

[44] Q. Yao, L. Chen, W. Zhang, S. Liufu and X. Chen, "Enhanced Thermoelectric Performance of Single-Walled Carbon Nanotubes/Polyaniline Hybrid Nanocomposites," *ACS Nano*, Vol. 4, No. 4, 2010, pp. 2445-2451.

[45] P. Ahadi, N. Haeri and A. Nazari, "The Use of Nanotechnology In Solar Systems," *Australian Journal of Basic and Applied Sciences*, Vol. 5, No. 11, 2011, pp. 1450-1456.

[46] R. Venkatasubramanian, C. Watkins, D. Stokes, J. Posthill and C. Caylor, "Energy Harvesting for Electronics with Thermoelectric Devices Using Nanoscale Materials," *IEEE International on Electron Devices Meeting*, Washington DC, 10-12 December 2007, pp. 367-370.

[47] A. I. Hochbaum and P. Yang, "Semiconductor Nanowires for Energy Conversion," *Chemical Reviews*, Vol. 110, No. 1, 2010, pp. 527-546.

12

Policy-Making for Households Appliances-Related Electricity Consumption in Indonesia— A Multicultural Country

Muhammad Ery Wijaya[*]**, Tetsuo Tezuka**

Department of Socio-Environmental Energy Science, Graduate School of Energy Science, Kyoto University, Kyoto, Japan

ABSTRACT

Household energy consumption is strongly influenced by culture. Therefore, the study of the influence of culture on energy consumption is important for designing the most suitable energy conservation policy to increase society's adaptation to policy. The present paper has the following aims: 1) to analyze and compare decision-making in the use of electrical appliances; and 2) to compile a strategy to improve the adoption of higher-efficiency appliances and the wise use of electricity in Indonesia to optimize households' energy conservation. All aims are in the framework analysis of the different cultural backgrounds and ethnicities represented by two cities—Yogyakarta and Bandung. The finding indicates that people in Yogyakarta show greater awareness of the benefits of adopting higher-efficiency appliances than people in Bandung. Therefore, the awareness rate of energy consumed by appliances and energy efficiency of appliances is significantly higher in Yogyakarta. This study shows the integration of intervention strategy in the purchase and use of electrical appliances with regard to the consumers' decision and behaviors within a framework of the local culture to manage electricity consumption in the household sector.

Keywords: Household; Energy Efficiency; Energy Saving; Decision-Making

1. Introduction

Household energy consumption has been studied in many countries, as it typically accounts for a large percentage of the total energy consumption [1]. Understanding the characteristics of household electricity consumption is important for researchers and policy-makers who are concerned with the impact of households on electricity use and the environment. Therefore, there is an urgent need to raise the awareness that energy use and its impact on the environment should concern all individuals in their daily activities.

Household energy consumption is strongly influenced by cultural factors [2-7]. Lutzenhiser [8] argued that there is a need to understand the relationships between human groups (culture, ethnics or races) and their technologies and that these relationships can be used to account the frequencies and magnitudes of energy flows. The study also compiled several energy research studies and found that the cultural approach in energy consump-

tion and policy-making has promising applications on three levels: the descriptive, explanatory and predictive analysis of specific applications of energy consumption and conservation. Energy-use patterns and the resulting energy policy implications vary by cultural background. Thus, the study of the influence of culture on energy consumption is important for designing the most suitable energy conservation policy to improve society's adaptation to policy. However, the development energy policy model based on cultural approaches has been hampered by a lack of empirical research.

The Indonesian society consists of different cultural backgrounds or ethnic groups, which can be distinguished into several regions. The variety of cultural backgrounds may affect the residents' decision-making behavior in the purchase and use of electrical appliances in Indonesia. Our previous study found that the differences in electricity consumption patterns between two different cultural backgrounds were influenced by several driving factors, such as income, duration of time at home, and family size [9]. However, further comprehensive study of

---

[*]Corresponding author.

the techno-socioeconomics of electricity consumption under different cultural backgrounds should be conducted. Therefore, we investigated the influence of cultures in the decision-making process in the purchase of electrical appliances in both cities [10]. The factors influencing the purchase of appliances indicated that people in Yogyakarta show a greater awareness of the benefits of adopting higher-efficiency appliances than do persons in Bandung. This variation stems from cultural aspects that reflect the way that people consider factors in the purchase of appliance.

The current study examines the influence of cultural backgrounds on Indonesian household electricity consumption within the perspective of electricity consumers' choice determinants in the purchase of electrical appliances and decision-making in consuming electricity. All aims in this study are in the framework analysis of different cultural backgrounds and ethnicities. Two cities are selected to show the influence of culture on electricity consumption, Yogyakarta is a center of Javanese culture and Bandung represents Sundanese culture. This paper has the following aims: 1) to analyse and compare decision-making in the use of electrical appliances; and 2) to compile a policy strategy in Indonesia to improve the implementation of energy conservation through the wise use of electricity and the adoption of higher-efficiency appliances as found in previous study [10]. The results of this study are expected to improve the strategy and effort to design an energy conservation policy based on local cultures to improve the success of policy implementation.

2. Overview of Indonesia

2.1. Indonesian Economy and Electricity

Indonesian economic development boosts the industrial and commercial sector and the people's welfare. During the years 2001 to 2009, the GDP per capita increased sharply, with an average of 15% annually. The GDP per capita amounted to US$ 748 in 2001 and US$ 2698 in 2009 [11,12]. This trend is predicted to continue over the coming years due to the revival of the world economy. The largest economic activity in Indonesia is centralized in the Jawa-Madura and Bali areas, known as the JaMaLi area. In 2010, the total population in this area was 141 million people, or nearly 60% of country's total population [13]. The total electricity consumption was 79 TWh in 2000 and increased to 135 TWh in 2009 [11]. Nearly 78% of this consumption was in the JaMali area. In 2009, the national electrification ratio reached 63.75%, and this value will continue to grow rapidly following national economic growth [14]. In the JaMaLi area, the industrial sector consumes the largest proportion of electricity (47%), followed by the household sector (39%) [15].

Although the industrial sector is the largest electricity consumer, the household electricity consumption tends to increase. Predictions suggest that by 2027, the household sector will be the largest electricity consumer in the JaMaLi area. According to the government's projection, the household sector will take 59% of the total electricity consumption share, whereas the commercial, industry and public sectors will consume 22%, 12% and 7%, respectively [15]. Thus, households are an important group when addressing energy conservation. An additional reason to focus on households is that electricity consumption in households continues to rise

2.2. Demography of Indonesia

Indonesia is a multi-ethnic society with more than 1000 ethnic/sub-ethnic groups. Nevertheless, only 15 groups have more than 1 million people. According to Lietar and De Meulenaere [16], definition of ethnic group refers to a cultural identity that involves language, beliefs, morals, laws, tradition and patterns of behavior. According to the 2000 Population Census, published by Statistics Indonesia, the two largest ethnic groups in Indonesia, the Javanese and Sundanese, accounted for 41.7% and 15.4% of the national population, respectively [17]. Therefore, Sundanese and Javanese people which the populations are located on Java Island were the largest electricity consumers in Indonesia.

3. Household's Behavior on Electricity Consumption

A high percentage of household energy consumption is associated with the use of major household appliances. Genjo [4] investigated the relationship between possession of home appliances and electricity consumption in Japanese households. The study found that the increase in the consumption of residential electricity was due to the use of a greater number of home appliances. Therefore, efforts to promote energy savings in the household sector are continuously increasing. The energy efficiency of electrical appliances has significantly improved in recent years. The improvement in energy efficiency will impact the overall energy demand and, subsequently, the environment [18]. However, these newly developed efficient appliances will not be widely used unless societies actively adopt them. The improvement in energy efficiency does not necessarily lead to an overall decline in physical consumption due to inappropriate consumption practices.

Guerin [19] summarized numerous studies of household energy consumption and found that the variables that most frequently affected energy behavior and consumption were occupant characteristics, occupant attitudes and occupant actions. However, a few studies have investigated energy consumption from the consumers' perspective. Yamamoto [20] investigated decision-making concerning electrical appliance use in Japan. The study found that the price did not function as a decision-making signal for electrical appliance users. Rather, decision-making was dependent on the characteristics of particular electrical appliances and the electricity payment system. Gaspar and Antunes [21] studied the consumer choice determinants in the purchase of appliances and analyzed the factors that drove the consideration of the energy efficiency class in Europe. Nevertheless, there is no comprehensive analysis that combines the consumer choice determinants in the purchase of appliances and behavior in the use of appliances in a framework to improve the energy conservation in the household sector.

Several researchers have studied household energy consumption as it relates to various population groups. Poyer [3] investigated differences in residential energy consumption between Latino and non-Latino households in the USA. The results showed significant variations in the patterns of energy consumption for Latinos and non-Latinos. Helbert [5] conducted a study on household energy consumption patterns in Guatemala and found that the energy portfolio of different ethnic groups significantly varied.

Nevertheless, there is no comprehensive analysis that combines the consumer choice determinants in the purchase of appliances and behavior in the use of appliances in a framework of different cultural backgrounds and ethnicities to improve the energy conservation in the household sector. To address these gaps, this study proposes to achieve energy conservation in the household sector with regards to its cultural backgrounds, both an increase in energy efficiency (through the adoption of higher-efficiency class appliance types) and the promotion of environmental consumer behaviors (through better use of electrical appliances) should be analyzed and promoted.

4. Methodology

Although an understanding of the technological use, social and economic characteristics of electricity consumers is important for improving energy conservation, it is not sufficient. Such knowledge should be complemented by the assessment of behavioral economics and human psychosocial variables such as attitudes, beliefs and perceived benefits in the purchase and use of electrical appliances. In fact, electricity consumption is affected not only by the use of electrical appliances but also by choice determinants of the efficiency of appliances upon purchase. However, few studies or projects have assessed electricity consumption based on cultural background. Particularly appliance attributes, psychosocial and socio-economic variables have not been examined in an inte-

grated manner with knowledge of technological choices and utilization. Hence, not all energy conservation policies that are currently practiced in the world may be applied in all societies. The present comprehensive study is designed to develop a policy framework that is suitable for society based on the influence of culture.

This research conducted a questionnaire survey in two cities. In Yogyakarta, only Javanese people were recruited as respondents. In Bandung, only Sundanese people were selected. The research was conducted from October to November 2011. The respondents were recruited through a door-to-door solicitation procedure in which they were asked to consent to a survey of their home appliances, review of their monthly electricity bill, and an in-depth interview regarding household members' typical electricity use behaviour. Approximately 30% of households from both cities in this survey used a prepaid system for their electricity payment.

With several barriers during data collection, such as time and financial limitations. The total sample size was rounded to 100 respondents in each city according to the income level distribution. The selected sample was validated with the statistics of both cities published by Statistics Indonesia to avoid sampling bias. Details of respondents' profile were similar as defined in [10].

Outline of the Questionnaire Survey

The study investigated the following: 1) choice determinants of appliance purchases; and 2) decision-making in appliance use. Meanwhile, the questions and analysis concerning the choice determinants of appliance purchases were presented in our previous study [10].

The questions related to decision-making in the use of electrical appliances were centered on two aspects, attitude and knowledge and technological perspectives (see details at **Table 1**). The content of the questionnaire

Table 1. Questions related to decision-making in the use of electrical appliances.

No.	Criteria	Question
1.	Attitude and knowledge	How do you understand the electricity prices?
		How do you remember the monthly electricity bill?
		How do you remain aware of changes in the monthly electricity bill?
		How do you remain aware of the change of the electricity prices?
		How do the electricity payment systems affect your awareness of the monthly bill?
2.	Technological perspective	How do you remain aware of the price of electricity consumed by appliances?
		How do you remain aware of the energy efficiency of appliances?

covered the following: 1) knowledge of electricity prices; 2) memory of electricity payment; 3) awareness of the changes of electricity bill; 4) awareness of electricity consumed by electrical appliances; 5) awareness of the energy efficiency; 6) response to the change of electricity price; and 7) effects of payment system.

Respondents were provided many options related to the questions and rated their level of agreement or disagreement on a Likert scale. After receiving the responses from respondents, the average scores for every question were calculated. For each question, the option with the higher average score was the first-ranked option and so on. Additionally, to compare decision-making in the purchase and use of electrical appliances in the two cities, a T-test was employed.

5. Results and Discussion

5.1. Analysis of Choice Determinants in the Purchase of Electrical Appliances

In order to understand the choice determinants of electricity consumers in the purchase of electrical appliances, the summary of our previous study [10] is presented in this section.

5.1.1. Use of Appliances

The replacement of an old electrical appliance with a new appliance allows a process of adoption of higher-efficiency appliances. In addition, improvements in the energy efficiency of appliances have the potential to decrease households' energy use. Therefore, the rate at which households replace various appliances has important implications for the realization of household energy demand saving in response to technological improvements [18]. The survey results indicate that the average lifetimes of the appliances were similar in the two cities (see **Table 2**). However, these values are lower than those of developed countries, such as the USA and Canada, where the average lifetime for refrigerators and washing machines is 16 and 12 years, respectively [18].

This difference, however, does not indicate that the adoption of more efficient appliances is more likely in Indonesian households than in the developed countries. Indeed, the values of the standard deviation for most appliances were quite large in both cities. This variation might stem from two sources, technical aspects and cultural aspects. The technical aspects of the problem include several issues. First, the quality of the electrical appliances sold in Indonesia is low. These appliances show low durability. Second, the reliability of the power supply in Indonesia is poor. Due to this quality problem, the electrical appliances are more easily damaged. The cultural aspect of the problem could reflect the way that people use electrical appliances and people's lack of

Table 2. Descriptive statistics of lifetime of appliances.

Appliance	City	N	M	SD
TV	Yogyakarta	100	9.86	4.85
	Bandung	100	10.16	4.76
Refrigerator	Yogyakarta	72	11.54	3.33
	Bandung	75	9.96	3.29
Air Conditioner	Yogyakarta	10	11.10	3.11
	Bandung	11	12.45	2.58
Electric Fan	Yogyakarta	75	5.71	3.38
	Bandung	75	7.95	2.84
Lighting	Yogyakarta	100	1.56	0.79
	Bandung	100	1.25	1.16
Rice Cooker	Yogyakarta	86	6.04	3.47
	Bandung	80	7.90	2.93
Water Pump	Yogyakarta	76	8.30	4.60
	Bandung	50	9.80	5.58
Washing Machine	Yogyakarta	41	7.71	3.73
	Bandung	38	7.68	2.31

Note: N stands for Number.

knowledge about operating the appliances. These arguments are strengthened by the respondents' reasons for replacing their appliances (see **Figure 1**). In the two cities, the most frequent reason to replace an appliance was because it was broken (92% in Yogyakarta and 97% in Bandung).

5.1.2. Required Information on Appliance Characteristics

The information prior to purchasing an appliance helps

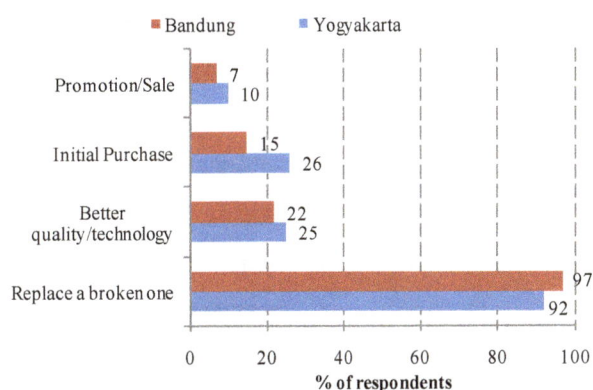

Figure 1. The reasons to replace electrical appliances (because some respondents selected two or more reasons, the totals may exceed 100%).

people make a decision about the purchase of the appliance. The characteristics of appliance including the price, quality, energy consumption, warranty, user friendliness, technology, safety, accessories, type, brand, and country of origin were assessed. The respondents could select more than one answer. The results are shown in **Figure 2**.

In Yogyakarta, the quality was the most frequently required information and followed by price and brand. Meanwhile, in Bandung, the price was the most frequently required information and followed by quality and warranty. Information related to energy consumption and the technology of the appliance was not a priority for the people in either city. This might be due to a lack of available information on these topics.

In addition, several media were used to access the information that people sought prior to purchasing an appliance (see **Figure 3**). In Yogyakarta, commercial advertisements in newspapers, magazines, on the Internet or in public spaces were the primary sources of information. In Bandung, most respondents sought information regarding appliance characteristics and used the store's sales staff as a primary source of information.

5.1.3. Factors that Influence the Purchase of Appliances

The analysis of the consumer's decision to select the electrical appliance that is most strongly preferred, is

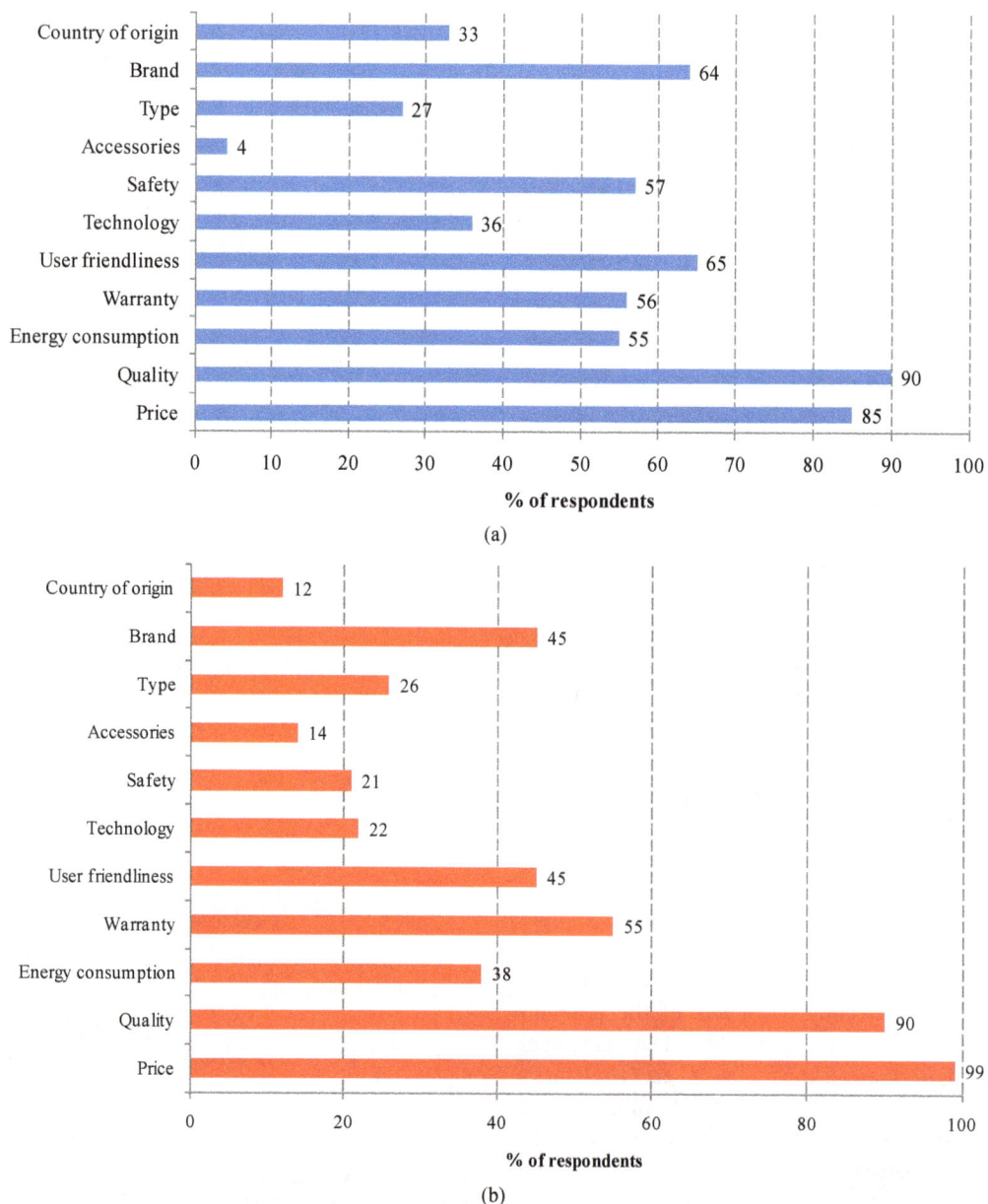

(a)

(b)

Figure 2. Required information prior to the purchase of an appliance in (a) Yogyakarta and (b) Bandung.

important because it reveals the relationship between the factors that influence the purchase of an appliance and the information required prior to purchasing the appliance. By this analysis, a further strategy can be formulated to improve Indonesian households' adoption of higher-efficiency appliances.

The response of respondents indicate that quality, price, safety and energy consumption were the most frequently considered factors in the purchase of an appliance by the respondents in Yogyakarta. In Bandung, the respondents primarily considered price, quality, brand, and user friendliness. **Figure 4** shows the respondents' responses on the factors that influence the purchase of appliances in Yogyakarta and Bandung. **Table 3** presents a comparison between the two cities on the factors that

influence the purchase of appliances. A detailed comparison of the responses shows that people in Yogyakarta devoted more attention to considerations of quality, energy consumption, user friendliness, technology, safety, and brand. These results differed significantly from those found in Bandung. People in Bandung devoted more attention to the accessories sold with the appliance.

5.2. Analysis of Decision-Making in Electrical Appliances Use

5.2.1. Understanding of Electricity Price

In daily life, behavioral economics tend to close relationship with price variable. Therefore, price is often considered to be an important signal in individuals' decision-making concerning electricity consumption [20]. In this survey, the behavioral economics of the respondents in reference to electrical consumption was investigated by questions concerning the following: 1) understanding of the government's electricity prices; 2) memory of their monthly electricity bill (both questions are scaled 1 (not at all) to 4 (highly)); and 3) awareness of the changes in their monthly electricity bill on a scale of 1 (not at all) to 5 (highly). The difference of the point scale in the assessment of awareness of the changes of their monthly electricity bill is more subjective than the two previous questions.

The results, as shown in **Table 4**, indicate that people in Bandung had a significantly better understanding of electricity prices than people in Yogyakarta. In terms of remembering the monthly electricity bill and awareness of changes in the monthly electricity bill, no significant differences were found between Bandung and Yogyakarta (p-value > 0.05).

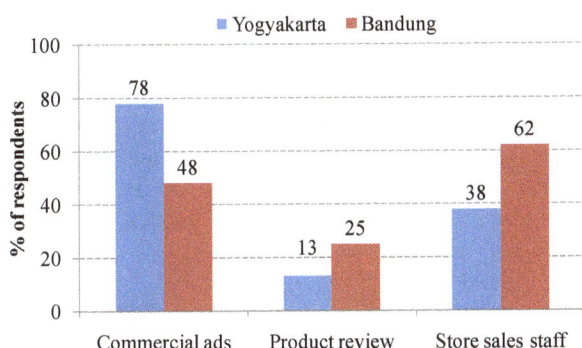

Figure 3. Sources of information prior to purchasing an appliance.

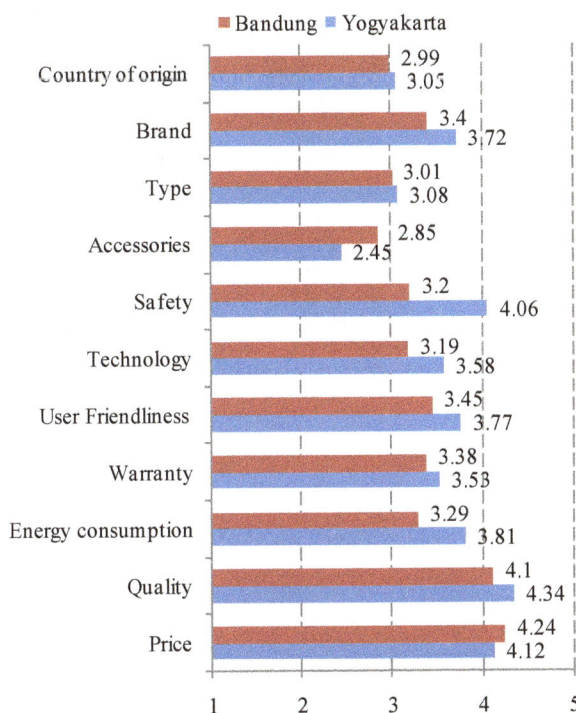

Figure 4. Respondents' response on factors that influence the purchase of appliances.

Table 3. Comparison of influencing factors in the purchase of appliances between Yogyakarta and Bandung.

Factor	p-value
Price	0.19
Quality	0.02
Energy Consumption	0.00
Warranty	0.20
User Friendliness	0.01
Technology	0.00
Safety	0.00
Accessories	0.00
Type	0.58
Brand	0.01
Country of Origin	0.68

Table 4. Comparison of respondents' understanding of the electricity price between Yogyakarta and Bandung.

Question	City	M	p-value
How do you understand the electricity prices	Yogyakarta	2.15	0.02
	Bandung	2.42	
How do you remember the monthly electricity bill	Yogyakarta	2.98	0.92
	Bandung	2.99	
How do you remain aware of changes in monthly electricity bill	Yogyakarta	3.69	0.09
	Bandung	3.50	

The results show that electricity prices were not well understood in Yogyakarta and Bandung. By contrast, in terms of remembering the monthly electricity bill, the average response of respondents indicate that people had a good memory of electricity expenditure in both cities (Yogyakarta, Mean = 2.98 and Bandung, Mean = 2.99, on a 4-point scale), even though the people in both cities had little knowledge of electricity prices. Similarly, people's awareness of changes in the monthly electricity bill was fair to high in both cities (Yogyakarta Mean = 3.69 and Bandung Mean = 3.50, on a 5-point scale). These results indicate that the respondents in Yogyakarta and Bandung devoted considerable attention to their monthly electricity expenditure, even though their knowledge of the prices of electricity set by the government was sparse.

These results, linked with the results obtained in sub-chapters 5.1.2 and 5.1.3, indicate that a lack of understanding of the prices of electricity could be the reason that energy consumption was not selected as the first factor considered in decision-making about the purchase of an appliance in either city. In fact, a component of price that is often overlooked in the purchase of an appliance is the price of the energy consumed by the appliance. If the overall price of the appliance includes an overview of the price of the energy consumed by the appliance during its lifetime, people's awareness of higher-efficiency appliances could be increased significantly. This argument is supported by the results in **Table 4**. These results show, for both cities, that respondents' awareness was higher for the overall price of the monthly consumption of electricity than the unit price of electricity.

5.2.2. Electricity Consumed by Appliances
Electrical appliances consume electricity during use, and electricity costs are calculated based on the amount of electricity consumed by appliances. Hence, knowledge of the power consumed by appliances can be an indication of decision-making in the use of appliances. It is important to note that the rate of power consumed by applian-

ces could be a consideration in electrical appliance purchases and use. From this point, respondents in the two cities assessed their awareness of the electricity consumed by the eight appliances (the appliances mentioned in sub-chapter 5.1.1) on a scale of 1 (not at all) to 5 (highly).

The survey results (see **Table 5**) show that the awareness rate of energy consumed by appliances was significantly higher in Yogyakarta than in Bandung for all appliances except air conditioners (p-value > 0.05). Regarding the frequency of respondents' responses, people in Yogyakarta were concerned with the electricity consumed by the water pump, rice cooker, washing machine and refrigerator (Mean 3.83, 3.76, 3.74 and 3.72, respectively). **Figure 5** presents the details of respondents' awareness of the price of electricity consumed by appliances. The people's awareness concerning the water pump and washing machine is because these appliances are considered necessary for daily life and consume a large amount of power. Meanwhile, the rice cooker and refrigerator are always switched on. Overall, the average frequency of respondents' responses to awareness of electricity consumed by appliances is moderate to fair to high for all appliances in both cities.

5.2.3. Awareness of Energy Efficiency of Appliances
The efficiency of an appliance influences its electricity consumption; higher rates of efficiency are associated with lower electricity consumption. Therefore, the knowledge and awareness of the energy efficiency of appliances are important in measuring the costs of electricity. The awareness of the energy efficiency of appliances was analyzed to determine whether people utilize this know-

Table 5. Comparison of respondents' responses to various questions between Bandung and Yogyakarta.

Appliances	T-test significance (p-value)		
	Awareness of rate of power consumed by appliances	Awareness of energy efficiency of appliances	Response of change of electricity price
TV	0.00	0.00	0.04
Refrigerator	0.00	0.00	0.00
Air conditioner	0.83	0.69	0.76
Electric fan	0.00	0.01	0.06
Lighting	0.00	0.00	0.67
Rice cooker	0.00	0.00	0.00
Water pump	0.00	0.00	0.01
Washing machine	0.00	0.00	0.00

ledge in using electrical appliances in their home to reduce the electricity consumption and monthly electricity bill. From the survey in both cities, respondents were asked to rate their awareness of the energy efficiency of eight appliances. The question was scaled from 1 (not at all) to 5 (highly).

The results indicate that, on average, the respondents' awareness was moderate to fair to high for all appliances, with averages above 3 (but less than 4) on the 5-point scale (see details in **Figure 6**). The survey finds significant differences between the two cities. The awareness rate of energy efficiency was significantly in Yogyakarta than in Bandung for all appliances except air conditioners (*p*-value > 0.05). The comparison of awareness of energy efficiency of appliances in Bandung and Yogyakarta is shown in **Table 5**.

From sub-chapter 5.2.2 and 5.2.3, the results indicate

that respondents in Yogyakarta give greater attention to electrical appliance use (power consumption and energy efficiency of appliances) than respondents in Bandung. In addition, the results show that awareness of energy efficiency and electricity consumed by appliances could be a decision-making signal for electricity consumption in the home. This result is surprising because energy consumption was not considered as the priority factor in the choice determinants of appliance purchases (as demonstrated in sub-chapter 5.1.3).

5.2.4. Response to the Change in Electricity Price

The price is suspected to be an important indicator in decision-making concerning electrical appliance use. Thus, consumers' response to the change in prices was assessed. Respondents in each city were asked how their use of each appliance would change if the current electricity price increased by 10%. The question was scaled from 1 (large decrease) to 5 (large increase).

The results indicate that respondents in Yogyakarta were significantly more likely to reduce their electrical appliance than those in Bandung for TV, refrigerator, rice cooker, water pump, and washing machine. However, no significant differences were found for air conditioner, electric fan and lighting (*p*-value > 0.05). Details of the comparison of respondents' response to change in electricity prices are presented in **Table 5**. In terms of the average frequency of respondents' response, if the electricity price increased by 10%, the people in Bandung tended not to change their appliance use (all average frequencies of respondents' responses are over 2.5), except for air conditioner use, which was slightly decreased (Mean = 2.19). Similarly, respondents in Yogyakarta tended to maintain their normal electricity use, except for air conditioner use (see **Figure 7**). According to these findings, change in electricity prices did not serve as a signal in decision-making about electricity consumption

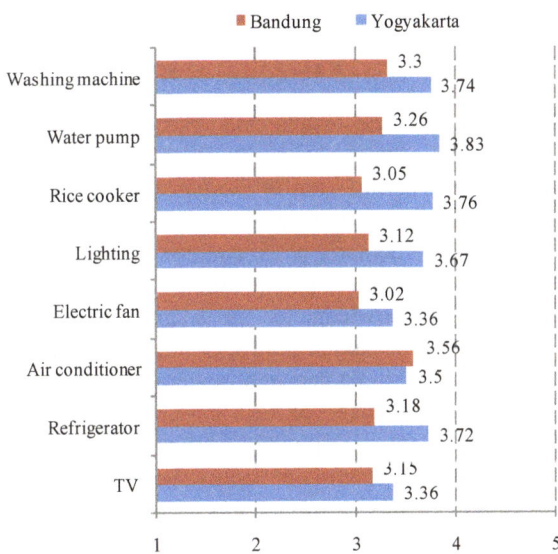

Figure 5. Respondents' awareness of the rate of power consumed by appliances.

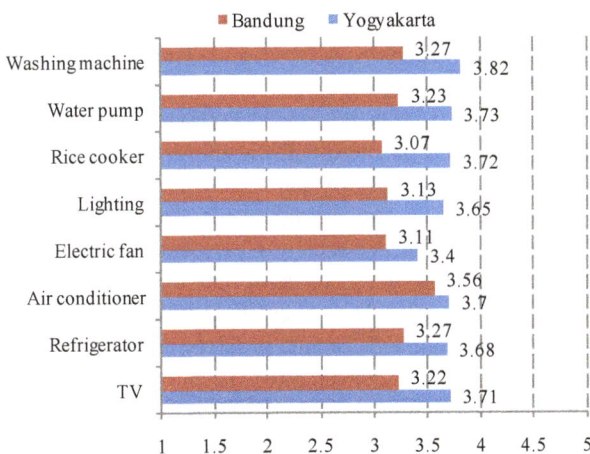

Figure 6. Respondents' awareness of appliances' energy efficiency.

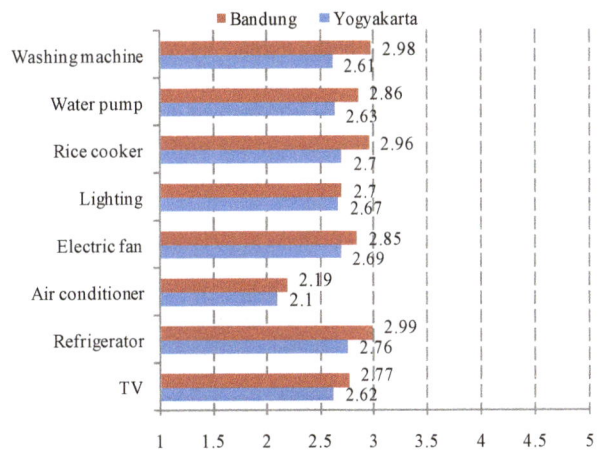

Figure 7. Respondents' responses to change in electricity price.

in the home.

5.2.5. Effects of Payment System

Yamamoto [20] argued that the payment system for home electricity consumption plays an important role in decision-making. Faruqui [22] reviewed 12 pilot studies that investigated the effect of in-home displays that showed electricity use on consumer behavior and found that prepayment metering increased awareness of electricity use. In Indonesia, two payment systems have been enacted, the post-paid system and the prepaid system. With the post-paid system, the consumers are charged a load fee and usage fee. The consumers can view the total amount of electricity use during a month on their bill. Meanwhile, with the prepaid system, the consumer does not pay a load fee, but must purchase a voucher through an Automated Teller Machine or specific designated kiosks for an amount that they select. The prepaid system shows the amount of electricity consumed in real time and the remaining electricity that can be used by the consumer.

The next assessment determined differences in the respondents' awareness of their monthly electricity use based on the prepaid and post-paid systems on a scale of 1 (not at all) to 5 (highly). The results are presented in **Table 6**. In Yogyakarta, there was no significant difference in respondents' awareness of their electricity payment between the two systems (p-value > 0.05); however, the respondents with the prepaid system displayed greater awareness of electricity payment than respondents with the post-paid system (Mean = 4.23 and Mean = 4.14, respectively). In Bandung, there was a significant difference between post-paid and prepaid systems (p-value = 0.01). Respondents with the post-paid system displayed greater awareness than respondents with the prepaid system; however, the prepaid system (Mean = 3.97) did not guide people to attain greater awareness of electricity consumption than did the post-paid system (Mean = 4.37). In addition, according to the average frequency of responses, both payment systems in Bandung and Yogyakarta served as a signal in decision-making about electricity consumption (Mean of all payment systems > 3.90). These results are in line with the findings in sub-chapter 5.2.1 that people in Bandung and Yogyakarta

Table 6. Comparison of effects of electricity payment.

City	Payment system	M	p-value
Yogyakarta	Post-paid	4.14	0.63
	Prepaid	4.23	
Bandung	Post-paid	4.37	0.01
	Prepaid	3.97	

give great attention to monthly electricity expenditure, although the knowledge of government's electricity prices is low.

6. Policy Implications

Most policies tend to be designed to affect appliance production and distribution or to influence consumer decision and behaviors [21]. This conventional approach in energy efficiency and energy saving policies should be replaced by a new paradigm that integrates consumer decisions and behaviors within a framework of the local cultures. In this section, a general conceptual policy framework is developed to promote energy conservation in two cities. In addition to the general framework, specific policy based on different indigenous responses is proposed.

6.1. Conceptual Household Energy Efficiency and Energy Saving Policy Framework for Yogyakarta and Bandung Cities

As presented in the results and discussion section, a choice determinant in the purchase of electrical appliances and decision-making in the purchase of electrical appliances originate from three internal factors: economic motives (particularly profit and loss reasons), human psychology (awareness, habit, attitude, and norm), and perspective or knowledge of the technology of appliances. Meanwhile, culture indirectly influences the internal factors in making a decision concerning the purchase of electrical appliances or consumption of electricity. Thus, when developing policy, local cultures should be considered as a source of sensitivity. Such consideration will result in the public's wider acceptance of policy

The findings of this study show that policy-makers should utilize differences approaches in two cities to successfully implement energy conservation policy. However, there is a general policy framework that could be implemented in both cities based on the analysis of the respondents' responses. A Conceptual Household Energy Efficiency and Energy Saving Policy Framework (CHE3SPF) has been developed for two cities to show how the policies of energy efficiency and energy saving can result in interventions that aim to manage electricity consumption in the household sector. The CHE3SPF is designed based on the society's perspectives; therefore, the policies input are considered based on the cultural characteristics only. The schematic of the CHE3SPF is presented in **Figure 8**. Details of the policies that aim to manage electricity consumption in the two cities are presented below:

1) *Financial incentives.* The price of appliances is the first factor considered prior to purchase; therefore, higher-efficiency appliances should be sold at an affordable

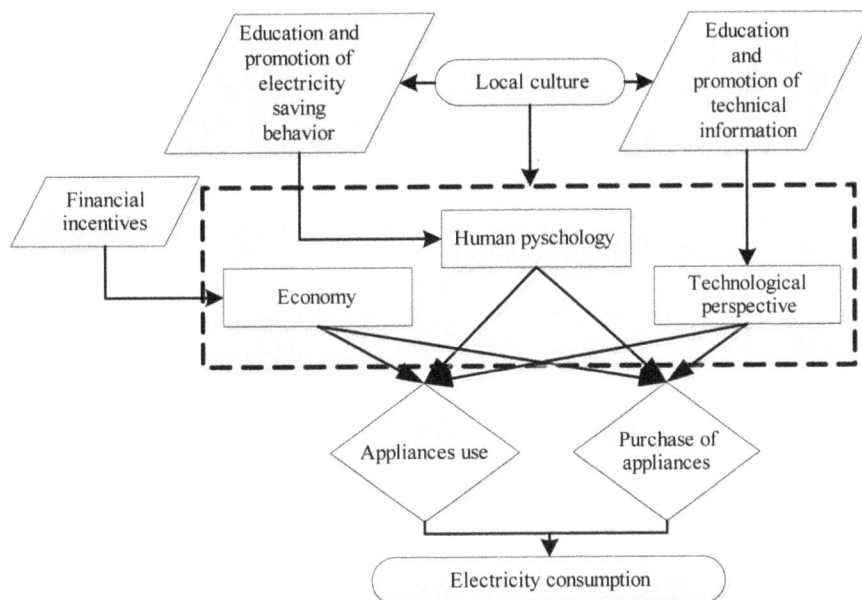

Figure 8. Conceptual Household Energy Efficiency and Energy Saving Policy Framework for both cities.

price. A financial incentive to consumers who are willing to purchase the higher-efficiency appliance class might be required. This policy will attract more people to these appliance types.

2) *Education and promotion of technical information.* The study found that people's knowledge of electrical appliance use (power consumption and energy efficiency of appliances) could be a signal in decision-making concerning electrical appliances. Therefore, promoting the adoption of high-efficiency appliances is an important way to improve awareness of electricity use.

3) *Education and promotion of electricity saving behavior.* Technology policies are one of the options available for the reduction of carbon emissions and the usage of energy. However, gains in the efficiency of energy consumption will result in an effective reduction in the per unit price of energy services. As a result, consumption of energy services should increase [23]. In the literature, this impact is often called the "rebound effect of energy efficiency". Doubts have recurrently been raised concerning the extent to which energy efficiency can reduce the demand for energy [24]. The level of the rebound effect for household sectors varies depending on consumer awareness during consumption of appliances. Economy, particularly motivation of profit and loss, is identified as having the greatest influence on the awareness of appliance use. However, based on the current findings, the change in electricity prices policy to increase people's awareness of their electricity consumption and reduce the level of the rebound effect will not result in a significant electricity saving. In the interest of decreasing the rebound effect, Oikonomou [25] integrated several theories in environmental psychology to iden-

tify parameters that affect end-use behavior in energy saving and energy efficiency. Although, on average, people in both cities displayed awareness of electricity consumption, the education and promotion of electricity saving behavior must be implemented to foster this habit in all people. This effort is important in combating and anticipating the rebound effect of energy efficiency to enforce the aims of energy conservation.

6.2. Proposal of Specific Policy Based on Indigenous Response

Based on the differences in driving factors in choice determinants in the purchase of electrical appliances and decision-making in the purchase of electrical appliances between the two cities, intervention should be tailored with an emphasis on local culture. Therefore, additional extended specific policies from CHE3SPF might be required. The implementation of energy labeling is essential for improving the adoption of higher-efficiency appliances. This scheme might be used in enriching the commercial advertisements with the information related to the energy consumption of appliances, as commercial advertisements play a key role in facilitating the consideration of the purchase of appliances in Yogyakarta. This scheme will increase people's knowledge of the technology within an appliance, particularly related to energy efficiency and energy consumption. The energy labeling scheme has been realized in many countries, such as European Union countries, US, Iran, Brazil, Thailand, Australia and India [26,27], as a key component of efforts to increase the diffusion of energy efficient household appliances. In addition, the energy labeling scheme is expected to indirectly improve the quality of appliances sold in domestic market, as the standard of energy labeling requires a minimal efficiency that should be obtained by an appliance.

The specific policies that might be applied for people in Bandung are as follows:

1) An increased implementation of the post-paid system would increase people's awareness of electrical appliance use.

2) In Bandung, the store's sales staff is a primary source of information about appliances. Thus, the sales staff has the greatest impact on people's choice of energy-efficient appliances. Therefore, the policy to provide a regular training for sales staff would result in better penetration of high-efficiency appliance purchases.

7. Conclusions

Electricity consumption in Indonesian households with different cultural backgrounds and ethnicities has been studied and analyzed with a case study of Yogyakarta city and Bandung city. The results indicate that in Yogyakarta and Bandung, the quality and price of an appliance are the most important factors considered prior to the purchase and decision-making concerning the purchase. In Yogyakarta, the third factor considered in the decision-making concerning the purchase is energy consumption. However, in Bandung, energy consumption is not a priority factor.

The decision-making concerning electrical appliance use in the home in two cities with different cultural backgrounds has been studied and analyzed. The results show that monthly electricity expenditure and knowledge of electrical appliance use (power consumption and energy efficiency of appliances) could be a signal in decision-making concerning electrical appliances in both cities regardless of the payment system, even though people have little knowledge of the government's electricity prices. The awareness rate of energy consumed by appliances and energy efficiency of appliances is significantly higher in Yogyakarta than in Bandung for all appliances except air conditioners. The results also indicate that if electricity prices increased by 10%, respondents in Yogyakarta are significantly more likely to reduce their electrical appliances use than those in Bandung. However, on average, respondents' would not change their appliance use. This indicates that the change in electricity prices does not serve as a signal in decision-making.

Finally, based on these results, a CHE3SPF for both cities has been developed. It illustrates the integration of intervention strategy in the purchase and use of electrical appliances with regard to the consumers' decisions and behaviors within a framework of the local culture to manage electricity consumption in the household sector.

The CHE3SPF aids in developing interventions of choice determinants in the purchase of electrical appliances and decision-making in the use of electrical appliances through economy motives, human psychology, and perspective on the technology of appliances. To address the sensitivity influenced by cultures, several additional specific policies are proposed to improve success in the implementation of energy saving and energy efficiency. These policies are an energy labeling scheme in Yogyakarta city and electricity payment with pre-paid system and training for store sales staff in Bandung city.

Overall, the results of this study have presented a strategy to improve the adoption of higher-efficiency appliances in Indonesia to manage the energy use of households with different cultural backgrounds and ethnicities. For this strategy to succeed, policy improvement and a strong willingness to carry out these strategies are required. A further study that integrates the consumer decision and behaviors in the purchase and use of electrical appliances in the electricity system in the framework of different cultures should be conducted to determine the effectiveness of intervention policies in the future long-term electricity consumption.

8. Acknowledgements

This study was funded by the Kyoto University Global Centre of Excellence (GCOE) Program "Energy Science in the Age of Global Warming". The first author conveys gratitude to the Ministry of Education, Culture, Sports, Science and Technology (MEXT) Japan for providing support during the study.

REFERENCES

[1] E. Ghisi, S. Gosch and R. Lamberts, "Electricity End-Uses in the Residential Sector of Brazil," *Energy Policy*, Vol. 35, No. 8, 2007, pp. 4107-4120.

[2] H. Wilhite, H. Nakagami, T. Masuda, Y. Yamaga and H. Haneda, "A Cross-Cultural Analysis of Household Energy Use Behavior in Japan and Norway," *Energy Policy*, Vol. 24, No. 9, 1996, pp. 795-803.

[3] A. D. Poyer, L. Henderson and A. P. S. Teotia, "Residential Energy Consumption across Different Population Groups: Comparative Analysis for Latino and Non-Latino Households in USA," *Energy Economics*, Vo. 19, No. 4, 1997, pp. 445-463.

[4] K. Genjo, S. Tanabe, S. Matsumoto, K. Hasegawa and H. Yoshino, "Relationship between Possession of Electric Appliances and Electricity for Lighting and Others in Japanese Households," *Energy and Buildings*, Vol. 37, No. 3, 2005, pp. 259-272.

[5] R. Helbert, "Factors Determining Household Fuel Choice

in Guatemala," *Environment and Development Economics*, Vol. 10, No. 3, 2005, pp. 337-361.

[6]   C. Wilson and H. Dowlatabadi, "Models of Decision Making and Residential Energy Use," *Annual Review of Environment and Resources*, Vol. 32, 2007, pp. 169-203.

[7]   R. Kowsari and H. Zerriffi, "Three Dimensional Energy Profiles: A Conceptual Framework for Assessing Household Energy Use," *Energy Policy*, Vol. 39, No. 12, 2011, pp. 7505-7517.

[8]   L. Lutzenhiser, "A Cultural Model of Household Energy Consumption," *Energy*, Vol. 19, No. 1, 1992, pp. 47-60.

[9]   M. E. Wijaya and T. Tezuka, "Understanding Socio-Economic Driving Factors of Indonesian Households Electricity Consumption in Two Urban Areas," In: T. Yao, Ed., *Zero-Carbon Energy Kyoto* 2011, Springer, Tokyo, 2012, pp. 56-60.

[10]   M. E. Wijaya and T. Tezuka, "Measures for Improving the Adoption of Higher Efficiency Appliances in Indonesian Households: An Analysis of Lifetime Use and Decision-Making in The Purchase of Electrical Appliances," *Applied Energy*, 2013, in Press.

[11]   Statistics Indonesia, "Strategic Data 2010," 2011. http://www.bps.go.id/ 65tahun/data_strategis.pdf

[12]   Center for Data and Information on Energy and Mineral Resources, "Handbook of Energy and Economic Statistics of Indonesia 2010," 2011. http://prokum.esdm.go.id/Publikasi/Handbook%20of%20 Energy%20&%20Economic%20Statistics%20of%20Indo- nesia%20/Handbook%202010.pdf

[13]   Statistics Indonesia, "Population Census 2010," 2011. http://www.bps.go.id/download_file   /SP2010_agregat_ data_perProvinsi.pdf

[14]   P. T. Perusahaan and L. Negara, "Annual Report 2009: Brightening the Nation in Harmony," 2010. http://www.pln.co.id/dataweb/AR/ARPLN2009.pdf

[15]   Ministry of Energy and Mineral Resources, "National Electricity Master Plan 2008-2027," 2008, http://prokum.esdm.go.id/kepmen/2008/Kepmen%20ESD M%202682%202008%20 ZRUKN.pdf

[16]   B. Lietaer and S. De Meulenaere, "Sustaining Cultural Vitality in a Globalizing World: The Balinese Example," *International Journal of Social Economics*, Vol. 30, No. 9, 2003, pp. 967-984.

[17]   L. Suryadunata, E. N. Arifin and A. Ananta. "Indonesia's Population: Ethnicity and Religion in Changing Political Landscape," Institute of Southeast Asian Studies, Singapore, 2003.

[18]   D. Young, "When Do Energy-Efficiency Appliances Generate Energy Savings? Some Evidence from Canada," *Energy Policy*, Vol. 36, No. 1, 2007, pp. 34-46.

[19]   D. A. Guerin, B. L. Yust and J. G. Coopet, "Occupant Predictors of Household Energy Behavior and Consumption Change as Found in Energy Studies Since 1975," *Family and Consumer Science Research Journal*, Vol. 19, No. 1, 2000, pp. 48-80.

[20]   Y. Yamamoto, A. Suzuki, Y. Fuwa and T. Satu, "Decision-making in Electrical Appliance Use in the Home," *Energy Policy*, Vol. 36, No. 5, 2008, pp. 1679-1686.

[21]   R. Gaspar and D. Antunes, "Energy Efficiency and Appliance Purchase in Europe: Consumer Profiles and Choice Determinants," *Energy Policy*, Vol. 39, No. 11, 2011, pp. 7335-7346.

[22]   A. Faruqui, S. Sergici and A. Sharif, "The Impact of Informational Feedback on Energy Consumption: A Survey of the Experimental Evidence," *Energy*, Vol. 35, No. 4, 2010, pp. 1598-1608.

[23]   L. A. Greening, D. L. Greene and C. Difiglio, "Energy Efficiency and Consumption: The Rebound Effect—A Survey," *Energy Policy*, Vol. 28, No. 6-7, 2000, pp. 389-340.

[24]   J. Nässén and J. Holmberg, "Quantifying the Rebound Effects of Energy Efficiency Improvements and Energy Conserving Behavior in Sweden," *Energy Efficiency*, Vol. 2, No. 3, 2009, pp. 221-231.

[25]   V. Oikonomou, F. Becchis, L. Steg and D. Russolillo, "Energy Saving and Energy Efficiency Concepts for Policy Making," *Energy Policy*, Vol. 37, No. 11, 2009, pp. 4787-4796.

[26]   L. Harrington and G. Wilkenfeld, "Appliance Efficiency Program in Australia: Labeling and Standards," *Energy and Buildings*, Vol. 26, No. 1, 1997, pp. 81-88.

[27]   B. Mills and J. Schleich, "What's Driving Energy Efficiency Appliance Label Awareness and Purchase Propensity?" *Energy Policy*, Vol. 38, No. 2, 2010, pp. 814-825.

13

Energy Efficient Hospitals Air Conditioning Systems

Essam E. Khalil

Mechanical Engineering, Faculty of Engineering, Cairo University, Cairo, Egypt

ABSTRACT

Energy Efficiency and Indoor Air Quality in the healthcare applications and particularly in surgical operating theatres are important features in modernized designs. The various reasons for deviation from obtaining optimum IAQ and energy efficient buildings are listed. The air conditioning systems serving the operating rooms require careful design to minimize the concentration of airborne organisms. Numerical approach is an appropriate tool to be utilized to adequately identify the airflow patterns temperatures and relative humidity distributions and hence energy efficient designs.

Keywords: CFD; Hospitals; Air Conditioning; Energy Efficiency

1. Introduction

The air quality of the indoor environment affects human comfort in a multitude of ways, depending on the contaminant. Airborne contaminants range from toxic substances such as carbon monoxide to nuisance matter such as large dust particles. The effects of airborne contaminant on humans vary greatly with the nature and type of contaminant. Hospitals and other healthcare facilities are complex environments that require ventilation for comfort and to control hazardous emissions. Ventilation and air distribution pattern have a great effect on the IAQ; also air movement in a room becomes of vital importance to comfort. To design and construct ventilation system that is capable of efficiently fulfilling all requirements, often even contradictory, is a great challenge [1-14].

2. IAQ Index, Factors and Hygiene

Comfort air conditioning is defined as "the process of treating air to control simultaneously its temperature, humidity, cleanliness, and distribution to meet the comfort requirements of the occupants of the conditioned space." The ASHRAE standard 55-targeted quoted. "Thermal comfort is that condition of mind that expresses satisfaction with the thermal environment."

2.1. Indoor Temperatures and Relative Humidity

Relative humidity affects our comfort in numerous ways both directly and indirectly. It is a thermal sensation, skin moisture, discomfort, and tactile sensation of fabrics, health and perception of air quality. Low humidity affects comfort and health. In hot climates, it is recommended to set the temperature from 16°C to 22°C at the working domain with relative humidity of 45% to 55%.

2.2. Airflow Velocity

The laminar airflow concept developed for hospitals, healthcare facilities and clean room use has advocated the need for both vertical and horizontal laminar airflow systems, around the surgical team. The unidirectional laminar airflow pattern is commonly attained at a discharge velocity of 0.45 ± 0.10 m/s.

2.3. Pressure Relationship

Ventilation recommendations for comfort, asepsis, and odor control in areas of acute care in hospitals that directly affect patient care are presented by the healthcare standards. Positive pressure is recommended and maintained in operating theatres with high air changes per hour up to 40 ACH

2.4. Air Movement Efficiency

In hospital facilities, the air movement takes an extra important role in the controlling of the healthy criteria. Undesirable airflow between rooms and floors is often difficult to control because of open doors, movement of staff and patients, temperature differentials, and stack effect. It should be emphasized that the air distribution and direction play the more important role in the airborne-infectious-disease management; good IAQ starts with building design.

3. Numerical Method

Three time averaged velocity components in X, Y, and Z

coordinate directions were obtained by solving the governing equations using a "SIMPLE Numerical Algorithm" [Semi Implicit Method for Pressure Linked Equation] described earlier in the work Spalding and Patankar, [14], Launder and Spalding, [15], Khalil [16]. The turbulence characteristics were represented by a two-equation k-ε model that accounts for normal and shear stresses and near-wall functions. Fluid properties such as densities, viscosity and thermal conductivity were obtained from references. The present work made use of the Computer Program 3DHVAC, which was developed, by Khalil [5] and modified later by Kameel [7,8], and by Kameel and Khalil [9-13]. The program solves the differential equations governing the transport of mass, three momentum components and energy in three-dimensional configurations under steady and unsteady conditions. It uses the SIMPLE algorithm with orthogonal three dimensional meshes of grid nodes that covers the calculation zone (**Figure 1**).

The different governing partial differential equations are typically expressed in a general form as:

$$\text{Div}\left(\rho V\Phi - \Gamma_{\Phi,\text{eff}} \cdot \text{grad } \Phi\right) = S_\Phi \qquad (1)$$

where:

ρ = Air density, kg/m^3;
Φ = Dependent variable;
V = Velocity vector;
$\Gamma_{\Phi,\text{eff}}$ = Effective diffusion coefficient;
S_Φ = Source term of Φ.

The effective diffusion coefficients and source terms for the various differential equations are listed in **Table 1**.

Table 1. Values of Φ, $\Gamma_{\Phi,\text{eff}}$, and S_Φ for partial differential equations.

	Φ	$\Gamma_{\Phi,\text{eff}}$	S_Φ
Continuity	1	0	0
X-momentum	U	μ	$-\partial P/\partial x + \rho g_x$
Y-momentum	V	μ	$-\partial P/\partial y + \rho g_y$
Z-momentum	W	μ	$-\partial P/\partial z + \rho g_z + \rho g\beta\Delta t$
Enthalpy	h	μ/σ_h	S_h
τ-age equation	τ	μ/σ_τ	ρ
k-equation	k	μ/σ_k	$G - \rho\varepsilon$
ε-equation	ε	μ/σ_ε	$C_1\varepsilon G/k - C_2\rho\varepsilon^2/k$

$$\mu = \mu_{\text{lam}} + \mu_t$$

$$\mu_t = \rho C_\mu k^2/\varepsilon$$

$$G = \mu\,[2\{(\partial U/\partial x)^2 + (\partial V/\partial y)^2 + (\partial W/\partial z)^2\} + (\partial U/\partial y + \partial V/\partial x)^2 + (\partial V/\partial z + \partial W/\partial y)^2 + (\partial U/\partial z + \partial W/\partial x)^2]$$

$$C_1 = 1.44,\ C_2 = 1.92,\ C_\mu = 0.09$$

$$\sigma_\tau = 0.9,\ \sigma_k = 1.0,\ \sigma_\varepsilon = 1.3$$

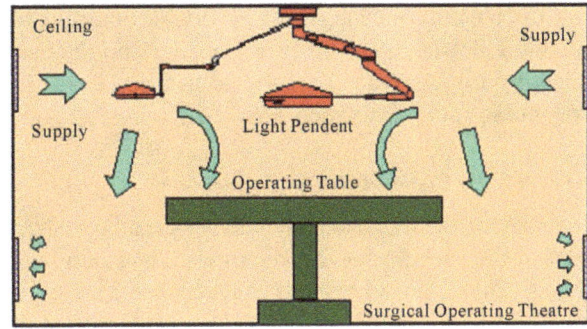

(a)

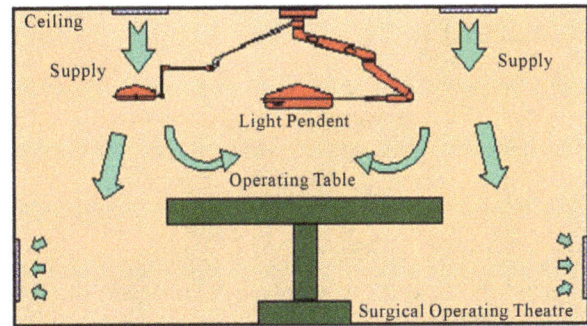

(b)

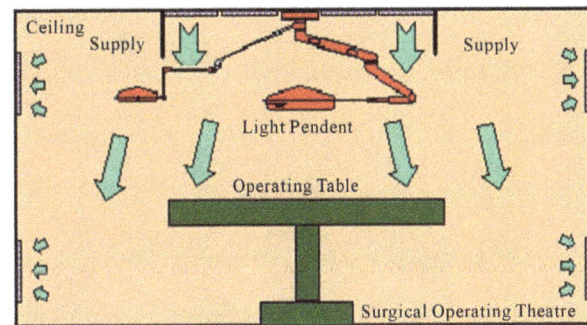

(c)

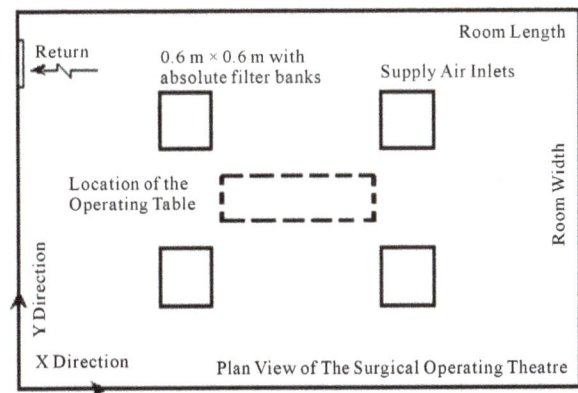

(d)

Figure 1. Proposed Airside Designs of HVAC Systems Serve the Surgical Operating Theatres. (a) Design A Configuration(X-Z plane); (b) Design B Configuration(X-Z plane); (c) Design C Configuration(X-Z plane); (d) X-Y Plan view of operating theatre.

The Computational Fluid Dynamics (CFD) model utilizes the following approximations in calculating the turbulence quantities, such as isotropic turbulence and the Boussinesq eddy viscosity concept.

3.1. Boundary Conditions

The solution of the governing equations can be realized through the specifications of appropriate boundary conditions. The values of velocity, temperature, kinetic energy, and its dissipation rate should be specified at all boundaries.

3.2. External Walls

A non-slip condition at all solid wall is applied to the velocities. The logarithmic law of the wall (wall function) of Launder and Spalding [15] was used here, for the near wall boundary layer.

3.3. Air Supply Inlets

At inlets, the air velocity was assumed to have a uniform distribution; inlet values of the temperature were assumed to be of a constant value and uniform distribution. The kinetic energy of turbulence and its dissipation rate are commonly estimated as follow.

$k_{in} = 3 \ (0.5 \ (I_{in} \ U_{in})^2)$,

$\varepsilon_{in} = C_\mu \ (k_{in})^{1.5}/l_e$,

where

I_{in} = Intensity of disturbance at air inlet.

l_e = Dissipation length at air inlet.

3.4. Initial Guessed Values

All velocity components were set as zeros initially, and temperatures were assumed to be equal to the steady state value of the comfort condition. The kinetic energy and its dissipation are estimated in the following manners.

$$k_{initial} = 11E-5, \ \varepsilon_{initial} = C_\mu(k_{initial})^{1.5}/c \cdot d,$$

where c = constant, and d = distance to nearest sidewall.

3.5. Numerical Procedure

The Computer Program, 3DHVAC is used to solve the time-independent (steady state) conservation equations together with the standard k-ε model as Launder and Spalding [15], Khalil [16] and Khalil et al. [17], and the corresponding boundary conditions. The numerical solution grid divided the space of the surgical operating theatre into discretized computational cells 500,000 grid nodes using the procedure of Kameel and Khalil [18]. The discrete finite difference equations were solved with the SIMPLE algorithm, Patankar, [19]. Solution convergence criteria, was applied at each iteration and ensured the summations of normalized residuals were less than

0.1% for flow, 1% for k and ε, and 0.1% for energy.

3.6. Model Validation

Previous comparisons between measured and predicted flow pattern, turbulence characteristics, and heat transfer were reported earlier in the open literature utilizing the present computational capabilities (3DHVAC), reference should be made to these for further details and assessments. A summary of the main assessment is expressed here as follows. The predictions of flow and turbulence characteristics are in general qualitative agreement with the corresponding experiments and numerical simulations published by others, Blum, [20], Neilsen, [21] were shown the work of Kameel [8]; the trends are in adequate agreement for engineering purposes. Nevertheless discrepancies exist and particularly in the vicinity of recirculation zone boundaries. More discrepancies were also observed in situation with heating flows than those of ventilation or cooling. For more details of validations for present application (Surgical Operating Theatres) review the work of Kameel and Khalil [13] and [22-24]. These papers describe the validation of the turbulence model to use it for the prediction of airflow characteristics in the surgical operating theatres. Also, these papers introduce a complete case study performed in a 1200 bed teaching hospital (New Kasr El-Aini Teaching Hospital, Cairo University).

4. Results and Discussion

Figure 2 shows prediction results of the proposed designs "A", "B", and "C". The three designs are simulated for the same ACH condition and different flow patterns were developed in the vicinity of the supply diffusers of each case. The present flow pattern, turbulence, and temperatures contours were predicted and represented as zonal areas. Each air characteristic, (velocity "V", turbulent kinetic energy "K", temperature "T", and local mean age "LMA"), is divided into two clear zone categories, acceptable and rejected zones. The acceptable zone category is the area that containing the desired design values of the considered characteristic (Φ), and the other space is the rejected category zone for the same air and flow characteristics. The areas that are allocated on the figures represent the rejected zones of each flow characteristic. If we suppose that the area of refused value for any characteristics (Φ) is R(Φ), then the area of acceptable values is A(Φ)= [1 − R(Φ)]. The identification and allocation of the unacceptable zones are aimed to highlight the final optimum clean and recommended areas of occupancy in the domain, which has all acceptable values for all air characteristics.

The optimum area of the operating theatre is the U(D), where U(D) = [A(Φ₁) ∩ A(Φ₂) ∩ ⋯ ∩ A(Φₙ)], and n is

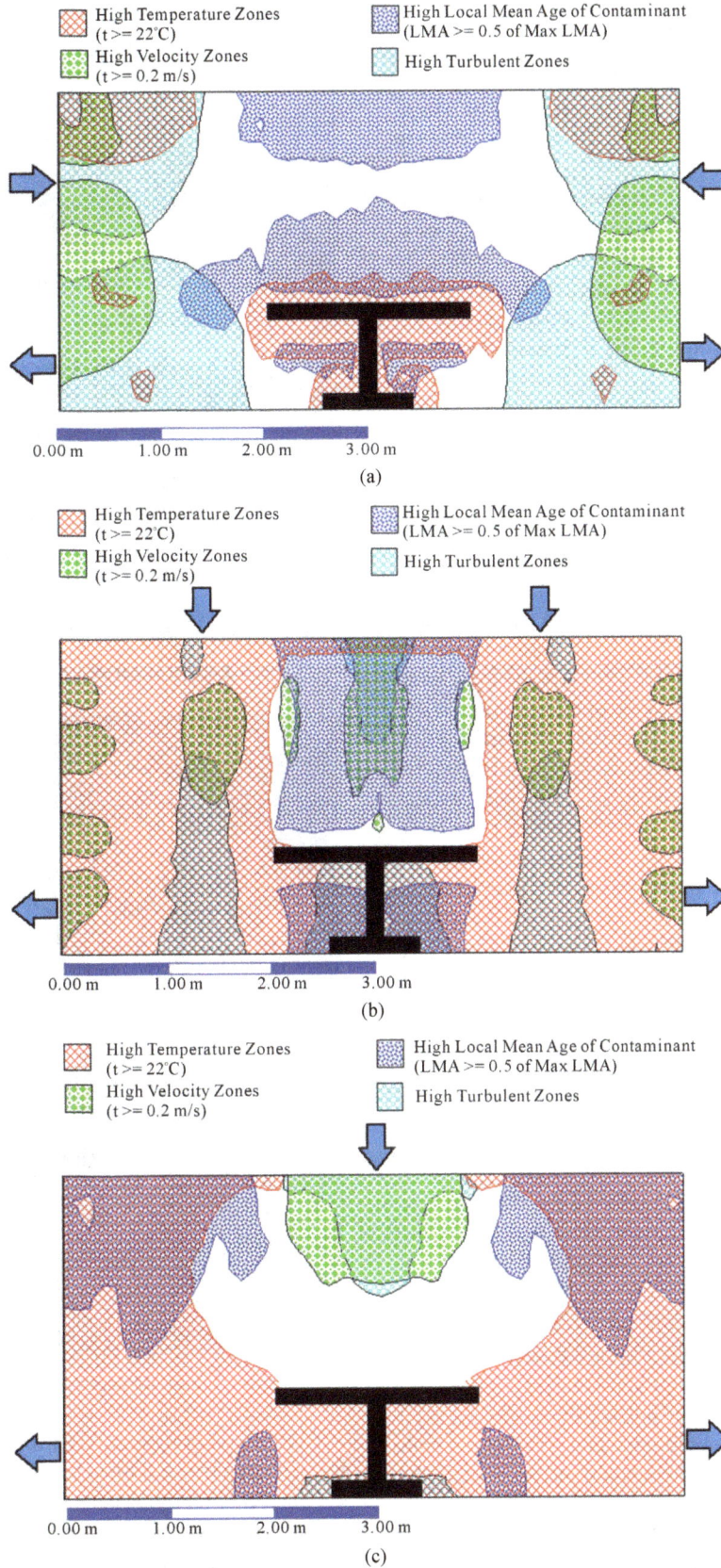

Figure 2. Prediction Results of Proposed Airside Designs of HVAC Systems, (X-Z plane). (a) Prediction Results of Design "A", (X-Z plane); (b) Prediction Results of Design "B" (X-Z Plane); (c) Prediction Results of Design "C".

the numbers of characteristics involved in the investigation. Indeed, it is so simple to calculate $U(D) = 1 − [R(\Phi_1)$ U $R(\Phi_2)$ U $\cdots$ U $R(\Phi_n)]$. For the present work, the calculation of $U(D) = 1 − [R(V)$ U $R(K)$ U $R(T)$ U $R(LMA)]$. So in the un-shaded area represents the optimum domain in the proposed designs.

From the results, one can observe that the proposed design "B" doesn't provide any optimum clean occupancy area or healthy area in the room according the proposed analytical previous equations. The proposed design "B" is completely unaccepted design model and should be excluded from any future recommendations of the HVAC designs.

According to the design parameter of supplying the same ACH, the supply air velocity in the case of design "A" is greater in magnitude than the corresponding velocity of design "C". That leads to decrease the maximum LMA value of design "A" than the corresponding value of design "C". But still, the unacceptable are of LMA values concentrated over the operating table. The LMA values in the vicinity of the extract ports is found to be less than the corresponding LMA values in the vicinity of the operating table, this indicates the increasing of contaminant concentration in the operating zone.

Indeed, the air distribution pattern is participating strongly in characterizing the LMA distribution pattern. The short circuit between the supply inlets and extract ports in case "A", not only, participated of creating a poor airflow movement over the operating table and correspondingly poorer scavenging, which decreased the hygiene level, but also, expensed the most of the conditioned air (paid for air) by throwing it without any useful utilization. The proper description of the design "A" is "inefficient sick design". On the other hand the proposed design "C" gave a perfect protection in the vicinity of the operating table and created the proper conditions in the operating area as a whole.

5. Analyses

The **Figures 3-6** represent the conventional flow patterns, temperature and relative humidity distributions in the typical operating theatres configurations in the vertical plane (X-Z directions).

These predictions were obtained for room configurations of 6 m × 5 m and height 3 m in the X, Y and Z coordinates directions, with a middle table of 2 m length; 0.5 m wide and 1.0 m above finished floor. The surgical team is distributed around the table and was treated both physically and as source of heat and moisture. The lighting pendent was ceiling mounted .The air velocity from the ceiling plenum was 0.287 m/s. The effects of the presence of the surgical team and lighting pendent are clear on the velocities, temperatures and relative humidity contour distributions, Khalil, [25-28].

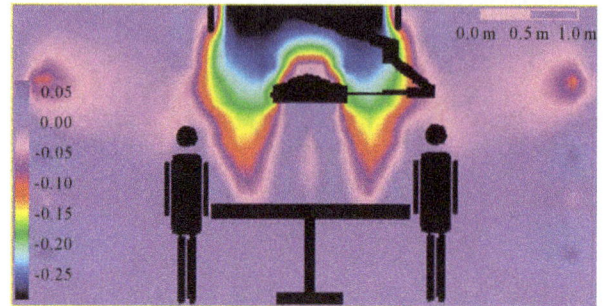

Figure 3. Predicted air velocity patterns.

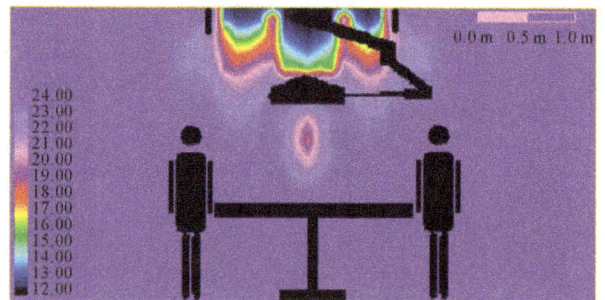

Figure 4. Air temperature patterns.

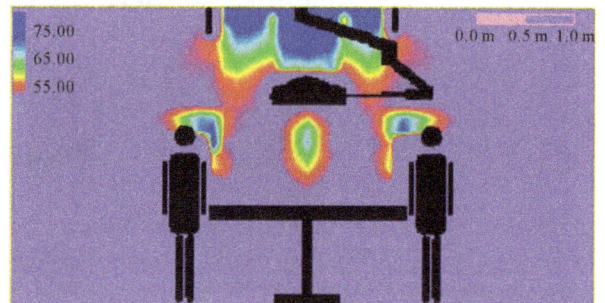

Figure 5. Relative humidity patterns.

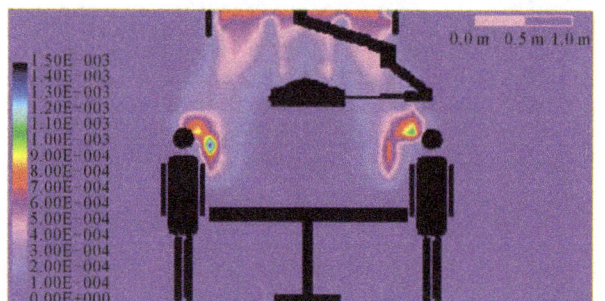

Figure 6. Kinetic energy of turbulence patterns.

The following tips represent the average more general and effective tips and hints to energy managers:

1) Define Energy Management through typical Operation and maintenance perceptions; hospital energy management is both mandatory and essential.

2) Determine available data such as metered Data, Data bases, reporting tools, and Energy costs, Equipment

Energy Efficiency/Performance, Direct Digital Control/ Energy Management Systems, available Operation Manuals and Building Energy Performance

3) Select your hospital energy management team who can easily appreciate management concepts and energy efficient operation and auditing. Develop lines of communication and evaluate available staff resources, capabilities, required training

4) Involve Operators and Maintenance Team on the location; solicit their ideas, they know their systems best and already often have good energy efficiency ideas; determine what would be required to implement an electrical efficiency measure. Apply low-to-no cost options and evaluate ability to not sacrifice comfort or productivity while implementing an energy efficiency measure then identify how to make an energy efficiency measure systematized.

5) Select priority for energy saving through HVAC control, Lighting control, Process systems, Chillers and boilers. For that use potential team members that may include energy, utilities, facilities, Operation and maintenance and end-user stakeholders. Involve someone experienced with implementing energy efficiency procedures to facilitate the process. Collect adequate information and data; optimize use or turn it off if you can. Establish a mechanism to facilitate communication; do not walk away; set-up a mechanism to continually apply energy efficiency measures

6) It is very imperative to involve operations and facility staff on the team and attempt to understand their perspective, make it clear that suggested improvements are not a reflection of their current performance.

7) Measure and estimate how much energy is being used by different equipment, utilizing available meters and measurement instruments. The energy efficiency performance of facilities is to be evaluated through the knowledge of energy use data per gross floor area or other means to establish a rating/comparison; careful analyses of simulation data or engineering calculations should be followed.

8) Implement the energy efficiency operational ideas such as the low-to-no cost ideas and maintain communication among people involved and measure performance of the operational changes.

9) Measure where needed to determine savings from operational changes, in case you already have meters and sufficient measurement tools; determine if the data is being properly evaluated .Evaluate energy data and energy efficiency operational changes and document any successes and savings.

10) Operational changes need to be continually implemented and properly documented and retained specially implementation plan to track key energy efficiency operational changes. Maintain communication between Energy Manager, operators, maintenance, and management and document savings to justify energy efficiency program.

6. Concluding Remarks

The complexity of HVAC systems in healthcare premises is increasing due to the additional functions. The HVAC system in this case is intended to provide the comfort and to remove any airborne contaminants that are produced in this application. These design criteria do not influence the HVAC system only, but may also require a special care in the architectural design and including the choice of the room furniture and its location. The designers of HVAC systems should consider, for energy optimization, the importance of the air distribution; the positioning of operating furniture and the using of partial walls may be useful to maintain the air environment in the surgical operating theatres.

It is believed that for the purpose of the energy-efficient operation Air Conditioning and other systems in large healthcare facilities, the energy manager should take into account the tips that are set here in this work. Different actors need different information. For giving relevant advice to the property owner which measures are cost-effective a very careful examination and calculation of the building's energy balance is necessary. A careful analysis is also necessary to give relevant information to the users how they can decrease their energy use without decreasing, under an acceptable level, the indoor air quality and thermal comfort.

7. Acknowledgements

The author would like to acknowledge the assistance given to him by his colleagues and students, particular thanks are due to Dr. R. Kameel and Dr. A. Medhat and Eng. Rana Khalil.

REFERENCES

[1] L. G. Berglund, "Comfort and Humidity," *ASHRAE Journal*, Vol. 40, No. 8, 1998, pp. 35-41.

[2] M. H. Hosni, K. Tsai and A. N. Hawkins, "Numerical Predictions of Room Air Motion," *ASME Fluids Engineering Division Conference*, Part 2, 1996, pp. 745-750.

[3] A. M. Medhat, "Air Conditioning Flow Patterns in Enclosures," M.Sc. Thesis, Cairo University, Cairo, 1993.

[4] E. E. Khalil, "Three-Dimensional Flow Pattern in Enclosures," Interim Report, Egyptalum, Egypt, 1994.

[5] E. E. Khalil, "Fluid Flow Regimes Interactions in Air Conditioned Spaces," *Proceedings of 3rd Jordanian Mechnical Engineering Conference*, Amman, May 1999.

[6] E. E. Khalil, "Computer Aided Design for Comfort in Healthy Air Conditioned Spaces," *Proceedings of Healthy*

Buildings 2000, Finland, Vol. 2, 2000, pp. 461-466.

[7] R. Kameel, "Computer Aided Design of Flow Regimes in Air Conditioned Spaces," M.Sc. Thesis, Cairo University, Cairo, 2000.

[8] R. Kameel, "Computer Aided Design of Flow Regimes in Air Conditioned Operating Theatres," Ph.D. Thesis Work, Cairo University, Cairo, 2002.

[9] R. Kameel and E. E. Khalil, "Computer Aided Design of Flow Regimes in Air Conditioned Spaces," *Proceedings of ESDA2000 ASME 5th Biennial Conference on Engineering Systems Design & Analysis*, Monteux, 2000.

[10] R. Kameel and E. E. Khalil, "Fluid Flow and Heat Transfer in Air Conditioned Spaces," *International Conference of Energy Systems* (*ICES*), Amman, September 2000, 2K, pp. 188-200.

[11] R. Kameel and E. E. Khalil, "Numerical Computations of the Fluid Flow and Heat Transfer in Air-Conditioned Spaces," NHTC2001-20084, *35th National Heat Transfer Conference*, Anaheim, 2001.

[12] R. Kameel and E. E. Khalil, "Air Quality Appraisal in Air Conditioned Spaces: Numerical Analyses," *Proceedings of 4th IAQVEC Conference*, Changsha, 2001, pp. 287-297.

[13] R. Kameel and E. E. Khalil, "Verification of Numerical Prediction of 3-D Air-Conditioned Flow Behavior in Full and Reduced Scale Room Models," *40th Aerospace Sciences Meeting & Exhibit*, Reno, Nevada, AIAA-2002-654, 12-15 January 2002.

[14] D. B. Spalding and S. V. Patankar, "A Calculation Procedure for Heat, Mass and Momentum Transfer in Three Dimensional Parabolic Flows," *International Journal of Heat and Mass Transfer*, Vol. 15, 1974, pp. 1787-1799.

[15] B. E. Launder and D. B. Spalding, "The Numerical Computation of Turbulent Flows," *Computer Methods in Applied Mechanics and Engineering*, Vol. 3, No. 2, 1974, pp. 269-275.

[16] E. E. Khalil, "Flow, Combustion & Heat Transfer in Axisymmetric Furnaces," Ph.D. Thesis, London University, London, 1977.

[17] E. E. Khalil, D. B. Spalding and J. H. Whitelaw, "The Calculation of Local Flow Properties in Two-Dimen-sional Furnaces," *International Journal of Heat and Mass Transfer*, Vol. 18, 1975, pp. 775-792.

[18] R. Kameel and E. E. Khalil, "Generation of the Grid Node Distribution Using Modified Hyperbolic Equations," *40th Aerospace Sciences Meeting & Exhibit*, Reno, Nevada, AIAA-2002-656, January 2002.

[19] S. V. Patankar, "Numerical Heat Transfer and Fluid Flow," Hemisphere Pub., WDC, 1980.

[20] H. M. Blum, "Experimental Verification of Turbulence Models," *ASHRAE Fundamentals*, Vol. 1, PT30, ASHRAE, Atlanta, 1956.

[21] P. V. Nielsen, "Numerical Prediction of Air Distribution in Rooms," ASHRAE, Building Systems: Room Air and Air Contaminant Distribution, 1989.

[22] R. A. Kameel and E. E. Khalil, "Numerical Computations of Thermal Comfort and Air Quality in Air-Conditioned Healthcare Applications," ASME Congress 2006, Paper IMECE-13354, November 2006.

[23] E. E. Khalil, "Flow Regimes and Thermal Patterns in Air Conditioned Operating Theatres," *Proceedings Climamed* 2006, Lyon, November 2006.

[24] R. A. Kameel and E. E. Khalil, "Numerical Investigation of the Airborne Contaminant Age in Surgical Operating Theatres," AIAA Paper, AIAA-2007-0807, January 2007.

[25] E. E. Khalil, "Numerical Computations of Air Flow Regimes in Healthcare Facilities and Their Experimental Verifications," IECEC Paper, AIAA-2009-4510, August 2009.

[26] E. E. Khalil, "Thermal Comfort and Air Quality in Sustainable Climate Controlled Healthcare Applications," AIAA-2010-0802, Orlando, January 2010.

[27] E. E. Khalil, "Holistic Approach to Green Buildings from Construction Material to Services," *Proceedings of International Conference on Air-Conditioning & Refrigeration* (*ICACR*2011), Korea, July 2011, pp. 1-7.

[28] E. E. Khalil, "Energy Efficiency, Air Flow Regime and Relative Humidity in Air-Conditioned Surgical Operating Theatres," *Proceedings ASHRAE*, Paper ASHRAE-2012-CH-12-C056, January 2012.

Nomenclature

C_μ　Turbulence model constant.

h　Enthalpy, Kj/kg

k　Turbulence kinetic energy, m^2/s^2.

U,V,W　Instantaneous components of velocity in three directions, m/s.

X,Y,Z　Coordinate directions.

L,W,H　Length, width, height of the theatre.

Lo　Length of the outlet air supply, m

δ_{ij}　Kroncker delta function.

ε　Turbulence dissipation rate.

Φ　General dependent variable.

Γ　Exchange coefficient.

μ　Absolute viscosity of air, kg/ms.

ρ　Density of air, kg/m^3.

σ　Effective Prandtl number.

Subscripts

I, j, k　Denoting Cartesian coordinate direction takes the values of axes X, Y, Z.

14

Diffusion of Sustainable Construction Practices. A Case of International Cooperation

Andrea Giachetta, Katia Perini, Adriano Magliocco
Department of Architectural Sciences, University of Genoa, Genoa, Italy

ABSTRACT

Data shows that buildings play an important role in the field of energy efficiency and environmental sustainability. The exchange of information among different countries is a key element to promote a wider diffusion of practices for sustainable development in the fields of architecture and urban planning, contributing to the improvement of new skills and economic and production activities, while also reducing the environmental impact of construction on the territory. An international cooperation can lead an exchange of experiences and practices, which could play a fundamental role since countries can have similar problems to deal with, as it happens for the specific identity of the Mediterranean territory in coastal and rural areas. This paper analyses a case of international cooperation, the project SCORE, "*Sustainable COnstruction in Rural and fragile areas for energy Efficiency*", financed under the European MED Programme with the purpose to promote sustainable energy policies in the construction sector on fragile coastal and rural Mediterranean areas.

Keywords: Sustainable Construction; Energy Efficiency; International Cooperation; Mediterranean Areas

1. Introduction

In the last few years the attention towards the environment seems to have become one of the primary objectives to follow. Data shows that constructions play an important role in the field of sustainability. In industrialized countries, buildings represent ±40% of the energy used [1]. Buildings consume a significant amount of energy over their life-time; the energy consumption of these in Europe is about 40% of the total energy demand [2]. They generate 40% - 50% of the total output of greenhouse gases [3]. The building sector has one of the greatest impacts on the environment. This is caused by several variables, which are mainly related to: exploitation of non renewable material resources, indiscriminate use of the territory, high-energy consumption connected to the whole life cycle, and air pollution production [4].

The growing importance of what could be called environmentalism is, nowadays, leading to the formation of contrasts with respect to different approaches to this field; while, up to a few years ago, the argument was between the supporters of more awareness actions on the environment and the ones uninterested in this field, which considered the topic not interesting. In this period the same environmentalists are against some projects defined as "sustainable" or to the installation of energy production systems from renewable source, for the territory eco-system and landscape preservation [5].

To promote a sustainable approach to the planning and implementation of construction activities and urban and building renovation, diffusion activities are needed. An international cooperation can lead an exchange of experiences and practices, which could play a fundamental role since countries can have similar problems to deal with. This happens for the specific identity of the Mediterranean territory in coastal and rural areas. These areas are of great interest for their history, culture, landscape and nature, but are also extremely fragile if we consider the possible impact of human activities and the creation of infrastructures and residential, touristic and production sites, but also the impact of ports and agricultural activities [6].

The exchange of information among different countries is a key element to promote a wider diffusion of practices for sustainable development in the fields of architecture and urban planning, contributing to the improvement of new skills and economic and production activities, while also reducing the environmental impact of construction on the territory [7,8].

This paper analyses a case of international cooperation: the objectives, development, and results of the project SCORE, "*Sustainable COnstruction in Rural and fragile areas for energy Efficiency*". This project has been financed under the European MED Programme with the purpose to promote sustainable energy policies in the construction sector on fragile coastal and rural Mediterranean areas. The MED Programme is a transnational programme of European territorial cooperation. It is financed by the European Union as an instrument of its regional policy. The transnational setup allows the programme to tackle territorial challenges beyond national boundaries [7]. Starting from a brief description of the main objectives and partners (from different countries) involved, the article discusses the main methods, tools, and results of the SCORE project.

2. The Project SCORE

The project SCORE, "*Sustainable COnstruction in Rural and fragile areas for energy Efficiency*", is a European project financed under the European MED Programme [8]. The main goal of the project is to implement sustainable energy policies in the construction sector on fragile coastal and rural MED areas with exceptional landscape values exploiting eco-innovative potential, using traditional building elements combined with innovative green technologies.

The Province of Savona (in Liguria Region, Italy) is the leader partner of the project working with other ten *partners* from seven different countries: Cyprus, France, Greece, Italy, Portugal, Slovenia, and Spain. The partners involved regional agencies for energy, chambers of commerce and their relevant bodies have previous experiences of territorial management aimed at adopting sustainable development strategies and research experiences in the field of environmental sustainability.

SCORE, thanks to the collaboration of partners from different countries with similar territorial characteristics, aims at recording, selecting, organising and spreading the knowledge of good practices which are common to the partners involved, when dealing with issues such as the architectural integration of technologies and systems for energy production out of renewable sources, the reduction of polluting emissions, the protection of the environment and the health of the population.

The project does not mean to simply focus the attention on well-known sustainable strategies for construction design and building activities, but to assess the best methods for an effective implementation of the mentioned strategies, with reference to the building tradition and the local normative and productive framework, considering the continuous education of operators and companies.

With the purpose to effectively identifying the best practices for sustainable development and promote their implementation, the project SCORE deals with the direct application with the different local stakeholders of the building sector and the citizens, as the ones who want to protect the environment and its resources, or with the associations who represent their claims. The Project is based both on the profitable exchange of knowledge among the participant countries and on organising local *focus groups* aimed at creating a *stakeholders* network in the respective territories. For an international exchange and also to involve local stakeholders the most effective tool is Internet; therefore the project has a website, www. scoremed.eu, which is one of the intended results of the Project itself and effectively presents SCORE and its initiatives (**Figure 1**).

3. Normative Framework and Case Studies

Within the SCORE project the different *Partners* identified the normative framework applicable to the project at national and local level, in order to compare it with other countries. If the framework of European Directives ensures a certain uniformity of the normative in different countries, local implementation regulations can significantly vary, providing a different effectiveness and flexibility for territorial management tools. Since these regulations play a fundamental role in the dissemination of technologies and systems for the planning and construction of sustainable buildings, it is important to compare the impact they can have on the different territories where they apply in order to asses the need for amendments or integrations; this has to be taken into account since SCORE *partners* are public entities with decisional power, if not direct competence, in the drafting of this regulations.

Another important action implemented by the different *partners* in order to compare the best results achieved at local level by current norms and regulations was the identification of case studies to highlight the best local practices concerning the architectural integration of sustainable technologies and systems. Several *Case Studies* were identified, included in a special section of the website and classified according to specific keywords (**Figure 2**). They mainly refer to building complexes and existing buildings, located in the different territories of the *partner* countries involved, where particular relevance is given to factors such as: the adoption of bioclimatic strategies, the integration of particular installations, the adoption of solutions to reduce energy consumption or, again, the use of natural materials, management strategies for environmental resources, correct integration with the existing natural and anthropic context. These case studies come from heterogeneous experiences referred to settlements of various sizes (from single houses to entire districts), to buildings with different destination for use (housing, tourism, production), in different contexts (urban areas, ports, agricultural areas, mountains), newly built or resulting from the renovation of existing buildings, both modern or with an historical value. The project also covers case studies of a different nature concerning, for instance, specific didactic experiences on

Figure 1. SCORE web-site home page [8].

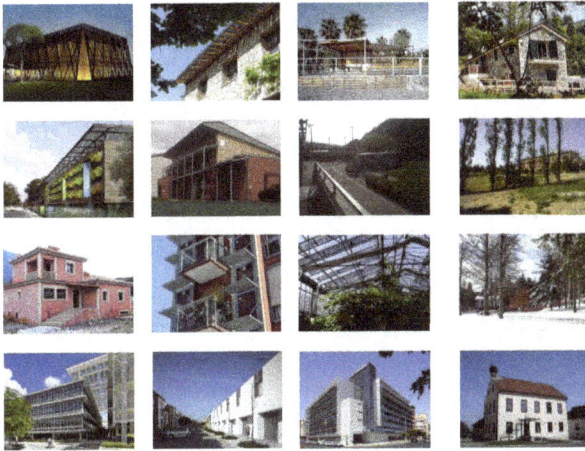

Figure 2. Some of the case studies analysed within SCORE project [8].

sustainability for future designers based on *learning by doing* approach (**Figure 2**, first line, third image), or local environmental certification systems for buildings. For example the first photograph (**Figure 2**) shows the new Mazan community centre. This French project is based on the integration of biocompatible materials and skin solutions, with a minimum amount of sophisticated effects (simple and natural) and materials. This is an eco-friendly project that meets demanding standards and provides an answer to the issues facing the environment and society. A strong feature of the project is the will to reduce energy consumption in the production and shipping of construction materials as well as the development of local production channels. This project makes considerable use of three local resources: wood, straw, and gypsum. Resorting to straw-bales as the main insulating material rather than the other industrially produced insulating materials has a number of benefits for the environment. Indeed, this is a material that enables the local value of a renewable natural resource, contributing also to the fight against greenhouse gases, as it is a material that stores CO_2 and has very useful thermal properties [8].

An interesting case study coming from Italy is the "Double purpose greenhouse", based on the use of renewable energy (**Figure 2**, third line, third photograph).

The work consists of an installation of semi-transparent photovoltaic panels on greenhouses. The study aims at testing the sustainability and the effectiveness of a "double purpose" greenhouse, which could produce more energy than the amount consumed, without impairing productive capacity and use flexibility. Results obtained show that no significant reduction of the productivity of plans grown is observed (both from a quantitative and a qualitative point of view) replacing 20% of the total surface originally covered by glasses with photovoltaic panels. The photovoltaic surface is almost 50 m^2 for a production of 4.1 KWp. The photovoltaic system fulfils about 40% of electric requirement of greenhouse (about 9.100 m^2 of indoor surface; [8]).

The third photograph in the last line of **Figure 2** is the Case study "Hotel Monte Malaga" (Spain). Monte Malaga is a modern and emblematic building due not only to its innovative design but also because it is an example of sustainable construction. According to the Andalusian Energy Agency, the big hotels with a proper use of energy saving measures can save up to 30% of primary energy. The Hotel Monte Málaga is the first hotel in Spain that integrates a cogeneration power center and bioclimatic principles. It is a huge generator of clean energy using solar thermal and solar photovoltaic, integrated into its architecture. The applications of current technologies for saving and clean energy generation are combined with traditional Mediterranean strategies as shade and natural ventilation. Bioclimatic criteria of sunlight, insulation and ventilation (energy saving), along with cleaner production of energy (photovoltaic and thermal), represent a real bet for a new and more respectful relationship between the hotel building and its environment [8].

One or more tags or categories (which can be assigned at same time to the projects) identify the case studies, these are:

- renewable energy;
- solar energy;
- photovoltaic;
- biocompatible materials;
- natural control strategy;
- thermal collector;
- skin solutions;
- biomass/biogas;
- passive;
- efficient management;
- geothermal gradient.

Several cards describe and critically analyse the case studies, highlighting the transfer potential of the adopted methodologies and strategies, which is a key element within the scope of the SCORE project.

This approach allowed the different *partners* to gain knowledge of interesting urban and construction activities in similar areas of other countries and to discover some situations of great interest within their own territory that were still relatively unknown and which deserved, however, to be promoted due to their innovative features.

The analysis and comparison of local norms and regulations and of the case studies allowed the identification of the main components, elements or construction materials, of design strategies and technological systems which become the focus for further phases of the project with regard to different possible implementation cases (ex-novo or on existing buildings) available in the respective territories.

4. Focus Groups as Tool to Elaborate Strategies of Governance

In the identification of case studies and of the issues connected to the implementation of norms and regulations by different stakeholders in the building process, described above, (designers, builders, suppliers of raw materials, building systems and plants, resellers and engineers working with these systems) local focus groups played a fundamental role. These are indeed a series of meetings (conferences, roundtables, workshops) held locally by different SCORE partners. These involved, in order to provide a positive opportunity for confrontation, the above-mentioned stakeholders of the construction sector together with local authorities, training and research institutes, universities and environmental associations. Thanks to the *focus groups* meetings SCORE and its several phases of development have been presented, thus providing the above-mentioned possibility to identify some case-studies, still unknown to the SCORE partners, but also to point out specific issues common interest highlighted by the SCORE project.

The leader partner province of Savona, for instance, through a series of interviews with market operators and trade associations could highlight some communication issues among manufacturer, designers and installers and local authorities and with respect to the interpretation of some legal requirements for promoting photovoltaic systems integrated into architectural structures in Liguria (Italian region on the Mediterranean seaside). This is one the regions with the lower number of photovoltaic installations in a country that, in the last few years, became a world *leader* in this branch. According to the "*Rapporto statistico* 2011 *Fotovoltaico*" of the GSE (the Energy Service Management of the Ministry for Economy and Finance) [9], Italy is the second nation in the world for installed capacity. Difficulties mainly concerned the lack of communication between operators. Therefore, one of the SCORE *focuses* ("Integrated Photovoltaic: opportunities and difficulties in spreading the application in Liguria") of the Province of Savona concentrate on this topic

to start setting up municipal strategies to tackle relevant problems. This involved representatives of the University of Genoa, of professional associations, manufacturer of photovoltaic modules and experts of the province and municipal authorities, of CeRSAA (Regional Center for Research and Agricultural Support, because of the prospective integration of photovoltaic solutions on greenhouses and agricultural land) and of Legambiente Liguria.

The *focus groups* represented an opportunity to effectively face specific problems and analyse the limits to the dissemination of these systems and technologies and find the best possible solutions. The creation of a network of stakeholders with shared interests and competences achieved through the *focus groups* plays also a fundamental role for disseminating project results.

5. An Eco-Construction Tool: SCORE Matrix

The analysis of norms and case-studies, as well as the local *focus groups*, allowed each SCORE partner to select building systems and technologies with high energy and environmental efficiency already in use or potentially available or suitable for development and diffusion in their own territory, while taking into account the historical-environmental value and fragile landscape of coastal and rural areas.

After highlighting these systems and technologies, a tool (defined as "*eco-construction too l*" in the project) has been developed; this has been realized to be user-friendly for the operators of the local network and to represent an effective model for the assessment, comparison and transfer of information among different *partners*. This process leaded to the development of the "Matrixes", which are developed by each *partner* starting from a common scheme to ensure a constant confrontation with each other.

The systems and technologies identified are: aggregation and exposure systems of the settlements for natural climatic control; passive solar Systems; thermal solar systems; photovoltaic systems; mini-micro wind-power systems; biomasses; geothermal systems, shading devices, natural ventilation; natural lighting; automatic control systems (smart buildings); coating systems (hyperinsulation and use of phase-change materials, PCM); eco-friendly materials; use of vegetation for microclimatic control; water management. The possible applications/declinations are: new constructions; renovation of recent buildings; recovery/renovation of historical buildings; ex-novo activities in historical contexts, as shown in **Figure 3**.

The various systems and technologies are cross-evaluated in each matrix together with possible implementations (*i.e.*: passive solar systems for the refitting of recent

Matrix - Evaluation model to assess the feasibility, sustainability and transferability of energy efficiency practices in MED territories.

| ITALY | GREECE | SPAIN | FRANCE | SLOVENIE | CYPRUS | PORTUGAL |

THEMES

This matrix is an eco-construction tool aimed to allow local planners and building practitioners to use criteria to make energy-efficient choices newbuild, conversion & renovation/retrofitting.

DECLINATIONS

Theme	New constructions	Requalifications of recent buildings	Renovation and refit works of historical buildings	Works "ex novo" in historical contexts
Aggregation/exposure for micro-climatic control	●	○	○	●
Passive solar	●	●	○	●
Solar thermal collector	○	○	●	●
Photovoltaic	○	○	●	●
Small wind turbine	●	○	○	●
Biomass	●	○	●	○
Geothermic	●	○	●	○
Different system of renewable energy	○	○	○	○
Sun screen control	○	○	○	○
Natural areation	●	○	○	○
Natural lighting	○	○	○	○
Automatic control system	●	○	○	●
Involucre (insulation, mass, PCM)	●	●	●	●
System (heating, conditioner)	○	○	○	○
Eco-compatibile materials	●	●	●	●
Microclimatic and environmental control through vegetation	●	●	○	●
Water resource (rain collection, etc)	○	○	○	○

The SCORE project is co-financed by the European Regional Development Fund in the framework of the MED Programme - Privacy policy - Contact

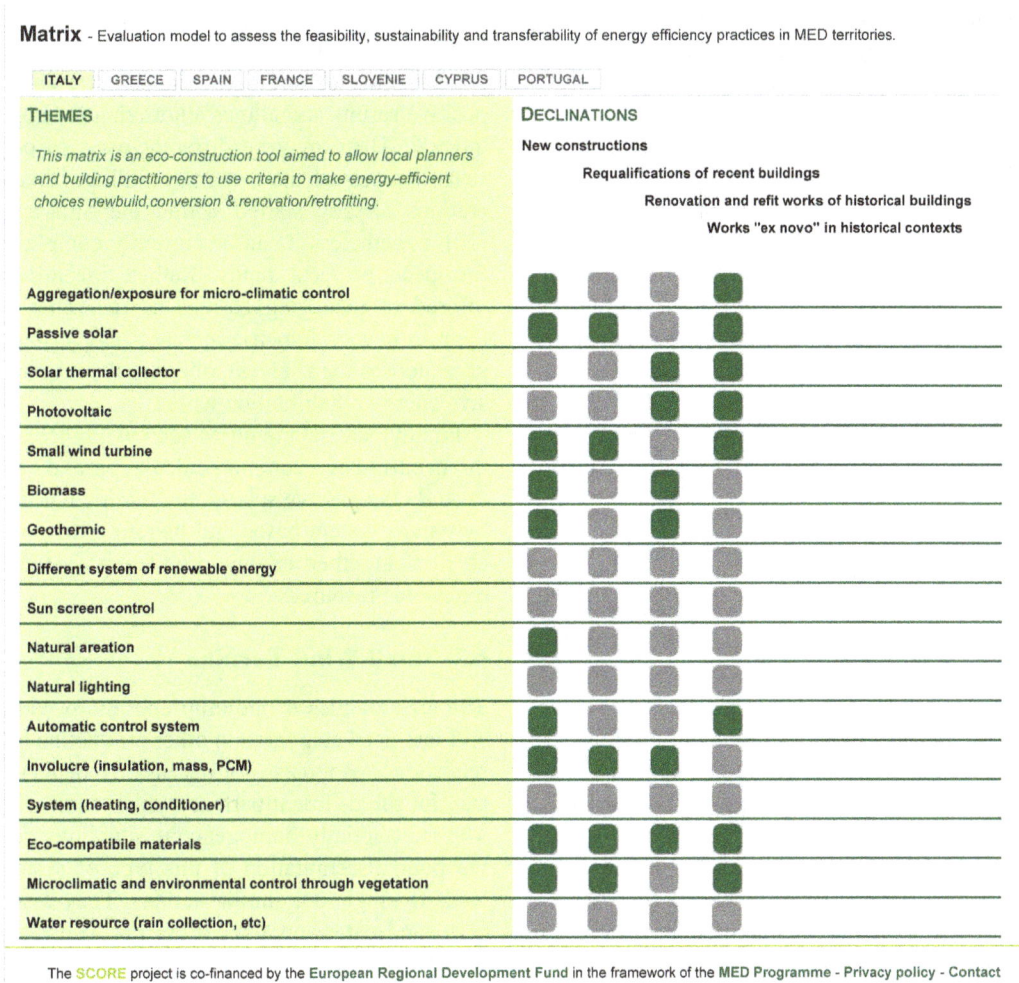

Figure 3. SCORE matrix [8].

buildings). Each partner then worked on specific documents containing analysis and proposals on the most-relevant cross-points, available for download from the project website in the language of each *partner* (for better understanding by local network operators) and with an English abstract (for international comparison). This provides a short description of the system and technology (with reference to the specific context and the state of implementation considered case by case), analysis of case-studies, relevant norms and regulations, with their possible critical evaluation in the annex, an evaluation of the relations with the local building history, the identification of strengths and advantages of the relevant system and technology (with reference to the specific context and nations in relation to the following aspects: Resource consumption reduction; reduction of the environmental burden; improvement of the quality of the internal environment; economical aspects; management; other factors) and of weaknesses/disadvantages of the relevant system and technology in relation to the following aspects: issues related to the architectonical integration, cultural differences (connected to the landscape perception of mainstream culture in the local context), differences in the legal framework (linked to local norms and to the presence of bureaucratic issues and similar), technical difficulties for installation/assembly linked to the local production context (lack of manufacturers, supply difficulties, economic and environmental difficulties/transport costs, lack of engineers with suitable qualification, etc.). Finally proposals to overcome the above mentioned weaknesses, also with reference to the results of the *focus groups*, to case-studies, to the solutions already implemented by other *partners*, are given.

The English abstract of each document is organised considering: the definition of guidelines according to the strengths and weaknesses highlighted in the system/technology and to the proposed solutions; the indications to develop an action plan ("*Bio-construction Action Plan*") to effectively implement the proposed solutions, also by means of further research programmes and pilot projects, and indications for developing an environmental quality certification, applicable in the partner countries involved, making reference to the possible weaknesses highlighted in the existing models of environmental certification for the relevant system and technology.

Guidelines for action plans and quality certifications have been defined thanks to the comparison of the elements highlighted for each system/technology and for the different applications, thanks to the information provided to the *Matrixes* by various *partners*. These plans and certifications could then be considered at local level and, therefore, be implemented by single *partners*. Moreover, they can be used to define which actions are common to two or more partners, with the possibility to adopt common action strategies.

6. Action Plans for Sustainable Development

The English abstracts included in the cards elaborated for the different Partner Countries Matrixes have been organized in specific sections dealing with various themes (for instance, passive solar, photovoltaics, biomass etc.) and applications (New Constructions, Refit Works of Recent Buildings, Renovation and refit works of historical buildings and Works ex novo in historical contexts), for setting up an action plan, defined "Bio-construction Action Plan", whose aim is to lay the foundations for concretely implementing the solutions suggested in the cards, through additional research programs and pilot projects. The collected information were then analysed and compared for the elaboration of the "Bio-construction Action Plans." It allows a fruitful exchange of information between the partners, which can be useful both for availing itself of the experience gained in the other countries, to find particularly effective solutions and to assess the opportunity to start common actions on a transnational and European level. Following some examples of the "Bio-construction Action Plan", organized by themes with information of transnational character, are provided with the aim to give an idea of the results achieved [8].

6.1. Passive Solar

The attitude of the various countries about this theme differs, thus making their proposals inhomogeneous. For instance, Italy deems it appropriate to analyse the existing properties, especially in the decayed peripheral contexts, in order to carry out large scale interventions; Greece believes that it would be useful to have demonstrative and dissemination projects about the necessary technological know-how for designing passive solar systems. The latter action suggested a proposal about the SCORE project, by selecting interesting case studies and innovative technologies in order to exchange know how among the different partners. Other projects of the same kind may be of great importance. With respect to the first possible action, the necessary economic resources are

high; nonetheless these studies are indispensable to assess where it is more appropriate to intervene; the energy rehabilitation of large real estates may even have very positive results and allows amortizing the costs relatively quickly. There may be different ways to perform large scale analysis of the existing buildings; this kind of operations already started within the Smartcity initiative [10], nonetheless it can be currently complex to resort to European projects funds. Rather, incentives could be offered to managing entities as tax relief or allowing them to have volumetric increases (for instance the solar greenhouses themselves), obliging them though to extensive energy rehabilitation works.

Finally, several countries such as France and Italy feel the need to have common calculation and software tools to study the passive solar systems which, actually, are not by nature system based and therefore they are less "reliable" than other energy production systems based on renewable resources.

6.2. Small Wind Turbine

The technologies of the mini and micro wind power are still less used compared to other production systems from renewable resources (not necessarily better); this is also true for the different partners which dealt with the topic. The substantially homogeneous situations deriving from the poor dissemination of this type of system solutions leads to single out similar actions. They essentially concern landscape and integration problems, finding environmental data (wind regime), defining a normative framework on this subject and suitable incentives. Besides the individual and important aspects, considering the reduced development of this promising technology and the substantial homogeneity of situations found in the MED countries, the theme of the mini and micro wind power seems strategic for starting specific lines of research on a European or transnational level.

6.3. Microclimatic and Environmental Control through Vegetation

As to this theme, there are very interesting proposals about training, normative reorganization, drafting technical standards, implementing pilot projects and operational test to assess benefits and risks. Databases about the types of employable plants with their possible effects on the microclimatic and environmental control are needed. If the shared interest for the creation of databases could imply common actions in the framework of transnational research projects, it is also true that the databases for the vegetation normally have a value as to the plants in the specific area of reference. Therefore, it would be necessary to start transnational projects in order to define the structure of the database in order then to implement the contents with investigations on a local scale.

7. Summary

The project presented demonstrates the considerable results, which can be obtained thanks to an international cooperation in the field sustainable energy policies in the construction sector on fragile coastal and rural MED areas with exceptional landscape values. Among the achieved results, special attention can be given in particular to the definition of guidelines and action plans for each partner and general action plans ("Bioconstruction action plans"). This has been possible thanks to what emerged from the work on the Matrix (**Figure 3**), and guarantees the diffusion and the application of energy saving solutions in the building sector and in town planning as for reference territories. Besides the definition of development and research projects that need to be implemented at local, transnational and European level through the identification of potential resources to foster such projects has been carried out, together with the identification of criteria for an environmental certification system applicable in the countries of partners involved. A key element for a wider diffusion of sustainable technology is the mutual exchange of good practices at procedural, regulatory, management and educational level, which comes from the broad communication of the results achieved by SCORE.

Another important result of the project is that a methodology has developed that seems very effective for local government interested in creating a network among themselves and with others in the area, in order to deal effectively and in a participatory way with the theme integration of sustainable building technologies and innovative solutions in areas of high landscape value.

In this historical moment the challenge of sustainability is no longer at least not only to identify the technologies and systems to be applied (there are already many, very advanced and effective), it is rather finding strategies for the concrete and widespread application of these technologies and systems, able to overcome the strong resistance and unfortunately still existing legal, regulatory, bureaucratic, productive, cultural and local resistances.

8. Acknowledgements

All the Partners of the SCORE project, the leader partner Provincia di Savona (for which the Department of Architectural Sciences of the University of Genoa has been the scientific coordination), and in particular the project manager Antonio Schizzi are acknowledged.

REFERENCES

[1] E. Wit and J. McClure, "Statistics for Microarrays: IEA-ECBCS," *Annual Report*, 2008, p. 44.

[2] C. Thormark, "A Low Energy Building in a Life Cycle. Its Embodied Energy, Energy Need for Operation And Recycling Potential," *Building and Environment*, Vol. 37, No. 4, 2002, pp. 429-435.

[3] D. Prasad and M. Hill, "The Construction Challenge: Sustainability in Developing Countries," London Royal Institution of Chartered Surveyors (RICS) Series 2004, Leading Edge Series.

[4] E. Nuzzo and E. Tomasinsig, "Recupero Ecoefficiente del costruito," Edicom Edizioni, Monfalcone (Gorizia) 2008.

[5] A. Giachetta "Il Progetto Ecologico Oggi: Visioni Contrapposte," Alinea Editrice, Firenze, 2012.

[6] G. Cassinelli and A. Magliocco, "Small Energy Production Plants in Ancient Settlements: The Case Study of Camogli," *3rd International Conference on Heritage and Sustainable Development*, Porto, 19-22 Jun 2012, pp. 433-442.

[7] www.programmemed.eu

[8] www.scoremed.eu

[9] Gse, "Rapporto Statistico 2011—Fotovoltaico" www.gse.it

[10] www.genovasmartcity.it

15

The Effectiveness of Energy Feedback for Conservation and Peak Demand: A Literature Review

Desley Vine, Laurie Buys, Peter Morris
Science & Engineering Faculty, Queensland University of Technology, Brisbane, Australia

ABSTRACT

This paper reviews electricity consumption feedback literature to explore the potential of electricity feedback to affect residential consumers' electricity usage patterns. The review highlights a substantial amount of literature covering the debate over the effectiveness of different feedback criteria to residential customer acceptance and overall conservation and peak demand reduction. Researchers studying the effects of feedback on everyday energy use have observed substantial variation in effect size, both within and between studies. Although researchers still continue to question the types of feedback that are most effective in encouraging conservation and peak load reduction, some trends have emerged. These include that feedback be received as quickly as possible to the time of consumption; be related to a standard; be clear and meaningful and where possible both direct and indirect feedback be customised to the customer. In general, the literature finds that feedback can reduce electricity consumption in homes by 5 to 20 percent, but that significant gaps remain in our knowledge of the effectiveness and cost benefit of feedback.

Keywords: Electricity; Energy Conservation; Feedback; Residential

1. Introduction

The only feedback received by many Australian households on their electricity consumption is their quarterly electricity bill [1]. The information (energy consumption per tariff, a comparison of consumption from previous billing periods and total owed for the current billing cycle) and the frequency of the Australian electricity bill have changed little in several decades [1]. This largely un-itemised, non-visual and infrequent feedback on their electricity consumption has been likened to driving cars without any information on the volume or price of fuel consumed and instead receiving a non-itemised invoice at some time in the future for the combined fuel consumption of all family vehicles [2]. This lack of information has become increasingly problematic in Australia, given forecasts for the price of domestic electricity increasing by over 37% between 2010.11 and 2013.14 [3]. This paper reviews electricity consumption feedback literature to explore the potential of electricity feedback to affect residential consumers' electricity usage patterns. "Feedback" in this context is household-specific information on electricity use.

There has been regular enquiry in the literature to improve electricity feedback to consumers since the 1970s [4]. Various methods of providing feedback have been explored including more detailed electricity bills [5], self-reading of meters [6], interactive tools [7] and in-home displays featuring various data including consumption comparisons and visualisations [2,8,9]. Feedback is also regularly studied in conjunction with additional instruments for electricity-saving or behaviour change such as time-of-use or real-time pricing [10] and critical peak pricing [2]. Encouraging the provision of feedback through subsidies, mandates, or other policies could be part of future utility demand-side management (DSM) programs [11]. However, consistent throughout the majority of feedback literature is the finding that feedback is linked to a conservation effect [12].

2. Functions of Feedback

Policies that provide feedback after consumption can be either "direct" or "indirect" [13]. Direct or real-time feedback is immediate and from a meter or other display monitor and has been found to provide greater energy savings than indirect feedback methods [14] which is information that has been processed in some way, e.g. more detailed electricity bills or household-specific advice for reducing electricity use [10,13]. Real-time or direct feedback has benefits over enhanced feedback. First, it can impact habitual, repetitive behaviour such as turning off lights or unplugging appliances [15,16]. People perform many everyday activities without reflection according to routines developed over time and this includes use of electricity [12]. Economists believe that full disclosure of information creates rational consumers [17]. With complete information people act rationally with the objective of maximising utility for dollars spent [18]. Without complete information, it is argued that people are imperfectly rational [17]. Direct feedback then should enhance other demand-response and DSM programs, including making users more responsive to real-time or time-of-use pricing programs and realising the conesquent benefits of load-shifting [2,11] thereby affecting peak consumption.

The second major benefit of direct feedback is the effect it can have on appliance purchasing decisions, as consumers notice from feedback that certain appliances are heavy energy consumers, they can consider replacing them with more efficient ones [12]. It has been argued that this could also lead to behavioural adjustment [16] with for example, people upon learning the real cost of leaving their television switched on decide to switch it off when no-one is in the room. Behavioural adjustment or product choice is more relevant when costs and the extent of energy use are made clearly apparent to the consumer [19]. The third major benefit is that real-time feedback can be more easily customised for individual households [15,20]. With the proper software to manipulate data, feedback could present usage patterns in formats most helpful to individual households [20,21]. Reviews of direct feedback experiments suggest direct feedback interventions yield between 5% and 15% energy savings for the time that they are installed, however their lasting impacts on behaviour are much less certain [2,13,22]. For example, a 15 month study undertaken by van Dam and colleagues [23] found that initial savings of 7.8% after four months could not be sustained in the medium to longer term.

3. Empirical Studies

Researchers studying the effects of feedback interventions on everyday energy use have observed substantial variation in effect size, both within and between studies with explanations often as varied as the results themselves. The variation is partly due to demographic, housing, and climate characteristics of the households. Studies have found that households with higher income, higher education levels, and higher electricity use show greater reductions when feedback is provided [5,24-26]. Other research has argued a positive correlation between income and household consumption levels [27,28], although some have found that the link is not always clear (Brandon & Lewis, 1999). In terms of income and consumption level reductions as a result of feedback (or other) interventions, again, researchers have found a positive correlation to exist [5,28], while others have not [25]. Climate will also impact reductions in use, as the same type of house would have a different demand in a

hot tropical or sub-tropical climate than it would in a cool, temperate climate. Households in more extreme climates (hotter in summer and colder in winter) would appear to have more potential for reducing electricity use. However, several studies have found that feedback has more of an impact when temperatures are more moderate [24,29]. Significant gaps remain on the effectiveness of feedback on different demographic groups and on households living in different climate regions.

Some studies have found that certain types of households respond better to feedback than others, but a considerable amount of uncertainty still exists over how different households will respond to increased level of information [30]. In a recent study of 21 households Wallenborn and colleagues [31] found that electricity feedback through smart meters could change electricity perception but only in households already interested or involved in energy savings or willing to understand the information provided. As a result of real-time feedback, the participants in another study reported taking action to reduce their energy consumption, however, the study found a statistically insignificant reduction in actual electricity consumption by the participants [22]. Alahmad and colleagues [22] suggest this could be due to the self-selection of participants and their already invested interest in electricity conservation prior to the study. In another recent study with 28 Australian households, Strengers [32] acknowledges in-home display feedback as an important visualisation tool illuminating what would otherwise be invisible. However, Strengers [32] argues that feedback has the potential to ignore practices considered non-negotiable and legitimise particular practices, for example, the routine use of clothes dryers by concentrating on what can be readily measured and saved rather than whether the practice is normal or necessary. Strengers [32] posits that failing to engage with the practices seen to be non-negotiable conditions of everyday living may cause householders to lose interest in this type of feedback over time. Indeed Ellegard and Palm [33] argue that a deeper knowledge of everyday energy consumption activities makes everyday life more sustainable.

Despite numerous studies on the effects of feedback, the potential impacts of feedback programs, especially large-scale, remain highly uncertain. In the majority of trials, feedback is designed and tested only in the context of its ability to facilitate a change in behaviour or to persuade consumers to use less power. While feedback provision generally results in consumers using less electricity for the period that it is installed, precisely why this is the case remains unclear [9,31]. As a result of testing feedback only for its energy saving potential, the scope of design and potential uses and interactions with regard to feedback, has been limited [34].

4. Goal Setting

One theme that is prevalent in the literature is the role of goal-setting with feedback. Several authors argue that feedback is only effective when it leads to the setting of a performance goal and that goal-setting is only effective in the presence of feedback that allows participants to evaluate their performance [35-38]. In Becker's [35] study, feedback alone led to a 4.5% decrease compared to results of energy savings of 5.7% when feedback was combined with the modest goal of a 2% reduction and 15.1% in electricity savings when feedback was combined with the difficult goal of a 20% reduction in electricity. More recently, Schultz [39] detailed examples of the energy savings that such strategies can achieve, including an assessment of OPower's increasingly popular Home Energy Report program which has achieved savings as high as 8% for those households that set personal conservation goals. Bonino and colleagues [40] in their online study of nearly 1000 participants found that energy goal setting is better for improving energy consumption.

The contribution of goal-setting to feedback programs, however, is not straight-forward. There are, according to Fischer [12] in her review of feedback programs, many studies where feedback alone appears to have worked. Fischer [12] also cites other studies involving commitment that delivered a small result with one study actually finding no effect on energy consumption. One possible explanation is that goals can be both implicit and explicit [36]. Studies where feedback alone appears to have worked could be the result of implicitly made commitments for goals that participants find meaningful and try to achieve for themselves.

5. Comparison Standards

The two main types of comparisons that have been investigated in the literature are "historic" and "normative" [12]. Historic feedback refers to consumption reported relative to the consumption of the same household from a similar time period in the past. Normative comparative feedback refers to consumption of a household reported in comparison to the consumption of some other similar group of households.

5.1. Historic Standards

Historic feedback provides residents with some frame of reference for their consumption levels and is generally perceived to be effective in this regard. In one example where it was implemented for the first time, treatment groups showed a 10% decrease in consumption which was maintained for more than three years [Wilhite and colleagues cited in 41]. The historic standard was found to be much preferred to a normative standard by focus group participants in the UK [42]. It has also been found to be the most readily recalled piece of information on an energy bill and what customers use to try and understand their consumption patterns [43]. In their study, Kempton and Layne [44] found that only 41% of their participants paid attention to the recent addition on the energy bill of a historic comparison standard. Despite Kempton and Layne's findings, historic feedback appears to be readily understandable, relevant, and useful for consumers. In her review, Fischer [45] found that an historic standard was one of the main features of some of the most effective studies for overall conservation.

5.2. Normative Standards

The effectiveness of normative comparative feedback is unclear. Comparing the consumption of one household to that of others is said to elicit social pressure to understand why consumption levels differ and to stimulate competition and ambition [45,46]. Cialdini [47] identified the importance of social proof in human decision making as people tend to imitate behaviour of others. Indeed, there are reports in which consumers have indicated that this sort of comparison based on similar demographics would be of interest to them [43,48]. In their study, Kempton and Layne [44] found that 70% of their participants had at some time discussed their bills with other people, including their neighbours. More recently, participants of another study indicated their interest in sharing energy-consumption feedback with family and friends [49]. It has been suggested that neighbour-based comparisons may be meaningful as neighbours tend to report similar attitudes and behaviours [Beaman and Vaske cited in 50]. Indeed, the effect of peers has been found to be more effective than incentives such as saving money, conserving resources, or being socially conscious [51]. Whilst the highest quality comparison combines various household attributes it has also been suggested that for practicality, individual streets in groups of 30 addresses is a good basis for geographical comparison [50]. An additional benefit of this type of comparative feedback is that there is no need for weather-adjusting.

Comparative standards are not universally popular, however. A UK study reported findings from their focus groups which suggested normative comparative standards to be very unpopular [42]. This preference may be cultural [45]. American [43,46,51,52] and Norwegian [48] studies have found that residents like normative comparison standards. Allcott [52] reviewed data from randomised natural field experiments of 600,000 treatment and control households in the United States that employed comparative electricity-use feedback, tips for energy conservation and an injunctive message of smiley face/s. Allcott [52] found that the effect of the intervention was equivalent to that of a short-run electricity price increase of 11% to 20% and that the cost effectiveness of the intervention compared favourably to traditional energy conservation programs.

As mentioned above, the effect of comparative standards on actual energy conservation is less clear. Bittle and colleagues [24] found in their study a result opposite to what was intended. They found that those who received the comparative standard feedback consumed more than those who did not. In ten studies reviewed by Fischer [45], there was no savings benefit with comparative standard feedback. Fischer [45] postulates that savings achieved by high users who were encouraged through the comparison to conserve energy may have been cancelled out by lower than average users being inadvertently encouraged to increase energy use because of the comparative standard. This phenomenon has been referred to as the "boomerang" effect and demonstrated in a study undertaken by Schultz and colleagues [46]. In their study, all households received comparative electricity-use feedback in which they were compared to their neighbours, but one group also received an injunctive message in the form of a hand-written smiley-face for households whose consumption was below the average level, and a sad-face for those whose consumption was above the average level. They found that those who consumed less than the average, but received the injunctive message of encouragement (the smiley face), maintained low consumption, whereas, those lower than average consumers who did not receive the injunctive message, increased their consumption. This study demonstrated that the "boomerang" effect can be mediated by not only providing descriptive norms but also including injunctive norms that somehow indicate what is commonly socially acceptable (or unacceptable) within a certain culture [46].

6. Criteria for Effective Feedback

While research is ongoing into the most effective types of feedback for encouraging conservation, some trends have emerged in the literature. It has been suggested that feedback must meet three characteristics for optimum effectiveness [53]. These characteristics include that: feedback must be received as close in time to the consumption event as possible; be related to some standard; and be presented in such a way that is meaningful to the consumer [53]. A fourth characteristic of customised and personalised feedback for individual households has also been suggested by Darby [54] and McMakin and colleagues [55]. According to Fischer [45], feedback should also be computerised, interactive, have appliance specific breakdown of consumption use and be provided over a prolonged period of time. A smaller body of research has explored detailed specifics of what should be included in feedback and while interesting and somewhat informa-

tive, Fischer [45] has argued that specific features may not always be generalizable across demographic groupings or cultures.

6.1. Trusted and Credible Feedback Source

The feedback information needs to be supplied by a trusted and credible source [19]. People can have an inherent distrust of social institutions and think that industry, business and government decisions and priorities are not aligned with energy efficiency objectives [56]. In an interesting study undertaken by Miller and Ford [57], they demonstrated this inherent distrust by sending a letter soliciting an energy conservation program using three different letterheads and they found that the letter that did not list affiliation with the utility received a significantly better response. In examining conservation programs, the use of community-based, non-profit contractors was effective [56,58].

6.2. Presentation of Energy Consumption Detail

For reporting consumption relative to a historic standard, Roberts, Humphries, and Hyldon [42] and Fitzpatrick and Smith [59] found that most of their participants preferred a bar graph representation. For comparative standards, Egan and colleagues [43] found that customers preferred a horizontal "sliding scale" bar chart that indicated on the scale with an arrow where the home's consumption lied. This was preferred over a distribution chart mimicking a bell curve. In general, Egan and colleagues [43] found that the comprehension of the graphics was relatively low, but that adding end-point labels to the charts helped. In marked contrast to these findings are those of Iyer and colleagues [50] who found that the distribution chart was easily understood by participants. In another study, Wilhite and colleagues [48] found their focus group participants were divided over the preference for linear representation and distribution charts for depicting energy consumption. In her review of feedback literature, Fischer [45] summarized her findings by suggesting that, for historical comparisons, vertical bar charts were preferred and for comparative feedback, the single bar graph is preferred. For information displays in general, Roberts and Baker [60] found that graphical displays such as pie charts were preferred, and that they required text labels for improved clarity. It appears that appliance usage information is also best represented in pie chart format [48,61].

6.3. Appliance Usage Charts

Often consumers believe that the appliances that are most visible to them (e.g. lights, dishwashers) are the ones that consume the most electricity [48]. It has been argued, therefore, that providing information on specific appliances and the home's appliance mix is desirable and beneficial to customers and to electricity conservation [59,61]. In her review, Fischer [45] found that some of the most effective studies often contained appliance specific detail. However, appliance-monitoring systems are expensive and require user configuration [49]. Sundramoorthy and colleagues [49] found in their study that participants were able to attribute dips and curves in the load to particular appliances and activities thereby possibly negating the need and therefore the cost of specific appliance-monitoring.

6.4. Consumption Metrics

The electricity consumption metric is also an important consideration. Dollar values of consumption are considered by some authors to be more desirable and useful to consumers [12,13]. Farhar and Fitzpatrick [25] found that their participants liked cost-based energy feedback and that it consistently resulted in reductions. However, Hutton and colleagues [62] and Fitzpatrick and Smith [59] found that feedback emphasising financial values did not have positive results across all their samples.

Environmental metrics have been used infrequently by researchers with Fischer [45] reporting only two in her review of feedback studies. The use of environmental metrics is perhaps one way of stimulating personal norms with regard to environmental concern [12,59] especially given current climate change issues. In their small study, Brandon and Lewis [25] found no significant impact on electricity conservation with the use of feedback containing environmental metrics. The link between environmental concerns and consumption may not always be obvious even to environmentally aware households [27].

6.5. Conservation Tips-Customising

Customisation seems to be important for conservation tips or advice to be effective [49]. In one study a customised newsletter including conservation tips was distributed with presumably customised consumption information and customers reported that the tips were the most useful in helping them to conserve [61]. Other focus group research recorded participants' dislike for a generic leaflet that would have been provided as an insert and their intention to discard such an insert [42].

6.6. Frequency

More immediate and frequent feedback is more likely to result in behaviour change [63]. Allen and Janda's [64] review of feedback studies recognised the primary benefit of real-time feedback as that of affecting customer awareness. The current state of the art for feedback devices is electric monitors that indicate how much electricity the household is using at any given moment. Elec-

tric monitors have the advantage over written feedback of being completely automated and likely being much more cost effective on a large-scale basis. While pilot projects testing continuous feedback have been more common recently [18], continuous feedback has been the subject of research since the 1970s. McClelland and Cook [65] was the first continuous feedback study and it found savings of 12 percent.

Ueno and colleagues [66,67] have conducted studies in Japan using continuous energy monitors. The 2005 study considered meters which disaggregated feedback by appliance and achieved savings of 17.8 percent. The 2006 study achieved savings of 9 percent with meters that did not disaggregate by appliance. Allen and Janda [64] reported on a study of 10 households that found no conservation effect from a continuous feedback device known as "The Energy Detective" (TED). The study participants reported that TED was not user-friendly and this seemed to cause the participants to ignore the device rather than explore and use the manual. Interestingly, in a more recent study with TED, Parker and colleagues [67] identified average savings among 17 households of 7.4 percent. However, the savings in the study ranged widely and the study participants were self-selected.

6.7. Delivery Medium

Fischer [45] found that the common feature in the "best" of 10 feedback studies she reviewed was interactivity in a computerised format and it could be argued that feedback delivery via email is an extension of this [13,68]. Gleerup and colleagues [68] in a recent Danish study found that timely information about a household's exceptional consumption communicated via email and sms messaging resulted in average reductions in total annual electricity use of about 3%. They argued that the type of feedback tested in their study could have a larger effect in other countries because Danish households are likely to be more efficient with electricity consumption as Denmark has the highest marginal electricity price in the world [68]. Email delivery also allows for feedback to be sent directly to the consumer and it can easily be linked to websites that are perhaps more interactive than the email feedback alone. Fischer [45] identified that effective feedback allows for multiple options that the user can choose interactively which is possible with internet-based feedback.

While email and internet-based feedback is generally more feasible for utilities [61] and there are some studies indicating that customers perceive paper-based feedback as wasteful [60], demographics and connectivity factors need to be considered as well. Martinez and Geltz [61] found in their study of 400 Californian residential customers that two-thirds preferred paper-based mail as their choice of medium for feedback, with a similar percentage of commercial customers indicating the same preference.

6.8. Layout, Appearance and Location

Fitzpatrick and Smith [59], Donnelly [69], Hargreaves *et al.* [9], Riche *et al.* [70], Karjalainen [34] Rodgers and Bartram [71] and Bonino and colleagues [40] explore design issues related to the integration of feedback into the household. Concerns over placement, aesthetic appeal and privacy considerations are found to be important considerations in the successful and long-term integration of a feedback medium in the home [70]. Bonino and colleagues [40] found that power visualisation should be in every room or on a portable device, e.g. a smart phone and that colour-based feedback was more easily understood and appreciated by their participants. Ambient and artistic visualisation was found to be a promising method of providing real-time feedback of residential energy use [71]. Preferences regarding functionality and aesthetic appeal vary widely both between and within households however, with results suggesting that regardless of the functionality of the feedback, devices which are aesthetically displeasing tended to become hidden from view and not utilised [9]. Confusingly, half of the participants in the Bonino and colleagues' study [40], wanted a less central "aesthetically acceptable" location while the other half wanted a visible location to track electricity use.

7. Conclusions

In general, the literature finds that feedback can reduce electricity consumption in homes by 5 to 20 percent [13], and that it works best when it is:

- delivered regularly;
- presented plainly and engagingly;
- tailored to the householder;
- interactive and digital;
- capable of providing information by appliance;
- accompanied by advice for reducing electricity use, and
- associated with a challenging goal for energy conservation.

However, there are key uncertainties from the literature and significant gaps still remain in our knowledge of the effectiveness and cost benefit of feedback. A number of research gaps identified by EPRI [18] and verified in this review include:

- The effect of feedback on consumers in different demographic groups;
- The effect of feedback on appliance purchasing decisions;
- The response effect on energy consumption from different formats of feedback;
- Whether feedback continues to work over time or whether it needs to be renewed/reshaped to keep householders engaged and maintain any conservation effects.

In addition, further gaps in research have been highlighted through this review and they include:

- The ability for feedback to facilitate the sharing of electricity information between households, friends or neighbours is almost entirely unexplored;
- The scope of design, potential uses and interactions with regard to feedback has been limited;
- The divergence of cost-benefit calculations for feedback with advanced metering infrastructure needs to be explored as does the conditions under which the costs of feedback outweigh the benefits.

This review has explored one potential solution, more detailed feedback, to help control the growth of residential energy and the expansion of electricity infrastructure. Finding such solutions could become increasingly important if demand for heating and cooling appliances continues at its current projection and/or hybrid and full-electric vehicles become a substantial portion of automobile sales. These are just two examples of situations which would add significant demand to the grid and where it would become more critical to try and control how and when consumers used their heating and cooling appliances and recharged their cars so as not to exceed peak capacity. Research shows that feedback does have the potential to positively affect residential electricity conservation. With increasing smart grid investment and improving feedback devices to be more user-friendly there will be greater opportunity to connect the consumer to the grid and therefore study the effect of feedback on large groups of consumers. Results from such studies would potentially be of interest to a diverse range of professional areas such as social science, computer science, power engineering and energy economics.

REFERENCES

[1] P. Simshauser, E. Molyneux and M. Shepherd, "The Entry Cost Shock and the Re-Rating of Power Prices in New South Wales, Australia," *Australian Economic Review*, Vol. 43, No. 2, 2010, pp. 114-135.

[2] A. Faruqui, S. Sergici and A. Sharif, "The Impact of Informational Feedback on Energy Consumption—A Survey of the Experimental Evidence," *Energy*, Vol. 35, No. 4, 2010, pp. 1598-1608.

[3] Australian Energy Market Commission, "Retail Electricity Price Estimates: Final Report for 2010-2011 to 2013-2014," Australian Energy Market Commission, Sydney, 2011.

[4] R. A. Winett, J. H. Kagel, R. C. Battalio and R. C. Winkler, "Effects of Monetary Rebates, Feedback, and Information on Residential Electricity Conservation," *Journal of Applied Psychology*, Vol. 63, No. 1, 1978, pp. 73-80.

[5]   H. Wilhite and R. Ling, "Measured Energy Savings from a More Informative Energy Bill," *Energy and Buildings*, Vol. 22, No. 2, 1995, pp. 145-155.

[6]   L. Nielsen, "How to Get the Birds in the Bush into Your Hand: Results from a Danish Research Project on Electricity Savings," *Energy Policy*, Vol. 21, No. 11, 1993, pp. 1133-1144.

[7]   T. Ueno, R. Inada, O. Saeki and K. Tsuji, "Effectiveness of an Energy-Consumption Information System for Residential Buildings," *Applied Energy*, Vol. 83, No. 8, 2006, pp. 868-883.

[8]   A. Gronhoj and J. Thogersen, "Feedback on Household Electricity Consumption: Learning and Social Influence Processes," *International Journal of Consumer Studies*, Vol. 35, No. 2, 2011, pp. 138-145.

[9]   T. Hargreaves, M. Nye and J. Burgess, "Making Energy Visible: A Qualitative Field Study of How Householders Interact with Feedback from Smart Energy Monitors," *Energy Policy*, Vol. 38, No. 10, 2010, pp. 6111-6119.

[10] S. Darby, "Smart Metering: What Potential for Householder Engagement?" *Building Research and Information*, Vol. 38, No. 5, 2010, pp. 442-457.

[11] M. Martiskainen and J. Coburn, "The Role of Information and Communication Technologies (ICTs) in Household Energy Consumption-Prospects for the UK," *Energy Efficiency*, Vol. 4, No. 2, 2011, pp. 209-221.

[12] C. Fischer, "Feedback on Household Electricity Consumption: A Tool for Saving Energy?" *Energy Efficiency*, Vol. 1, No. 1, 2008, pp. 79-104.

[13] S. Darby, "The Effectiveness of Feedback on Energy Consumption: A Review for DEFRA of the Literature on Metering, Billing and Direct Displays," Environmental Change Institute, University of Oxford, Oxford, 2006.

[14] G. Wood and M. Newborough, "Dynamic Energy-Consumption Indicators for Domestic Appliances: Environment, Behaviour and Design," *Energy and Buildings*, Vol. 35, No. 8, 2003, pp. 821-841.

[15] G. Jacucci, A. Spagnolli, L. Gamberini, A. Chalambalakis, C. Björksog, M. Bertoncini, C. Torstensson and P. Monti, "Designing Effective Feedback of Electricity Consumption for Mobile User Interfaces," *PsychNology Journal*, Vol. 7, No. 3, 2009, pp. 265-289.

[16] J. H. van Houwelingen and W. F. van Raaij, "The Effect of Goal-Setting and Daily Electronic Feedback on In-Home Energy Use," *Journal of Consumer Research*, Vol. 16, No. 1, 1989, pp. 98-105.

[17] E. Kamenica, S. Mullainathan and R. Thaler, "Helping Consumers Know Themselves," *American Economic Review*, Vol. 101, No. 3, 2011, pp. 417-422.

[18] Electric Power Research Institute, "Residential Electricity Use Feedback: A Research Synthesis and Economic Framework 1016844," Electric Power Research Institute, Palo Alto, 2009.

[19] J. Burgess and M. Nye, "Re-Materialising Energy Use through Transparent Monitoring Systems," *Energy Policy*, Vol. 36, No. 12, 2008, p. 4454.

[20] A. Spagnolli, N. Corradi, L. Gamberini, E. Hoggan, G. Jacucci, C. Katzeff, L. Broms and L. Jonsson, "Eco-Feedback on the Go: Motivating Energy Awareness," *Computer*, Vol. 44, No. 5, 2011, pp. 38-45.

[21] D. J. Kerrigan, L. Gamberini, A. Spagnolli and G. Jacucci, "Smart Meters: A Users' View," *PsychNology Journal*, Vol. 9, No. 1, 2011, pp. 55-72.

[22] M. A. Alahmad, P. G. Wheeler, A. Schwer, J. Eiden and A. Brumbaugh, "A Comparative Study of Three Feedback Devices for Residential Real-Time Energy Monitoring," *IEEE Transactions on Industrial Electronics*, Vol. 59, No. 4, 2012, pp. 2002-2013.

[23] S. S. van Dam, C. A. Bakker and J. D. M. van Hal, "Home Energy Monitors: Impact over the Medium-Term," *Building Research and Information*, Vol. 38, No. 5, 2010, pp. 458-469.

[24] R. G. Bittle, R. Valesano and G. Thaler, "The Effects of Daily Cost Feedback on Residential Electric Consumption," *Behaviour Modification*, Vol. 3, No. 2, 1979, pp. 187-202.

[25] G. Brandon and A. Lewis, "Reducing Household Energy Consumption: A Qualitative and Quantitative Field Study," *Journal of Environmental Psychology*, Vol. 19, No. 1, 1999, pp. 75-85.

[26] I. Ayres, S. Raseman and A. Shih, "Evidence from Two Large Field Experiments that Peer Comparison Feedback can Reduce Residential Energy Usage," Rochester, 2009.

[27] B. Gatersleben, L. Steg and C. Vlek, "Measurement and Determinants of Environmentally Significant Consumer Behavior," *Environment and Behavior*, Vol. 34, No. 3, 2002, pp. 335-362.

[28] D. A. Guerin, B. L. Yust and J. G. Coopet, "Occupant Predictors of Household Energy Behavior and Consumption Change as Found in Energy Studies Since 1975," *Family and Consumer Sciences Research Journal*, Vol. 29, No. 1, 2000, pp. 48-80.

[29] D. C. Mountain, "Real-Time Feedback and Residential Electricity Consumption: British Columbia and Newfoundland and Labrador Pilots," 2007.

[30] K. Ehrhardt-Martinez, J. A. S. Laitner and K. A. Donnelly, "Chapter 10—Beyond the Meter: Enabling Better Home Energy Management," In: S. Fereidoon Perry, Ed., *Energy, Sustainability and the Environment*, Butterworth-Heinemann, Boston, 2011, pp. 273-303.

[31] G. Wallenborn, M. Orsini and J. Vanhaverbeke, "Household Appropriation of Electricity Monitors," *International Journal of Consumer Studies*, Vol. 35, No. 2, 2011,

pp. 146-152.

[32] Y. Strengers, "Negotiating Everyday Life: The Role of Energy and Water Consumption Feedback," *Journal of Consumer Culture*, Vol. 11, No. 3, 2011, pp. 319-338.

[33] K. Ellegård and J. Palm, "Visualizing Energy Consumption Activities as a Tool for Making Everyday Life More Sustainable," *Applied Energy*, Vol. 88, No. 5, 2011, pp. 1920-1926.

[34] S. Karjalainen, "Consumer Preferences for Feedback on Household Electricity Consumption," *Energy and Buildings*, Vol. 43, No. 2-3, 2011, pp. 458-467.

[35] L. J. Becker, "Joint Effect of Feedback and Goal Setting on Performance: A Field Study of Residential Energy Conservation," *Journal of Applied Psychology*, Vol. 63, No. 4, 1978, pp. 428-433.

[36] C. Seligman, L. J. Becker and J. M. Darley, "Encouraging Residential Energy Conservation through Feedback," *Advances in Environmental Psychology*, Vol. 3, 1981, pp. 93-113.

[37] L. T. McCalley and C. J. H. Midden, "Energy Conservation through Product-Integrated Feedback: The Roles of Goal-Setting and Social Orientation," *Journal of Economic Psychology*, Vol. 23, No. 5, 2002, pp. 589-603.

[38] L. T. McCalley and C. J. H. Midden, "Making Energy Feedback Work," In: P.-P. Verbeek and A. Slob, Eds., *User Behavior and Technology Development*, Springer Netherlands, Vol. 20, 2006, pp. 127-137.

[39] P. W. Schultz, "Making Energy Conservation the Norm," In: K. Ehrhardt-Martinez and J. A. S. Laitner, Eds., *People-Centered Initiatives for Increasing Energy Savings*, ACEEE, Colorado, 2010.

[40] D. Bonino, F. Corno and L. De Russis, "Home Energy Consumption Feedback: A User Survey," *Energy and Buildings*, Vol. 47, 2012, pp. 383-393.

[41] G. Wood and M. Newborough, "Energy-Use Information Transfer for Intelligent Homes: Enabling Energy Conservation with Central and Local Displays," Vol. 39, No. 4, 2007, pp. 495-503.

[42] S. Roberts, H. Humphries and V. Hyldon, "Consumer Preferences for Improving Energy Consumption Feedback: Report to Ofgem," Centre for Sustainable Energy, Bristol, 2004.

[43] C. Egan, W. Kempton, A. Eide, D. Lord and C. Payne, "How Customers Interpret and Use Comparative Graphics of Their Energy Use," ACEEE, Washington D.C., 1996, pp. 39-46.

[44] W. Kempton and L. L. Layne, "The Consumer's Energy Analysis Environment," *Energy Policy*, Vol. 22, No. 10, 1994, pp. 857-866.

[45] C. Fischer, "Discussion Paper 8: Influencing Electricity Consumption Via Consumer Feedback: A Review of Experience," ECEEE 2007 Summer Study, Berlin, 2007.

[46] P. W. Schultz, J. M. Nolan, R. B. Cialdini, N. J. Goldstein and V. Griskevicius, "The Constructive, Destructive, and Reconstructive Power of Social Norms," *Psychological Science*, Vol. 18, No. 5, 2007, pp. 429-434.

[47] R. B. Cialdini, "Influence: The Psychology of Persuasion," William Morrow & Co., New York, 1993.

[48] H. Wilhite, A. Hoivik and J. Olsen, "In Advances in the Use of Consumption Feedback Information in Energy Billing: The Experiences of a Norwegian Utility," ECEEE 1999 Summer Study Proceedings, 1999.

[49] V. Sundramoorthy, G. Cooper, N. Linge and L. Qi, "Domesticating Energy-Monitoring Systems: Challenges and Design Concerns," *IEEE on Pervasive Computing*, Vol. 10, No. 1, 2011, pp. 20-27.

[50] M. Iyer, W. Kempton and C. Payne, "Comparison Groups on Bills: Automated, Personalized Energy Information," *Energy and Buildings*, Vol. 38, No. 8, 2006, pp. 988-996.

[51] J. M. Nolan, P. W. Schultz, R. B. Cialdini, N. J. Goldstein and V. Griskevicius, "Normative Social Influence Is Underdetected," *Personality and Social Psychology Bulletin*, Vol. 34, No. 7, 2008, pp. 913-923.

[52] H. Allcott, "Social Norms and Energy Conservation," *Journal of Public Economics*, Vol. 95, No. 9-10, 2011, pp. 1082-1095.

[53] C. J. H. Midden, J. E. Meter, M. H. Weenig and H. J. A. Zieverink, "Using Feedback, Reinforcement and Information to Reduce Energy Consumption in Households: A Field-Experiment," *Journal of Economic Psychology*, Vol. 3, No. 1, 1983, pp. 65-86.

[54] S. Darby, "Making It Obvious: Designing Feedback into Energy Consumption," In Second International Conference on Energy Efficiency in Household Appliances and Lighting, Naples, 2000.

[55] A. H. McMakin, E. L. Malone and R. E. Lundgren, "Motivating Residents to Conserve Energy without Financial Incentives," *Environment and Behavior*, Vol. 34, No. 6, 2002, pp. 848-863.

[56] L. Lutzenhiser, "Social and Behavioral Aspects of Energy use," *Annual Review of Energy and the Environment*, Vol. 18, No. 1, 1993, pp. 247-289.

[57] R. D. Miller and J. M. Ford, "Shared Savings in the Residential Market: A Public/Private Partnership for Energy Conservation Urban Consortium for Technology Initiatives," Energy Task Force, Baltimore, 1985.

[58] R. B. Cialdini, "Basic Social Influence Is Underestimated," *Psychological Inquiry*, Vol. 16, No. 4, 2005, pp. 158-161.

[59] G. Fitzpatrick and G. Smith, "Technology-Enabled Feedback on Domestic Energy Consumption: Articulating a Set of Design Concerns," *IEEE on Pervasive Computing*, Vol. 8, No. 1, 2009, pp. 37-44.

[60] S. Roberts and W. Baker, "Towards Effective Energy

Information. Improving Consumer Feedback on Energy Consumption. A Report to OFGEM," Centre for Sustainable Energy, Bristol, 2003.

[61] M. S. Martinez and C. R. Geltz, "Utilizing a Pre-Attentive Technology for Modifying Customer Energy Usage European Council for an Energy-Efficient Economy," 2005.

[62] R. B. Hutton, G. A. Mauser, P. Filiatrault and O. T. Ahtola, "Effects of Cost-Related Feedback on Consumer Knowledge and Consumption Behavior: A Field Experimental Approach," *Journal of Consumer Research*, Vol. 13, No. 3, 1986, pp. 327-336.

[63] M. J. Bekker, T. D. Cumming, N. K. P. Osborne, A. M. Bruining, J. I. McClean and L. S. Leland Jr., "Encouraging Electricity Savings in a University Residential Hall through a Combination of Feedback, Visual Prompts, and Incentives," *Journal of Applied Behavior Analysis*, Vol. 43, No. 2, 2010, pp. 327-331.

[64] D. Allen and K. B. Janda, "In the Effects of Household Characteristics and Energy Use Consciousness on the Effectiveness of Real-Time Energy Use Feedback: A Pilot Study American Council for an Energy-Efficient Economy 2006 Summer Study, Washington D.C., 2006," ACEEE: Washington D.C., 2006.

[65] W. Abrahamse, L. Steg, C. Vlek and T. Rothengatter, "A Review of Intervention Studies Aimed at Household Energy Conservation," *Journal of Environmental Psychology*, Vol. 25, No. 3, 2005, pp. 273-291.

[66] T. Ueno, R. Inada, O. Saeki and K. Tsuji, "In Effectiveness of Displaying Energy Consumption Data in Residential Houses: Analysis on How the Residents Respond, 2005 Summer Study of the European Council for an Energy Efficient Economy, Stockholm, 2005," ECEEE, Stockholm, 2005, pp. 1289-1299.

[67] D. Parker, D. Hoak and J. Cummings, "Pilot Evaluation of Energy Savings and Persistence from Residential Energy Demand Feedback Devices in a Hot Climate. In Summer Study on Energy Efficiency in Buildings—The Climate for Efficiency is Now," ACEEE, Pacific Grove, CA, 2010.

[68] M. Gleerup, A. Larsen, S. Leth-Petersen and M. Togeby, "The Effect of Feedback by Text Message (SMS) and Email on Household Electricity Consumption: Experimental Evidence," *The Energy Journal*, Vol. 31, No. 3, 2010, pp. 113-132.

[69] K. A. Donnelly, "The Technological and Human Dimensions of Residential Feedback: An Introduction to the Broad Range of Today's Feedback Strategies," In: K. Ehrhardt-Martinez and J. A. S. Laitner, Eds., *People-Centered Initiatives for Increasing Energy Savings*, ACEEE, Colorado, 2010.

[70] Y. Riche, J. Dodge and R. A. Metoyer, "Studying Always-On Electricity Feedback in the Home," *Proceedings of the SIGCHI Conference on Human Factors in Computing Systems*, ACM, Atlanta, 2010, pp. 1995-1998.

[71] J. Rodgers and L. Bartram, "Exploring Ambient and Artistic Visualization for Residential Energy Use Feedback," *IEEE Transactions on Visualization and Computer Graphics*, Vol. 17, No. 12, 2011, pp. 2489-2497.

16

The Effect of Transition Hysteresis Width in Thermochromic Glazing Systems

Michael E. A. Warwick[1,2], Ian Ridley[3], Russell Binions[4*]
[1]Department of Chemistry, University College London, Christopher Ingold Laboratories, London, UK
[2]UCL Energy Institute, Central House, London, UK
[3]School of Property, Construction and Project Management, RMIT University, Melbourne, Australia
[4]School of Engineering and Materials Science, Queen Mary University of London, London, UK

ABSTRACT

Thermochromic glazing theoretically has the potential to lead to a large reduction in energy demand in modern buildings by allowing the transmission of visible light for day lighting whilst reducing unwanted solar gain during the cooling season, but allowing useful solar gain in the heating season. In this study building simulation is used to examine the effect of the thermochromic transition hysteresis width on the energy demand characteristics of a model system in a variety of climates. The results are also compared against current industry standard glazing products. The results suggest that in a warm climate with a low transition temperature and hysteresis width energy demand can be reduced by up to 54% compared to standard double glazing.

Keywords: Energy Simulation; Energy Demand Reduction; Thermochromic Glazing

1. Introduction

Thermochromic glazing systems have often been proposed as possible candidates for use as intelligent glazing systems for use in energy saving windows. There has been work done on potential thin film candidates, such as VO₂, and how the composition and structure can be modified to fit the best properties for a thermochromic glazing. These properties are assumed to be transition temperature for the material needs to be near that of room temperature (20˚C - 25˚C). The transition should occur quickly meaning that the gradient of the hysteresis should be steep and that the hysteresis loop width should be as narrow as possible. While these parameters seem to reasonable assumptions there has been very little work done [1-3] to see what effects each of these variables has on the potential efficiency of the material as an energy saving product. It is also worthwhile investigating the magnitude of the effect each of these factors has on the efficiency and whether they vary in effect with environmental factors.

Thin films of vanadium (IV) oxide have been the subject of significant research efforts in recent years due to their proposed application as an "intelligent" energy effi-

cient window coating [4,5]. These technologies are based on a temperature modulated structural phase change, which occurs in the pure material at 68˚C where the low temperature monoclinic phase (VO_2 M) converts to the higher temperature rutile phase (VO_2 R) [6]. Significant changes in electrical conductivity and infra-red optical properties occur in the material as a result of this structural change. The rutile material is metallic in nature and reflects a wide range of solar radiation, whereas the monoclinic phase is a semiconductor and transmissive across the same range of solar radiation. This dynamic switching behaviour is in contrast to existing commercial approaches which rely on glazing with static behaviour such as heat mirrors, absorbing glass or Low-E coatings [7]. For vanadium dioxide to be effective as an intelligent window coating it is highly desirable to lower the transition temperature from 68˚C to nearer room temperature in order to maximize the energy demand reduction properties [1]. Doping studies have shown that the transition temperature can be altered by the incorporation of metal ions into the vanadium dioxide lattice [8,9]. It has been demonstrated that the most effective metal ion dopant is tungsten, which is found to lower the transition temperature of the material by 25˚C for every atomic percent incorporated of the dopant [10]. The transition temperature has also been shown to be affected by film strain [11] and it has been demonstrated that strain can be introduced by careful choice of deposition conditions [12]. Tungsten doped vanadium dioxide films have been prepared by a variety of methods including physical vapour deposition [13], sol-gel [14], and chemical vapour deposition (CVD) methodologies [15-17]. CVD routes to the production of VO_2 films are generally considered more attractive because of the ability to integrate CVD processes with high volume float glass manufacture and the physical properties of CVD produced films, which are generally adherent and durable [18].

The synthesis of thermochromic VO_2 films is by now largely well known and understood. However there is a major gap in knowledge of the energy demand reduction properties that these films may confer to glazing systems. Such advantages in the use of these coatings have not been significantly investigated with only a few reports existing in the literature [1-3,18]. Indeed very little attention at all has been paid to the various parameters of the transition such as hysteresis width, hysteresis gradient and change in transmission/reflectance with respect to their energy saving performance.

Several studies have been performed on electrochromic [19-22] and thermotropic [23-25] systems suggesting that such systems can significantly reduce the overall energy demand and enhance overall building energy performance. The advantage of a thermochromic system based on VO_2 compared to these systems is that there is no change in the visible portion of the spectrum, only in the infrared portion [7].

In this paper we use energy-modeling studies of some idealized spectra to examine the effect of thermochromic transition hysteresis width on the energy demand reduction properties of thermochromic glazing systems based on VO_2 in several different climates. They are assessed with reference to some existing commercial products. This study, the first of this kind, is crucial to evaluate and quantify the performance of thermochromic glazing and the various transition parameters.

2. Experimental

Energy Plus software developed by the Lawrence Berkeley National Laboratory [26] and US Department of Energy was used to perform energy simulations and analysis. Energy Plus TM is an energy analysis and thermal load simulation program based on a user's description of a building from the perspective of the building's physical make-up, associated mechanical systems, etc. A series of simulations with different configurations and settings were run in order to evaluate the performance of the thermochromic coatings in different climates. The simulation set period is one year, with data points gathered every hour. A very simple model of a room in a build-

ing was constructed in Energy Plus TM. The room has external dimensions 6 m × 5 m × 3 m (length × width × height) and it is placed so that the axis of every wall is perpendicular to one of the orientation north, south, west and east. We consider the room to represent the facade of a generic building so that just one wall is exposed to the external environment (weather, sun, wind, etc.); the remaining three walls are not affected by external conditions. The building is located in the northern hemisphere and the external wall is supposed to be exposed to the southern side. The modeled zone is a mid floor office, of a multi-story block, buffered both above and below by conditioned spaces. The ground temperature would therefore have no effect on the performance of the studied zone. The choice of ground temperature was set not to reflect the real local ground temperature but rather the temperature of a further buffering zone, below the modeled spaces, and was taken to be 18°C throughout the year.

One glazing possibility was considered comprising the whole of the southern face (100%)—a glazing wall, representing a modern commercial building. Further details governing the materials used for walls, etc., have been previously reported [1]. In both cases the window is double glazed with a 12 mm air cavity, the coating was always modeled on the inside face of the outer pane. The only difference between each simulation was the glazing or coating used

The internal conditions were chosen to be air-conditioned between 19°C and 26°C to maintain a comfortable working/living environment. The required illuminance level in an office building is taken to be 500 lux, this corresponds to a lighting load of 400 W. The lights are fully dimmable: lowering their output when there is an adequate illuminance from the sun, in order to save energy. It is considered that they can be dimmed in the whole range from 0% to 100%. The dimming control is automatic and zoned. The casual heat gain (persons + equipment) is taken to be 500 W in total and the ventilation rate used is 0.025 m^3/s. Building occupancy was set as occupied from 8:00 till 18:00, five days a week, as is normal for an office. The simulations were run for three different cities representing heating dominated (Helsinki), cooling dominated (Palermo) and mixed (London) environments.

The thermochromic properties of the glazing were modeled in version 5.0.0 of Energy Plus by entering the spectral data of the glazing in the hot and cold states. The thermochromic hysteresis effect was controlled by use of a set of schedules and the thermochromic model built into the energy plus software. The thermochromic model allows for individual spectral properties to be assigned a specific temperature range in which they will be utilized according to the glazing surface temperature. This means that as the surface temperature changes, the spectral data used to determine the amount of solar radiation entering the building will change, thus simulating the thermochromic effect. By varying the temperature range over which each spectrum is used allows for the creation of a suitable thermochromic hysteresis gradient model. The window surface temperature was correlated against the incident solar radiation.

To model the hysteresis width into the systems the model was run then analyzed to establish when the glazing system was "heating" or "cooling" according to the surface temperature of the window, using this data a series of individual daily schedules was produced and these could be used to control which glazing was to be used a which times.

Two thermochromic hysteresis gradient models; one "heating", for when the system is in the heating portion of hysteresis loop, one "cooling", for the cooling system of the hysteresis loop were set up in the thermochromic model. The schedule produced from the heating and cooling data could then be used to assign which set of spectra was to be used at which times there for creating a thermochromic system model.

The model is clearly limited because the building is not ideal for all climates. Insulation layers, as well as the materials chosen here, in warmer and cooler climates would be different from that used in the model depending not only on local climate conditions but also on the constructive techniques and materials available in loco. Likewise the assumption that a constant ground temperature of 18°C throughout the year is significant. However, by using the results obtained from the plain glass simulations as a baseline we aim to isolate the change in energy performance caused by the use of different glazings.

2.1. Choice of Spectra Used

The results of the energy plus out puts for of the theoretical gradients are shown below. Blank glass and two already commercially available static coatings, absorbing glass (AG) and silver sputtered glass (SSG), spectra shown in **Figure 1**, were also modelled as to be used as references by which the effective energy savings could be compared.

It is defined that any plausible window glazing candidate would require a transmission of at least 60% in the visible region so as to still allow enough light in to effectively provide light source, and that to be effective energy efficient thermochromic glazing there would be a large change in transmission in the infra red region. Several transmission/reflectance UV-vis spectra (**Figure 2**) were generated with an overall change of 65% in the infrared region. These spectra could then be entered in to the thermochromic model of the Energy plus TM pro-

gramme to simulate variations in the thermochromic properties of a glazing system.

Figure 2 shows the maximum and minimum spectra for the transmission/reflectance, *i.e.* the spectra for the hottest and coldest states of the materials. In order to recreate a transition between these two states a series of spectra were produced at increments of 2.5% transmission/reflectance. Each of the pairs of spectra could then be assigned to a temperature and a hysteresis gradient created. By changing the temperature gap (ΔT) between each spectra the temperature range during which the thermochromic transition occurred could be varied and therefore the hysteresis gradient changed. In total 27 spectra were made so if ΔT was set at 1°C the thermochromic transition of 65% would occur of a 27°C range around transition temperature (T_c). Transition temperatures were chosen to be 20°C, 25°C, 30°C and 35°C.

2.2. Hysteresis Width Variation

The other factor thought to effect the efficiency of thermochromic glazing is the width of the hysteresis loop. This affects the temperature at which the thermochromic transition occurs on heating and cooling the window. To model the effect that this has on the performance three hysteresis widths were investigated corresponding to widths of 5°C, 10°C and 15°C these theoretical widths/gradients are shown in **Figure 3**.

The hysteresis used all had a gradient of $-2.5 \Delta T\%/°C$ and transition temperatures of 20°C, 25°C, 30°C and 35°C as the change in transition temperature can effect the efficiency of the systems with respect to the other transition parameters.

3. Results

3.1. Climates Investigated

Three cities were chosen to represent climates that are: cooling dominated—Palermo, heating dominated-Helsinki and mixed—London, in order to determine the type of climate that thermochromic glazing is able to have the greatest influence in.

3.2. Palermo

Figure 4 shows the daytime highs and night-time lows for a one-year period in Palermo, Italy. The figure shows that for the majority of the year the temperature during the day is greater than the desired room temperature of 22°C.

The average temperature throughout the year in Palermo is 29°C. The environment is therefore a hot environment and energy demand is cooling dominated.

3.3. London

Figure 5 shows the temperature profiles for the period of

Blank Glass

Absorbing Glass

Silver Sputtered Glass

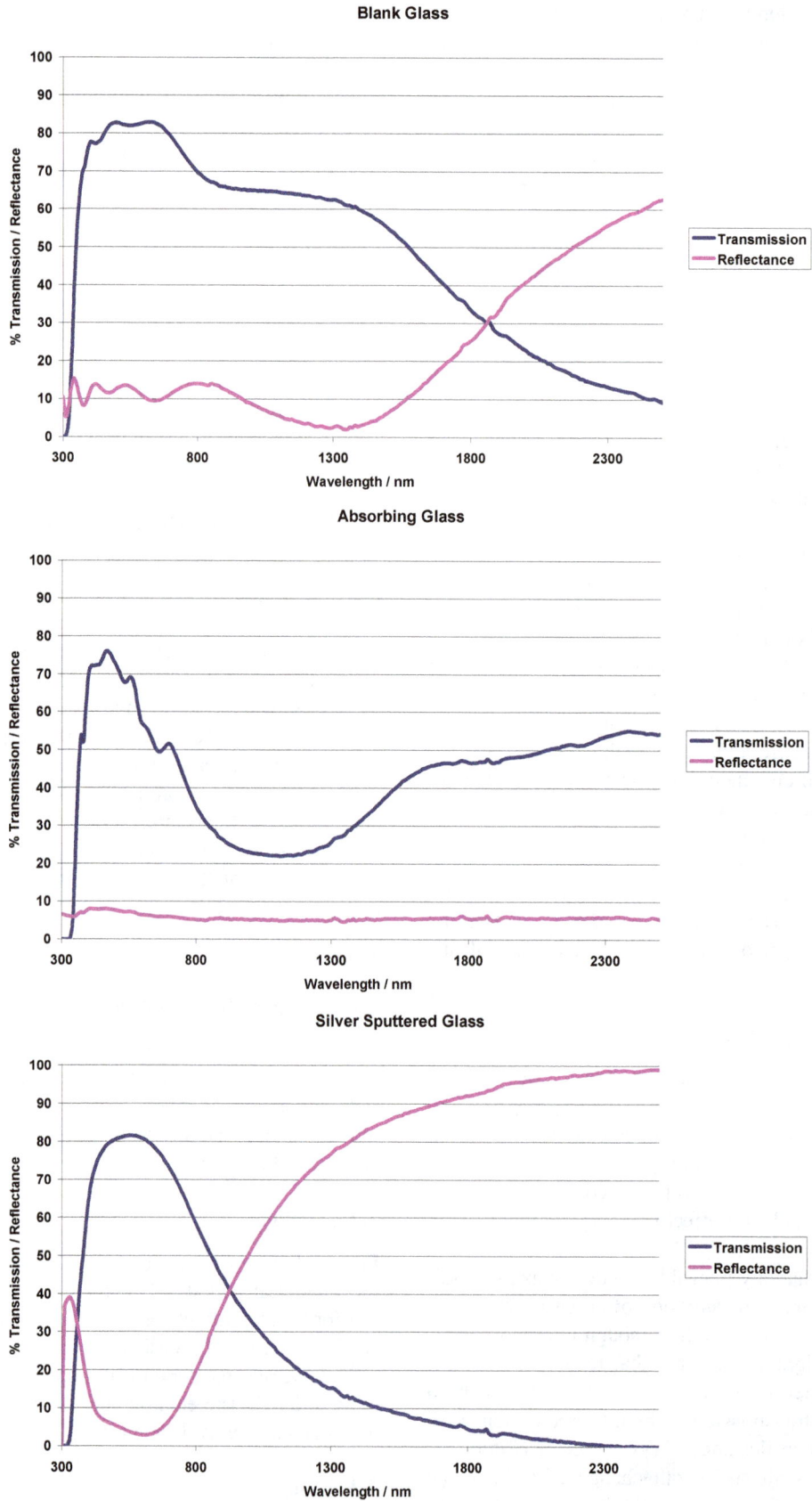

Figure 1. Transmission/reflectance spectral data for industrial standards and blank glass.

Theoretical Transmission Spectra

Theoretical Reflectance Spectra

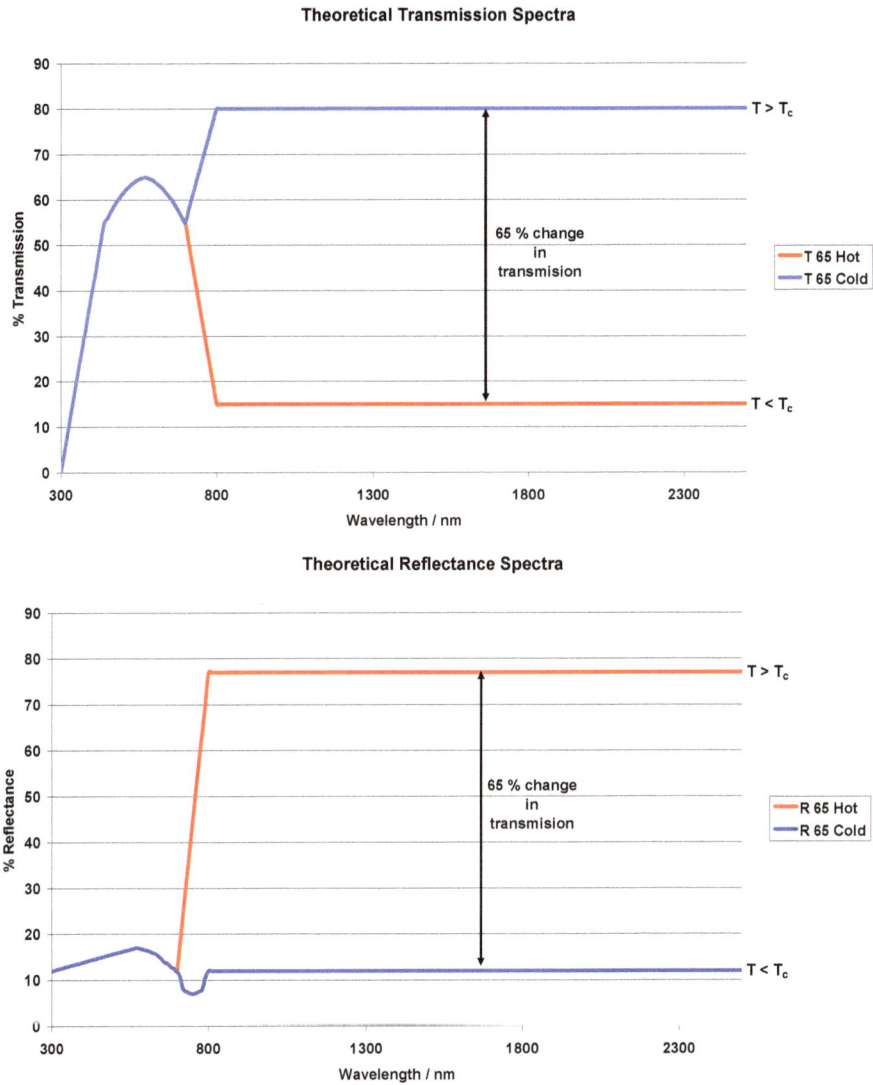

Figure 2. Theoretical transmission/reflectance data.

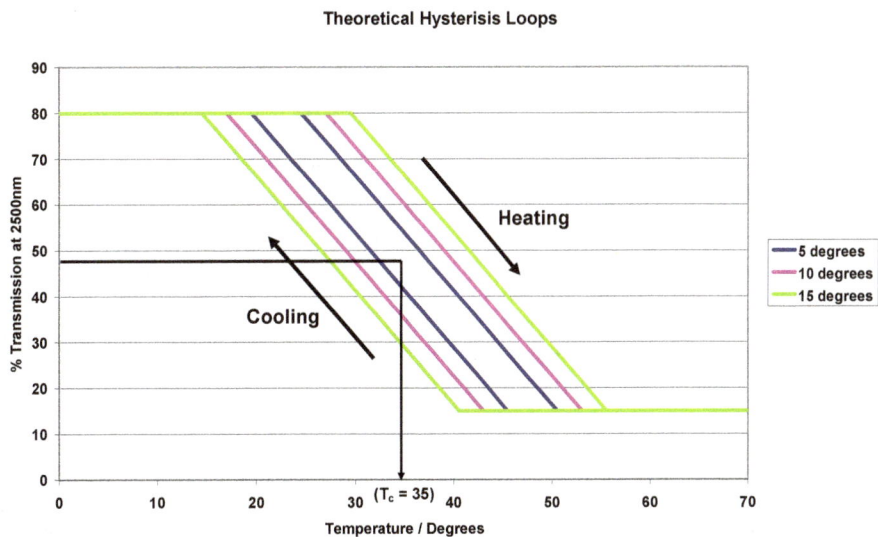

Theoretical Hysterisis Loops

Figure 3. Diagram showing the theoretical hysteresis loops modelled for $T_c = 35°C$.

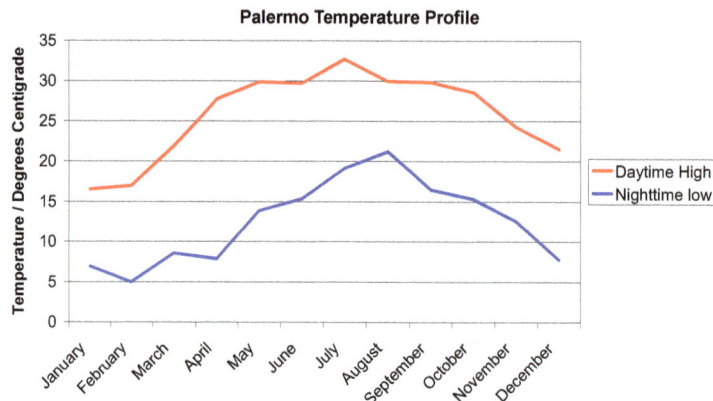

Figure 4. The daytime highs and night-time lows of the Palermo environment though out the year.

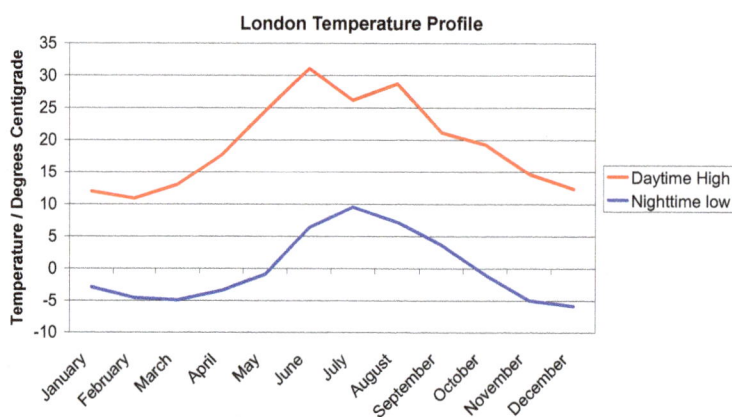

Figure 5. The daytime highs and night-time lows of the London environment throughout one year.

a year in London, UK.

We can see from **Figure 5** that for around one quarter of the year, May-August, the environment will be cooling dominated. For the winter and autumn period, November to March the environment will heating dominated. For the remainder of the year there will be a mixture of heating and cooling required.

3.4. Helsinki

In **Figure 6** the yearly temperature profiles for the daytime highs and the night-time lows for the Helsinki environment are shown.

It can be seen from **Figure 6** that the climate requires heating for the majority of the year; making Helsinki a heating dominated climate. The amount of time spent above a comfortable room temperature (22°C) is largely negligible, even in the summer months when the temperature may rise above this for one or two hours a day.

3.5. Intrinsic Energy Saving Properties of Glazing

The thermochromic systems modelled have an intrinsic energy reduction due to the absorbing properties of the systems, which will cause energy reduction even when the systems are in the cold state. To quantify the effect that this absorbing property would have the simulations were run using only the cold spectra. These results, shown in **Figure 7**, could then be used to see what portion of the energy reduction can be assigned to variable heat mirror properties and what is due to the absorbing properties.

4. Results

4.1. Palermo

Figure 8 shows the energy usage required for different modelled systems for the environment in Palermo. Blank glass creates an energy demand of 3655 KWh and the two industry standards of SSG and AG require 2930 and 2547 KWh respectively. This means the AG leads to a 1109 kWh reduction in energy usage when compared to blank glass and SSG has a reduction of 726 kWh compared to blank glass.

Figure 8(a) shows the results for the thermochromic systems when the T_c value was set to 35°C, in this set of systems the worst performing is the system where the hysteresis width is greatest (width = 15°C) in this case

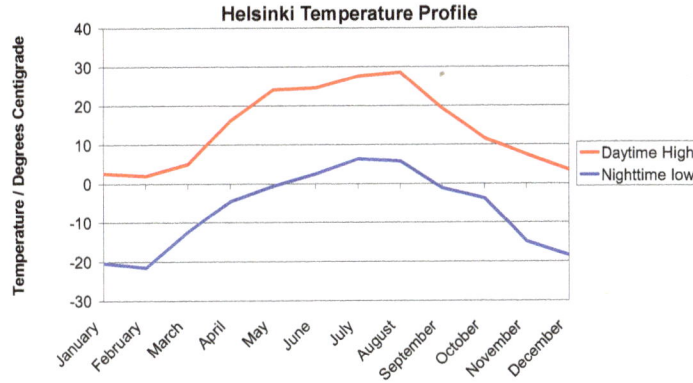

Figure 6. The daytime highs and night-time lows of the Helsinki environment throughout one year.

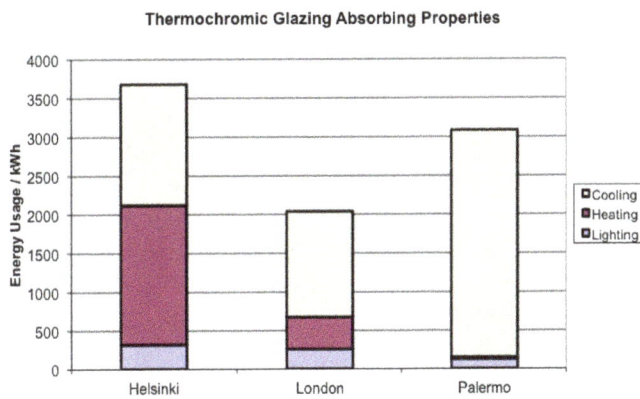

Figure 7. The total energy requirement in each environment due to absorbing properties of the thermochromic glazing.

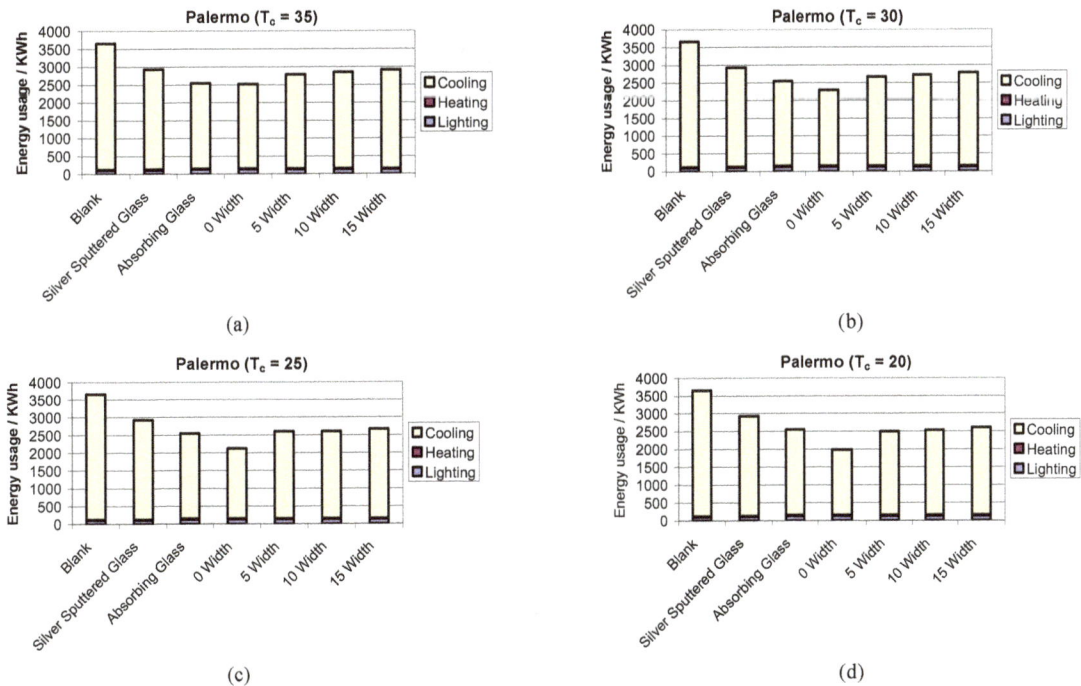

(a)

(b)

(c)

(d)

Figure 8. (a)-(d) show the breakdown of energy usage in a commercial building model based in Palermo for varying values of T_c ((a) = 35˚C, (b) = 30˚C, (c) = 25˚C, (d) = 20˚C) and varying hysteresis width (0˚C, 5˚C, 10˚C, 15˚C). Each Graph shows all modelled glazing systems. The colours correspond to the different energy requirements; yellow = cooling, maroon = heating and blue = lighting.

the energy demand is 2914 kWh a saving of 741 kWh compared to blank glass. As the width decreases the energy requirement also decreases when the width equals 10°C the energy requirement is 2851 kWh, 5°C the requirement is 2786 kWh and 0°C where it drops to 2519 kWh. These energy requirements correspond to an improvement of 804, 870 and 1136 kWh respectively when compared to blank glass.

Figure 8(b) shows the thermochromic systems energy usage when the T_c value is decreased to 30°C. When the hysteresis width is set to 0°C (**Figure 8(b)**, 0 width) the energy requirement is 2239 kWh corresponding to a reduction of 1362 kWh compared to blank glass. As the hysteresis width is increased the energy demand is increased. 5°C width requires 2605 kWh, 10°C width requires 2724 kWh and 15°C width requires 2786 kWh. These values, when compared to blank glass correspond to energy demand reduction values of 991, 932, and 869 kWh.

When the value of T_c is equal to 25°C (**Figure 8(c)**) the worst performing thermochromic system is the system with the largest hysteresis width (15°C width,) for this system the energy demand is 2666 kWh a reduction of 991 kWh compared to blank glass. 10°C width system requires 2600 kWh of power a reduction of 1056 kWh compared to blank glass, the 5°C width system requires 2599 kWh a saving of 1057 kWh. The best performing system, 0 Width, requires 2124 kWh corresponding to a saving of 1531 kWh compared to blank glass.

Figure 8(d) shows the energy demands for the systems when T_c is set to 20°C. For this set of simulations the least well performing system is the system with the greatest hysteresis width (15°C width). The widest hys-

teresis width had an energy load of 2599 kWh this is a reduction of 1057 kWh compared to blank glass. As the hysteresis width is decreased the energy load decreases, a 10°C width has an energy load of 2527 kWh a reduction of 1128 kWh compared to blank glass. 5°C width has a 2492 kWh energy load, 1163 kWh less than blank glass. 0°C width has the lowest energy demand of all the systems requiring 1983 kWh a reduction of 1672 kWh compared to blank glass.

The narrowing of the hysteresis width has greater affect at lower values of T_c, this can be seen by comparison of the energy load difference between the best and worst thermochromic for each value of T_c; $T_c = 35$°C gap = 395 kWh, $T_c = 30$°C gap = 493 kWh, $T_c = 25$°C gap = 540 kWh, $T_c = 20$°C gap = 615 kWh.

The results show that for the Palermo environment the thermochromic glazing with the lowest energy demand is one with a low value of T_c and as narrow a hysteresis width as possible (**Figure 9**). When comparing the thermochromics to the industry standards, AG and SSG, it can be seen that for all values of T_c the narrowest hysteresis width have a lower energy load. With low values of T_c it is still possible for the thermochromic glazing to out perform both industry standards. In all cases the thermochromic glazing out performs the SSG sample. It is clear from these results that the hysteresis width of the thermochromic transition has important consequences for the energy demand reduction properties of the glazing.

4.2. London

The energy load demand for the blank glass scenario is

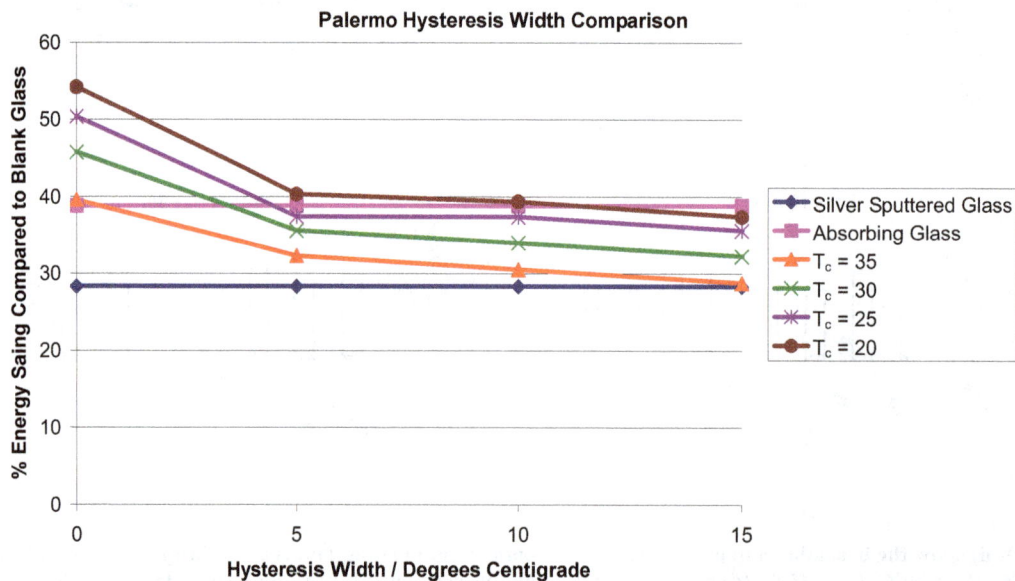

Figure 9. Shows the percentage energy saving of each system when compared to the blank glass system with variation of hysteresis width.

shown in **Figure 10** for the London environment. The blank glass energy load corresponds to a value of 2373 kWh. Also shown in all of the graphs is the energy requirement for two industry standards AG (absorbing glass) and SSG (a heat mirror) these require energy loads of 1707 and 1845 kWh respectively. This means that AG corresponds to a saving of 666 kWh and SSG a saving of 527 kWh compared to blank glass. This indicates that the absorbing glass AG is slightly more effective for this environment than the SSG case.

Figure 10(a) shows the systems modelled where the T_c is set to 35°C for the thermochromic systems. For these systems the best performing is the one with the thinnest hysteresis width, width = 0°C. As the hysteresis width is increased the energy demand is also in creased giving the following energy requirements; width = 0°C energy load = 1769 kWh, width = 5°C energy load = 1910 kWh, width = 10°C energy load 1946 kWh and width = 15°C energy load = 1978 kWh. These give energy demand reductions of 585, 429, 385 and 349 kWh compared to the blank glass system respectively.

For the systems where T_c = 30°C, shown in **Figure 10** b, the worst performing system is the one with the widest hysteresis width, width = 15°C. The energy requirements decrease for each environment as the hysteresis width is narrowed with the 0°C width requiring the lowest annual energy load. 15°C width requires 1910 kWh, 10°C width requires 1870 kWh, 5°C width requires 1830 kWh, 0°C width requires 1616 kWh when compared to a blank glass systems these systems gave reductions in energy usage of 462, 502, 542 and 757 kWh respectively.

Figure 10(c) shows the systems where the value of T_c is set to 25°C, as with the previous systems the energy requirement increases with the increase in hysteresis width but the lower T_c results in lower overall energy requirements. 15°C width has a 1830 kWh energy load, 10°C width has a 1789 kWh energy load, 5°C width has a 5°C width energy load, 0°C has a 1484 kWh energy load. These energy loads result in respective reductions of 542, 583, 612 and 888 kWh when compared to the system modelled for blank glass.

The final graph, **Figure 10(d)**, shows the systems when the T_c value is reduced to 20°C this yields thermochromic system with the lowest energy usage but still follow the trend as in the previous systems where the hysteresis width reduction corresponds to reduction in energy usage. 15°C width requires 161 kWh, 10°C width requires 1733 kWh, 5°C requires 1723 kWh and 0°C width requires an energy usage of 1388 kWh. These energy loads correspond to reductions in energy usage of 612, 639, 649 and 984 kWh respectively when compared to the blank glass system.

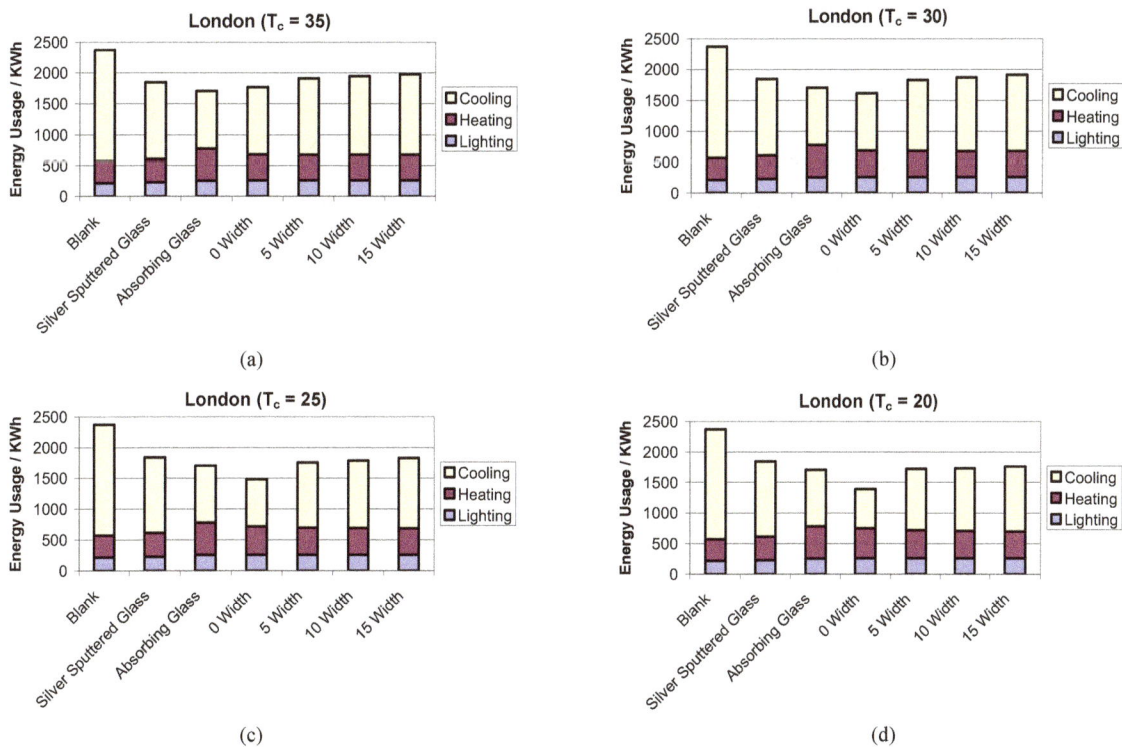

(a)

(b)

(c)

(d)

Figure 10. (a)-(d) show the breakdown of energy usage in a commercial building model based in London for varying values of T_c ((a) = 35°C, (b) = 30°C, (c) = 25°C, (d) = 20°C) and varying hysteresis width (0°C, 5°C, 10°C, 15°C). Each graph shows all modelled glazing systems. The colours correspond to the different energy requirements; yellow = cooling, maroon = heating, blue = lighting.

When comparing the results to the two industry standards (**Figure 11**) it can be seen that in the case where no hysteresis width is introduced, 0°C width, the systems out perform both industry standards for all values of T_c (except where $T_c = 35$°C). When hysteresis width is introduced there is a drop in the glazing improvement and the lower T_c cases (20°C & 25°C) perform comparably to the AG case and the higher T_c cases (30°C & 35°C) perform comparably to the SSG instance.

4.3. Helsinki

Figure 12 shows the different systems modelled for the Helsinki environment, each of the graphs, **(a)-(d)**, show the systems using blank glass and two industry standards SSG (heat mirror) and AG (absorbing glass). The blank glass system has an energy load requirement of 4088 kWh. SSG has an energy load of 3369 kWh which is a decrease of 719 kWh compared to blank glass. AG has an energy load of 3306 kWh, which is a 782 kWh reduction in energy demand compared to the blank glass case. This indicates that for this environment AG is slightly more efficient than SSG.

Figure 12(a) shows the thermochromic systems where the value of T_c has been set to 35°C. For the thermochromics in this set of simulations the best performing is the one with the thinnest hysteresis and as hysteresis width is increased the energy load requirement also increases. Then energy requirements are as follows. 0°C width 3417 kWh, 5°C width 3554 kWh, 10°C width 3597 kWh and 15°C width 3630 kWh. These energy loads, when compared to blank glass, correspond to reductions in energy of: 670, 533, 490 and 458 kWh respectively.

In **Figure 12(b)** the annual energy loads for systems with T_c value of 30°C is shown. These systems show the same relationship between hysteresis width and energy

load as was observable in the $T_c = 35$°C systems where increased hysteresis width leads to increased energy load requirement. 0°C width requires 3226 kWh, 5°C width requires 3449 kWh, 10°C width requires 3505 kWh and 15°C width requires 3554 kWh of energy per year. These energy values correspond to reductions of 862, 638, 583 and 533 kWh respectively compared to the blank glass system.

Figure 12(c) shows the system where the value of T_c is set to 25°C this system shows the same trend as the previous two systems with hysteresis width decrease leading to a decrease in energy load. 15°C width 3449 kWh energy load, 10°C width 3394 kWh energy load, 5°C width 3346 kWh energy load and 0°C width 3052 kWh energy load. These energy loads correspond to reductions, when compared to the blank glass system, of 638, 694, 741 and 1035 kWh respectively.

The last graph, **Figure 12(d)**, shows the system with the lowest T_c value. This system gives the lowest value for each thermochromic when compared to the higher T_c value systems and like the other systems it shows the same trend of decreased hysteresis width leading to lower energy demands. 15°C width requires 3346, 10°C width requires 3305 kWh, 5°C width requires 3276 kWh and 0°C width requires 2908 kWh. When compared to the blank glass system these energy requirements correspond to reductions of 741, 783, 812 and 1180 kWh respectively.

When comparing the systems to the Industry standards it can be seen from **Figure 13** that the thermochromic systems in which there has been a hysteresis width introduced all perform comparably or worse than the both the AG and the SSG systems. For this environment the best performing thermochromic is one with a low T_c value and a narrow hysteresis width, **Figure 13**, 0°C width.

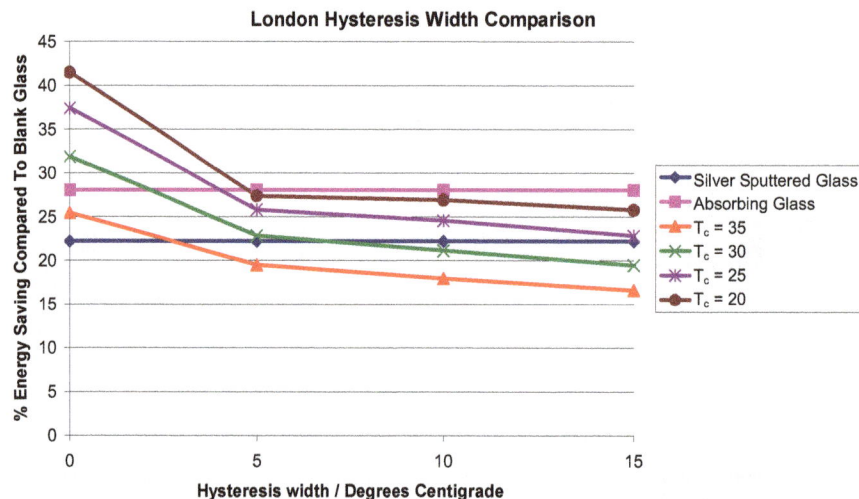

Figure 11. It shows the percentage energy saving of each system when compared to the blank glass system with variation of hysteresis width.

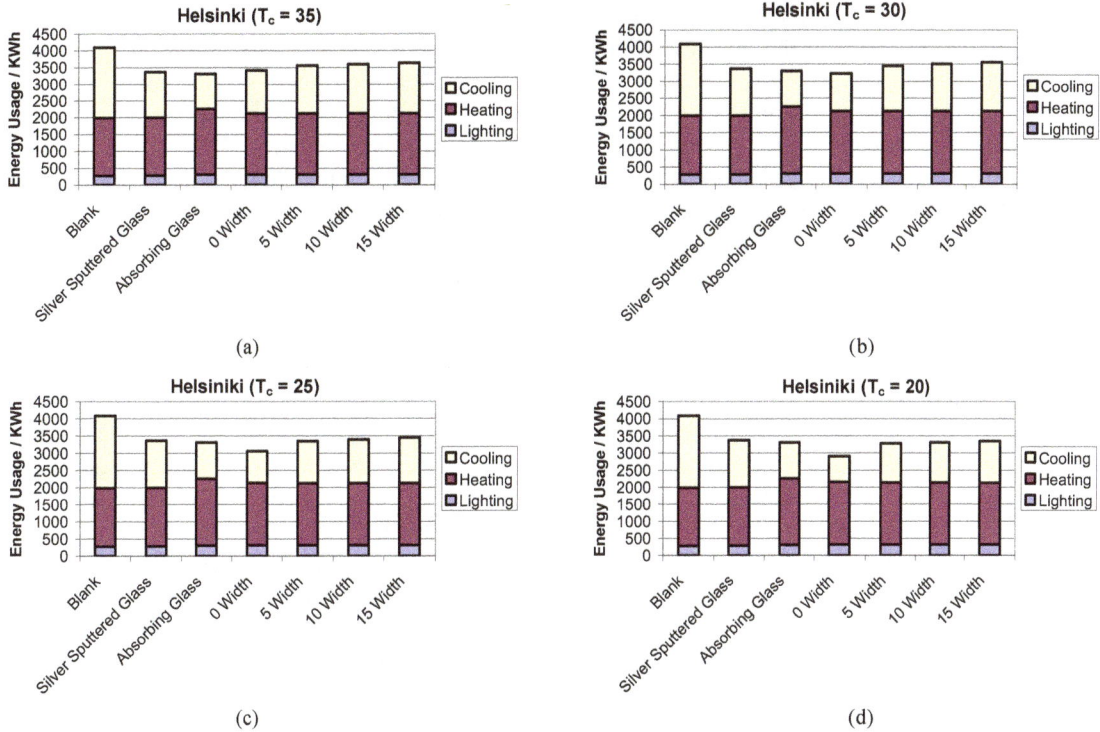

Figure 12. (a)-(d) show the breakdown of energy usage in a commercial building model based in Helsinki for varying values of T_c ((a) = 35°C, (b) = 30°C, (c) = 25°C, (d) = 20°C) and varying hysteresis width (0°C, 5°C, 10°C, 15°C). Each graph shows all modelled glazing systems. The colours correspond to the different energy requirements; yellow = cooling, maroon = heating, blue = lighting.

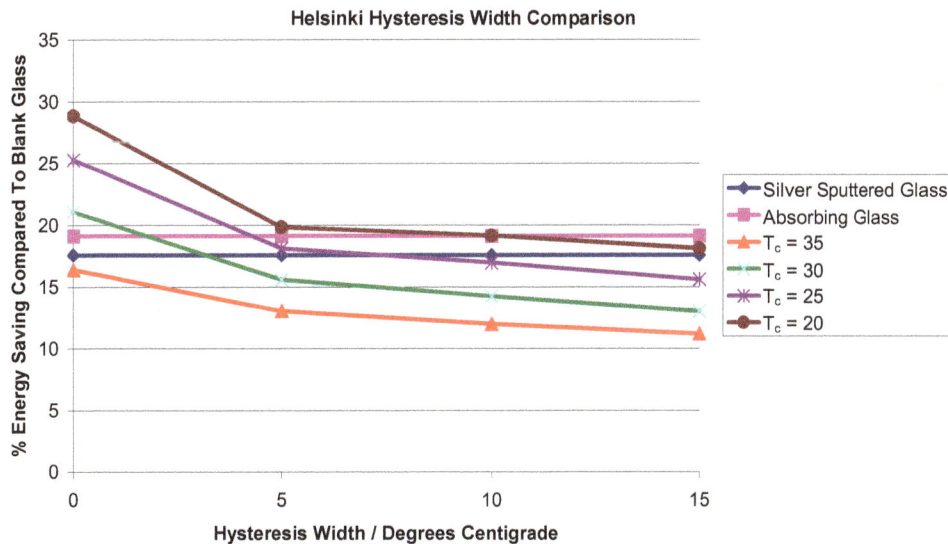

Figure 13. It shows the percentage energy saving of each system when compared to the blank glass system with variation of hysteresis width.

5. Discussion

There are some noticeable general trends in all of the data. Firstly and foremost a lower transition temperature leads to a bigger energy saving. Secondly a lower transition hysteresis width leads to a bigger energy saving. It is also noticeable that the transition temperature has a bigger effect than the hysteresis transition width, *i.e.* an improvement of 1% per degree T_c reduction as opposed to an improvement of ~0.5% per degree reduction in hysteresis width (except at very low width).

The overall effect in each of the climates investigated was to reduce the energy demand when compared to the clear-clear glazing system. In the hot environment, Palermo, the best performance was for the lowest transition temperature investigated (20°C) and the lowest hysteresis width (0°C) with a total energy load of 1983 kWh, representing a reduction in demand of 1672 kWh or 54% compared to the clear-clear glazing system. This was also improvement compared to the industry standards which gave energy demand reductions of 39% for the absorbing glass and 28% for the silver sputtered glass compared to the clear-clear glazing system.

In the mixed environment, London, the best performance was also given by the scenario with the lowest transition temperature investigated (20°C) and the lowest hysteresis width (0°C) with a total energy load of 1388 kWh, representing a reduction in demand of 984 kWh or 43% compared to the clear-clear glazing system. This was also improvement compared to the industry standards which gave energy demand reductions of 27% for the absorbing glass and 22% for the silver sputtered glass compared to the clear-clear glazing system.

In the cool environment, Helsinki, the scenario with the lowest transition temperature investigated (20°C) and the lowest hysteresis width (0°C) also gave the best energy demand reduction performance. A total energy demand of 2908 kWh was required, representing a reduction in demand of 1180 kWh or 27% compared to the clear-clear glazing system. This was also improvement compared to the industry standards which gave energy demand reductions of 18% for the absorbing glass and 17% for the silver sputtered glass compared to the clear-clear glazing system.

Biggest energy improvement is in the hot climate (Palermo), although there is still significant energy demand reduction in the mixed and cool climates investigated (London and Helsinki)

It is readily apparent that the switching of the glazing is having a significant effect. In the best case in each climate the inherent energy demand of the low temperature thermochromic glazing (*i.e.* no switching) is 3086 kWh for Palermo, 2028 kWh for London and 3676 kWh for Helsinki. In comparison the best thermochromic performance in each climate leads to an overall energy demand of 1983 kWh for Palermo, 1388 kWh for London and 2908 kWh for Helsinki, which represents an energy demand reduction of 36%, 32%, 21% for Palermo, London and Helsinki respectively.

One notable phenomenon is that typically an extra 10% - 15% in energy demand improvement over blank glass can be gained by reducing the hysteresis width from 5°C to 0°C. The thinner the hysteresis width the more quickly the hot state is entered, thus allowing for maximum efficiency in allowing most appropriate glazing characteristics to be used. However, the production of thin vanadium dioxide thin films with a 0°C hysteresis

width is not enormously realistic. It is not well understood how to obtain low hysteresis widths, although uniform particle size has been suggested as one possible route [12]. In any case, the vast majority of samples in the literature have widths of 5°C - 10 °C [16,17,26-30]. Which gives a comparable or moderately enhanced performance when compared to the absorbing glass, and a significantly improved performance over the sputtered silver glass.

None the less thermochromic glazing would still provide an additional energy benefit of 1% - 3% compared to absorbing glass and there are a number of practical reasons to prefer a coating of this type. Sputtered silver glasses are typically produced by PVD methodologies and have a limited life span, typically around 10 years much less than the average life span of a modern commercial building. Absorbing glasses are produced by body tinting, obtaining a stable glass melt in the float glass production time is a laborious process, typically taking around 100 hours, during which time any glass produced at the factory is not saleable this represent 100% loss during this time. Producing films via an integrated CVD process allows for deposition to be turned on and off at will, drastically reducing the float line down time and improving factory efficiency

6. Conclusion

The energy demand associated with the use of a theoretical thermochromic coating in architectural glazing has been evaluated, using the energy simulation package EnergyPlus. The study, the first of this kind, examines the effect of different hysteresis properties of the thermochromic transition. The results demonstrate that such thermochromic glazing could have an additional energy benefit in excess of those approaches currently used in warmer climates. This arises from a combination of absorbing and heat mirror behaviour. The best energy performance occurred in the warm climate examined and where the thermochromic transition temperature was lowest and the hysteresis width at it's thinnest. In this instance the energy demand of the model could be reduced by 54% compared to a standard double glazed system. The results also indicate that transition temperature has a much larger effect on the energy demand reduction properties of the glazing than the hysteresis width—an improvement of 1% per degree T_c reduction as opposed to an improvement of ~0.5% per degree reduction in hysteresis width. It does however, highlight that hysteresis width does have an important role to play in enhancing the energy demand reduction characteristics of thermochromic glazing systems and that this needs to be considered when thermochromic thin films are synthesised.

7. Acknowledgements

RB thanks the Royal Society for a Dorothy Hodgkin fellowship and the EPSRC for financial support (grant number EP/H005803/1). Mr. Kevin Reeves is thanked for invaluable assistance with electron microscopy.

REFERENCES

[1] M. Saeli, C. Piccirillo, I. P. Parkin, R. Binions and I. Ridley, "Energy Modelling Studies of Thermochromic Glazing," *Energy and Buildings*, Vol. 42, No. 10, 2010, pp. 1666-1673.

[2] H. Ye, X. Meng and B. Xu, "Theoretical Discussions of Perfect Window, Ideal Near Infrared Solar Spectrum Regulating Window and Current Thermochromic Window," *Energy and Buildings*, Vol. 49, 2012, pp. 164-172.

[3] X. Ye, Y. Luo, X. Gao and S. Zhu, "Design and Evaluation of a Thermochromic Roof System for Energy Saving Based on Poly(N-isopropylacrylamide) Aqueous Solution," *Energy and Buildings*, Vol. 48, 2012, pp. 175-179.

[4] C. G. Granqvist, "Window Coatings for the Future," *Thin Solid Films*, Vol. 193-194, 1990, pp. 730-741.

[5] C. G. Granqvist, "Solar Energy Materials," *Advanced Materials*, Vol. 15, No. 21, 2003, pp. 1789-1803.

[6] K. D. Rogers, "An X-Ray Diffraction Study of Semiconductor and Metallic Vanadium Dioxide," *Powder Diffraction*, Vol. 8, No. 4, 1993, pp. 240-244.

[7] S. S. Kanu and R. Binions, "Thin Films for Solar Control Coatings," *Proceedings of the Royal Society A: Mathematical, Physical and Engineering Science*, Vol. 466, 2009, pp. 19-44.

[8] F. Béteille, R. Morineau, J. Livage and M. Nagano, "Switching Properties of $V_{1-x}TixO_2$ Thin Films Deposited from Alkoxides," *Materials Research Bulletin*, Vol. 32, No. 8, 1997, pp. 1109-1117.

[9] T. E. Phillips, R. A. Murphy and T. O. Poehler, "Electrical Studies of Reactively Sputtered Fe-Doped VO_2 Thin Films," *Materials Research Bulletin*, Vol. 22, No. 8, 1987, pp. 1113-1123.

[10] T. D. Manning, I. P. Parkin, M. E. Pemble, D. Sheel and D. Vernardou, "Intelligent Window Coatings: Atmospheric Pressure Chemical Vapor Deposition of Tungsten-Doped Vanadium Dioxide," *Chemistry of Materials*, Vol. 16, No. 4, 2004, pp. 744-749.

[11] G. Xu, P. Jin, M. Tazawa and K. Yoshimura, "Thickness Dependence of Optical Properties of VO_2 Thin Films Epitaxially Grown on Sapphire (0 0 0 1)," *Applied Surface Science*, Vol. 244, No. 1-4, 2005, pp. 449-452.

[12] R. Binions, G. Hyett, C. Piccirillo and I. P. Parkin, "Doped and Un-Doped Vanadium Dioxide Thin Films Prepared by Atmospheric Pressure Chemical Vapour Deposition from Vanadyl Acetylacetonate and Tungsten Hexachloride: The Effects of Thickness and Crystallographic Orientation on Thermochromic Properties," *Journal of Materials Chemistry*, Vol. 17, No. 44, 2007, pp. 4652-4660.

[13] W. Burkhardt, T. Christmann, B. K. Meyer, W. Niessner, D. Schalch and A. Scharmann, "W- and F-doped VO_2 Films Studied by Photoelectron Spectrometry," *Thin Solid Films*, Vol. 345, No. 2, 1999, pp. 229-235.

[14] I. Takahashi, M. Hibino and T. Kudo, "Thermochromic Properties of Double-Doped VO_2 Thin Films Prepared by a Wet Coating Method Using Polyvanadate-Based Sols Containing W and Mo or W and Ti," *Japanese Journal of Applied Physics*, Vol. 40, 2001, p. 1391.

[15] D. Barreca, L. E. Depero, E. Franzato, G. A. Rizzi, L. Sangaletti, E. Tondello and U. Vettori, "Vanadyl Precursors Used to Modify the Properties of Vanadium Oxide Thin Films Obtained by Chemical Vapor Deposition," *Journal of The Electrochemical Society*, Vol. 146, 1999, p. 551.

[16] R. Binions, C. S. Blackman, T. D. Manning, C. Piccirillo and I. P. Parkin, "Thermochromic Coatings for Intelligent Architectural Glazing," *Journal of Nano Research*, Vol. 2, 2008, pp. 1-20.

[17] D. Vernardou, M. E. Pemble and D. W. Sheel, "Vanadium Oxides Prepared by Liquid Injection MOCVD Using Vanadyl Acetylacetonate," *Surface and Coatings Technology*, Vol. 188-189, 2004, pp. 250-254.

[18] M. Saeli, C. Piccirillo, I. P. Parkin, I. Ridley and R. Binions, "Nano-Composite Thermochromic Thin Films and Their Application in Energy-Efficient Glazing," *Solar Energy Materials and Solar Cells*, Vol. 94, No. 2, 2010, pp. 141-151.

[19] J. A. Clarke, M. Janak and P. Ruyssevelt, "Assessing the Overall Performance of Advanced Glazing Systems," *Solar Energy*, Vol. 63, No. 4, 1998, pp. 231-241.

[20] C. G. Granqvist and V. Wittwer, "Materials for Solar Energy Conversion: An Overview," *Solar Energy Materials and Solar Cells*, Vol. 54, No. 1-4, 1998, pp. 39-48.

[21] J. H. Klems, "Materials for Solar Energy Conversion: An Overview," *Energy and Buildings*, Vol. 33, No. 2, 2001, pp. 93-102.

[22] M. S. Reilly, F. C. Winkelmann, D. K. Arasteh and W. L. Carroll, "Modeling Windows in DOE-2.1E," *Energy and Buildings*, Vol. 22, No. 1, 1995, pp. 59-66.

[23] H. Feustel, A. de Almeida and C. Blumstein, "Alternatives to Compressor Cooling in Residences," *Energy and Buildings*, Vol. 18, No. 3-4, 1992, pp. 269-286.

[24] P. Nitz and H. Hartwig, "Solar Control with Thermo-

tropic Layers," *Solar Energy*, Vol. 79, No. 6, 2005, pp. 573-582.

[25] A. Raicu, H. R. Wilson, P. Nitz, W. Platzer, V. Wittwer and E. Jahns, "Facade Systems with Variable Solar Control Using Thermotropic Polymer Blends," *Solar Energy*, Vo. 72, No. 1, 2002, pp. 31-42.

[26] F. Béteille and J. Livage, "Optical Switching in VO$_2$ Thin Films," *Journal of Sol-Gel Science and Technology*, Vol. 13, No. 1-3, 1998, pp. 915-921.

[27] R. Binions, G. Hyett and P. Kiri, "Solid State Thermochromic Materials," *Advanced Materials Letters*, Vol. 1, 2010, pp. 86-105.

[28] W. Burkhardt, T. Christmann, S. Franke, W. Kriegseis, D. Meister, B. K. Meyer, W. Niessner, D. Schalch and A. Scharmann, "Tungsten and Fluorine Co-Doping of VO$_2$ Films," *Thin Solid Films*, Vol. 402, No. 1-2, 2002, pp. 226-231.

[29] C. G. Granqvist, "Transparent Conductors as Solar Energy Materials: A Panoramic Review," *Solar Energy Materials and Solar Cells*, Vol. 91, No. 17, 2007, pp. 1529-1598.

[30] N. Joyeeta and J. R. F. Haglund, "Synthesis of Vanadium Dioxide Thin Films and Nanoparticles," *Journal of Physics: Condensed Matter*, Vol. 20, No. 26, 2008, Article ID: 264016.

17

Case Studies of Energy Saving and CO$_2$ Reduction by Cogeneration and Heat Pump Systems

Satoru Okamoto

Department of Mathematics and Computer Science, Shimane University, Matsue, Japan

ABSTRACT

This paper describes two case studies: 1) a cogeneration system of a hospital and 2) a heat pump system installed in an aquarium that uses seawater for latent heat storage. The cogeneration system is an autonomous system that combines the generation of electrical, heating, and cooling energies in a hospital. Cogeneration systems can provide simultaneous heating and cooling. No technical obstacles were identified for implementing the cogeneration system. The average ratio between electric and thermal loads in the hospital was suitable for the cogeneration system operation. An analysis performed for a non-optimized cogeneration system predicted large potential for energy savings and CO$_2$ reduction. The heat pump system using a low-temperature unutilized heat source is introduced on a heat source load responsive heat pump system, which combines a load variation responsive heat pump utilizing seawater with a latent heat-storage system (ice and water slurry), using nighttime electric power to level the electric power load. The experimental coefficient of performance (COP) of the proposed heat exchanger from the heat pump system, assisted by using seawater as latent heat storage for cooling, is discussed in detail.

Keywords: Cogeneration System; Heat Pump System; Energy Saving; CO$_2$ Reduction; Hospital; Aquarium

1. Introduction

Following the crisis of the Great Eastern Japan Earthquake and the nuclear power plant accidents on March 11, 2011, the future energy balance flow for Japan was quantitatively estimated. Cogeneration systems and heat pump systems are especially becoming increasingly important technologies after the nuclear power plant accidents. This paper describes two case studies of energy systems: 1) a cogeneration system of a hospital; and 2) a heat pump system installed in an aquarium that uses seawater as latent heat storage. These systems have several advantages, including lower consumption of primary energy, reductions in the levels of air pollution, and less expenditure. Simultaneous production of heating, cooling, and power provides higher overall system efficiency. Depending on the conditions, these combined systems can be the most economical solution for a building. The requirement is that the system should be located where there is high consumption of electrical, heating, and cooling energy throughout the year. A hospital is a perfect example of the type of consumer with those conditions. This system might not be profitable during a certain period of the year because of the relative costs of gas

and electricity. Therefore, it is important to make a detailed analysis of any planned system and examine the various possible operating regimes [1-3].

The basis of a cogeneration system is its electrical, heating, and cooling device. However, different kinds of cogeneration systems are distinguished by the types of driving units and cooling devices that make up the systems. The driving unit of a cogeneration module can be a steam turbine, a gas turbine, a reciprocating engine, or a fuel cell. The cooling energy from a cogeneration system is mainly produced with either a turbo-chiller or an absorption chiller; the choice depends on the required output power and the operating regime [4-7]. This work examines the technical viability of the systems to determine the better choice for an operating pattern for a cogeneration system supplying energy to a hospital, starting from an analysis of energy consumption data available for the hospital; no obstacles were identified in the technical feasibility of the cogeneration [8].

On the other hand, to reduce the emissions of carbon dioxide that contributes to global warming, the effective use of energy, such as the efficient use of various types of waste heat and renewable energy, should be promoted. A heat pump system can produce more heat energy than the energy that is used to run the heat pump system. Therefore, a heat pump system is considered to be one representation of a machine system that can efficiently use energy, and a load leveling air-conditioning system that uses unutilized energies at high levels [9-13].

This paper also discusses a heat pump system installed in an aquarium that uses seawater as latent heat storage [14]. To maintain indoor temperatures, such as using air conditioner to maintain the indoor conditions at a constant temperature and a constant relative humidity and to cool the water supply to the fish tanks in the aquarium, heat from seawater is collected as the heat source for the heat pump system. The pump system using low-temperature unutilized heat sources is introduced on a heat source load responsive heat pump system, which combines a load variation responsive heat pump utilizing seawater with a latent heat-storage system (ice and water slurry), using nighttime electric power serving as electric power load leveling. The experimental coefficient of performance (COP) of the proposed heat pump with the latent heat-storage cooling system is discussed in detail.

2. Cogeneration System in a Hospital

It is very important to determine the optimal patterns for the driving units of a cogeneration system in a hospital. However, it has been very difficult to solve this problem under the actual conditions. Therefore, establishing the optimal controls to accommodate the changing energy demands of a hospital have never been technically accomplished because of the following reasons: 1) the monthly energy consumption of a hospital is affected by seasonal changes of energy demands throughout the year, 2) the daily energy consumption is affected by changes during weekdays or holidays, and 3) the hourly energy consumption is affected by differences in daytime or nighttime energy usage. This study will add to the challenges of obtaining a new solution. The cogeneration system is conceived as an autonomous system that combines the generation of electrical, heating, and cooling energy in the hospital. The gas engine generators of the adopted cogeneration units have higher efficiency (40.7%) than other conventional driving units; they are Miller cycle gas engines, which use a novel technology that drives cogeneration systems to efficiently produce electrical and heat energy. The average ratio between the electric and thermal loads in the hospital is suitable for the cogeneration system operation [1,2,5]. A case study performed for a non-optimized cogeneration system predicted large potential for energy savings and CO_2 reduction [15-18].

2.1. Energy Demand in the Hospital

The hospital examined in this study belongs to the Shi-

mane University Faculty of Medicine in Japan. The hospital is an eight-story building and was completed in 1977; therefore, its facilities are superannuated. The area that is heated and air-conditioned is 42,203 m^2, and there are 616 beds. Despite this quantitative analysis of energy consumption, a qualitative analysis is needed to assess whether a cogeneration system can meet the energy demand. The hospital has a central system for hot or warm water production (heavy oil-fueled boilers; 16 t/h × 2 and 5 t/h × 1) and cooling (three absorption chillers, 600 RT; and a turbo-chiller, 400 RT), as shown in **Figure 1**. Power is usually purchased from the electric utility. Mainly, all-air systems have been adopted using air-treatment units that allow the adjustment of temperature and humidity; a typical temperature range for the heat exchanger in such units is 70°C - 85°C.

The typical daily consumption in 2005, shown in **Figure 2**, for electrical, heating, and cooling loads, is nearly constant only for the summer, autumn, and winter, because the typical day energy consumption profiles in the spring is almost same that as the autumn. The typical hourly consumption profiles in 2005, shown in **Figure 3**, for heating and cooling are quite regular, with a small increase during the morning or afternoon, depending on whether it is a weekday or a holiday. The hourly electric load profile is characterized by regular power requests during the day with a leap in demand for the lighting system and the elevators. Obviously, certain electric loads require a very high level of supply security; therefore, dedicated engines or inverter groups are usually provided for energy supply in case of power-grid failures.

2.2. Description of the Cogeneration System

This study sights at the possibility of installing a cogeneration system in a hospital, *i.e.*, the simultaneous generation of heating, cooling, and electrical energy. **Figure 4** shows the concept of an autonomous system for the combined generation of electrical, heating, and cooling energy. The driving cogeneration units are two high-efficiency Miller cycle (40.7%) gas engines (GE-1 and GE-2).

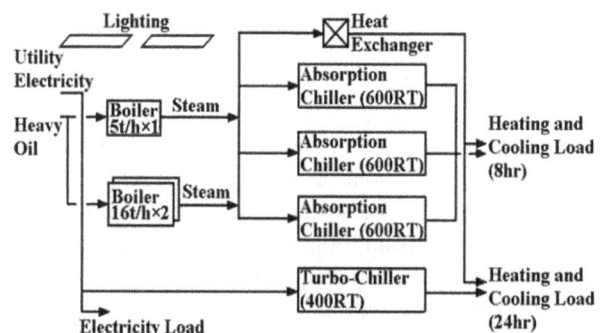

Figure 1. The conventional hospital system.

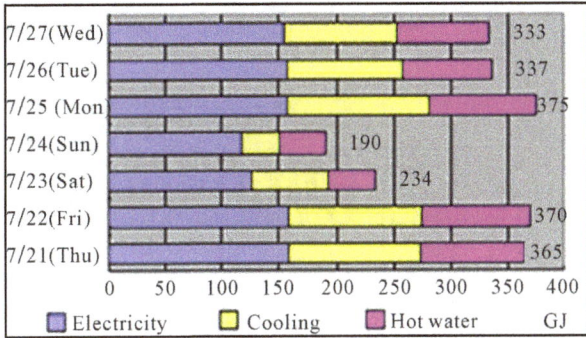

(a)

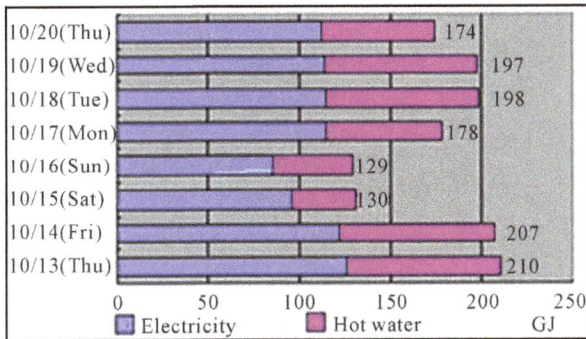

(b)

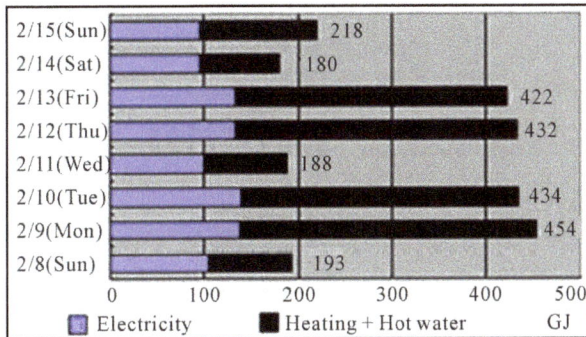

(c)

Figure 2. Daily energy consumption of the hospital. (a) Consumption during a typical week in summer; (b) Consumption during a typical week in autumn; (c) Consumption during a typical week in winter.

A gas engine is used as a driving unit because of the high demand for electrical and heating energy. The natural gas-fueled reciprocating engine generates 735 kW. Table 1 shows the energy consumption of the conventional system and the cogeneration system throughout the year. For our study, electrical energy will be used only in the hospital. Electricity can also be purchased from the public network to cover a deficit, as shown in Table 1. Table 1 shows the consumption of electricity from the public utility by the conventional system and the amount of natural gas consumed by the cogeneration system. Generated steam drives three steam-fired absorption chillers of the same conventional type (600 RT × 3) and use the present storage (1000 m^3), which is delivered to

individual heat consumers, as shown in **Figure 4**, which shows the amount of generated steam throughout the year. An additional peak-time waste heat boiler provides additional heat during the winter period. This system can provide simultaneous heating and cooling, as shown in **Figure 4**. **Figure 5** shows the heating load during the winter and the cooling load during the summer of 2009. It is necessary to install an additional absorption chiller

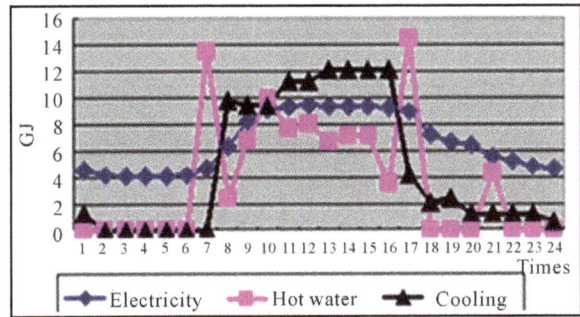

(a)

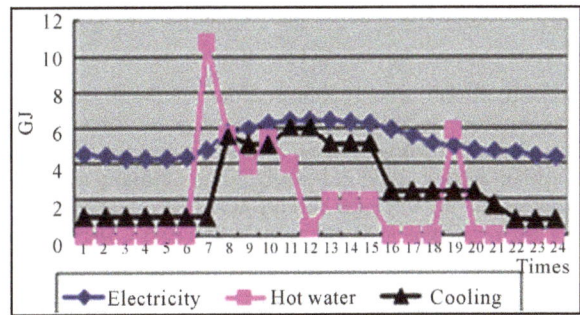

(b)

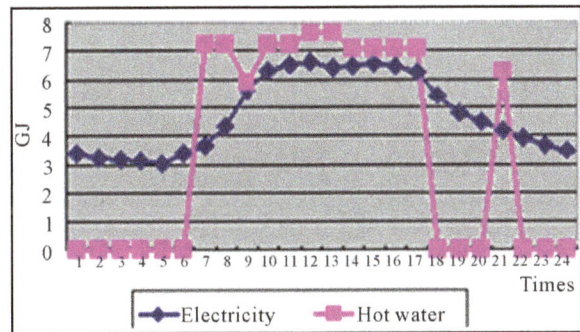

(c)

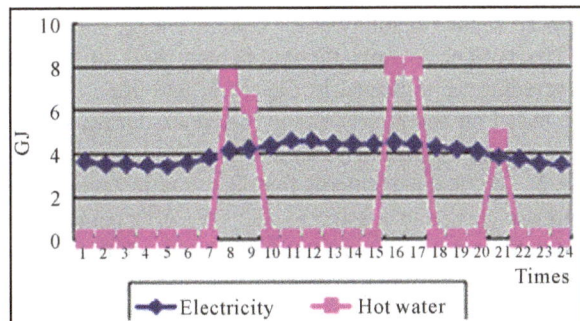

(d)

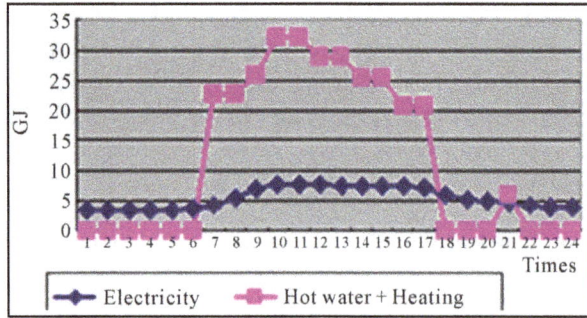

(e)

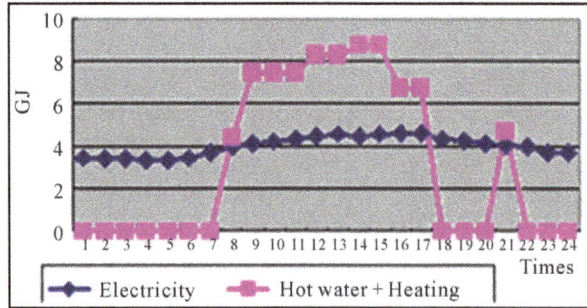

(f)

Figure 3. Hourly energy consumption of the hospital. (a) Consumption during a typical weekday in summer; (b) Consumption during a typical holiday in summer; (c) Consumption during a typical weekday in autumn; (d) Consumption during a typical holiday in autumn; (e) Consumption during a typical weekday in winter; (f) Consumption during a typical holiday in winter.

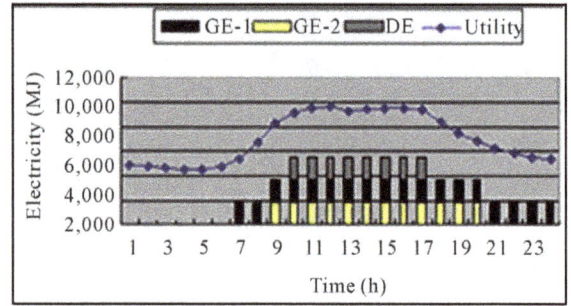

Figure 4. The cogeneration system of the hospital.

to increase the reliability of the cooling energy supply.

The typical patterns for the driving units of the cogeneration are indicated in **Figure 5**, and these patterns are based on the hourly energy demands during several seasons and on weekdays or holidays, as shown in **Figure 3**. The planned patterns for electricity in several seasons are shown in **Figure 5**. The choice of patterns for generated electricity, indicated in **Figures 5(a)** and **(b)**, are based on the hourly heating load and cooling loads for a summer weekday and holiday, respectively. The deficit in electricity is supplied by an emergency diesel

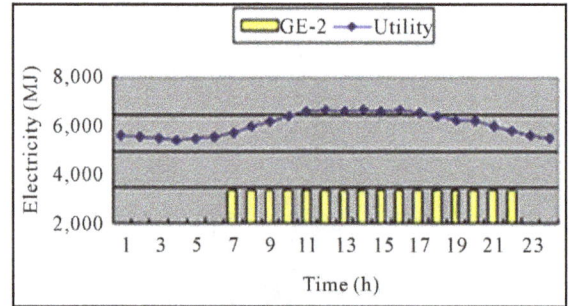

(a)

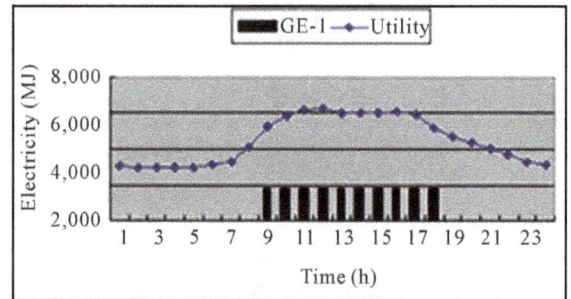

(b)

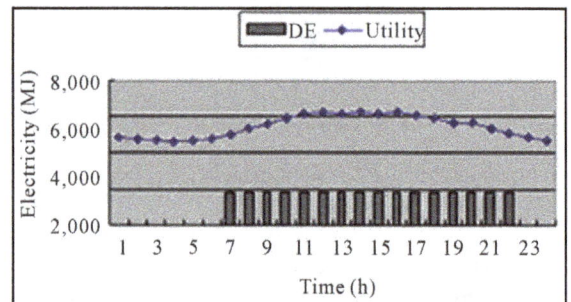

(c)

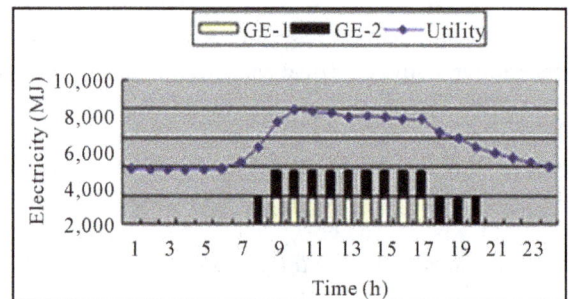

(d)

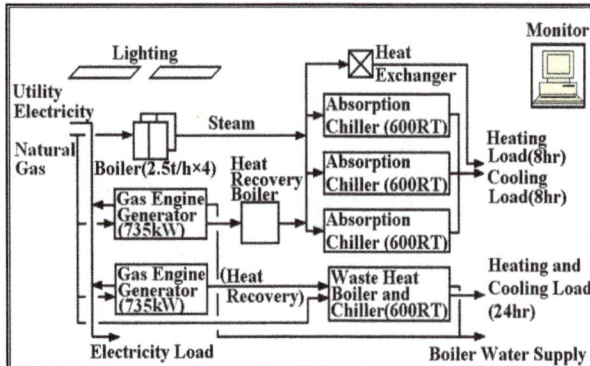

(e)

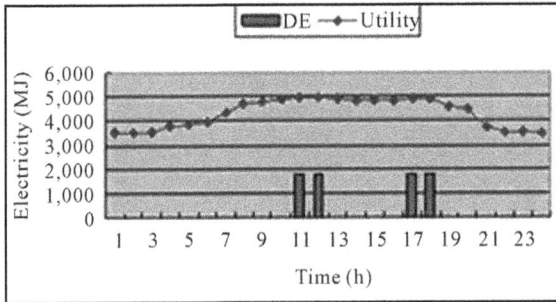

(f)

Figure 5. Energy consumption of the cogeneration system. (a) Energy use pattern during a typical summer weekday; (b) Energy use pattern during a typical summer holiday; (c) Energy use pattern during a typical autumn weekday; (d) Energy use pattern during a typical autumn holiday; (e) Energy use pattern during a typical winter weekday; (f) Energy use pattern during a typical winter holiday.

generator (DE). The patterns in autumn, indicated in **Figures 5(c)** and **(d)**, depend on the temporal hot water requirements for weekdays and holidays, as shown in **Figures 3(c)** and **(d)**, respectively. The patterns in winter indicated in **Figures 5(e)** and **(f)** are based on the hourly heating load (hot water and heating) on weekdays and holidays, as shown in **Figures 3(e)** and **(f)**, respectively. Normally, the heat energy needs in winter are too high to use as the basis of the maximum amount of heating energy required when planning a cogeneration system. The requirements for electrical, heating, and cooling energy vary within certain limits. These are the energy consumption estimates for the cogeneration system described in **Table 1**. We chose a cogeneration module on the basis of a peak cooling load. For cooling purposes, we chose three absorption chillers; one of them having cooling power of 600 RT.

2.3. Energy Saving and CO_2 Reduction of the Cogeneration System Installed in a Hospital

When retrofitting a conventional plant, as shown in **Figure 1**, with cogeneration technology, the existing components and equipment must be integrated with the new ones. This section examines a typical hospital energy system and how it would be integrated with the required system. **Figure 4** shows the new components or major changes that include two gas engine generators and a heat-recovery boiler. When the conventional system is converted to a cogeneration system, the existing boilers and chillers can be used as auxiliary systems. The same piping used for the existing centralized all-air system for heating and cooling can be used.

To confirm all these considerations, a brief analysis was performed for the hospital with the obtained monthly electricity, heavy oil, and natural gas consumption indicated in **Table 1**. This analysis is much more general; the

purpose of this section is to show the enormous potential for energy saving with a cogeneration system in the hospital, where interesting results might be achieved even by a non-optimized plant design and operation. The calculations for cogeneration system are performed using two gas engines with a heat-recovery boiler and three absorption chillers; it strictly follows the sum of electrical demand for direct users and for feeding the absorption chiller.

The energy-saving ratio was calculated by:

$$\frac{T-C}{T} \times 100 = 12.3\%$$

The energy provided by a conventional system and cogeneration system is described in **Table 1**. Here T represents the total energy consumption of the conventional system, and C represents the energy consumption of the cogeneration using electricity from a public utility. T is the sum of the fuel consumption in the local energy system and in a power plant that supplies electricity. The energy rates used in **Table 1** are explained in **Appendix**. The energy-saving ratio calculated by simulation is 16.5% in the feasibility study [19]. Because both the total energy consumption of the conventional systems in the feasibility study [19] and the verification study are the same, the actual energy consumption of the cogeneration system are overestimated in the verification study. The prime reason to adopt a cogeneration system cannot be entirely based on the hourly consumption profiles for heating and cooling with a small increase during the morning or afternoon depending on whether it is a weekday or holiday.

The CO_2 reduction ratio is calculated by:

$$\frac{X-Y}{X} \times 100 = 20.7\%$$

The CO_2 emitted by the conventional system and the cogeneration system is described in **Table 1**. X represents the total amount of CO_2 from the conventional system, and Y represents the amount of CO_2 from cogeneration with natural gas and electricity from public utilities. X is the sum of the CO_2 amount emitted from the utility electricity and the heavy oil plant. The detailed emission rates in **Table 1** are given in **Appendix**. The amounts of CO_2 emitted from the cogeneration system driven by natural gas are much smaller than those from the conventional system, because much of CO_2 amount from the conventional system is emitted from the heavy oil plant.

3. The Heat Pump System in the Aquarium

Before Energy consumption tests were ran in the aquarium for two years. Overall performance characteristics that were of particular interest were the integrated COP along with other instantaneous comparisons of power,

Table 1. Energy consumption and CO_2 emissions of the conventional system and the cogeneration system.

Month	Conventional System			Cogeneration System	
	Electricity (kWh)	Heavy Oil (L)	Natural Gas (m³)	Electricity (kWh)	Natural Gas (m³)
April	1,291,580	93,358	7034	825,840	160,104
May	1,306,100	50,258	7034	906,660	104,189
June	1,458,060	97,087	7034	946,320	145,528
July	1,657,000	154,286	7034	994,260	249,380
August	1,661,820	166,932	7034	1,004,520	225,876
September	1,504,160	114,955	7034	975,600	177,445
October	1,367,580	59,345	7034	972,240	111,506
November	1,321,180	110,781	7034	963,060	138,195
December	1,423,180	171,110	7034	958,260	202,818
January	1,489,120	212,179	7034	998,400	204,811
February	1,344,760	186,068	7034	888,600	173,531
March	1,449,840	176,325	7034	948,660	167,121
Total	17,274,380	1,592,684	84,408	11,382,420	2,060,504
Total (MJ)	169,807,155	62,273,944	3,891,209	111,889,189	94,989,234
CO_2 (kg)	9,587,281	4,316,174	199,625	6,317,243	4,873,092
Total (MJ)	235,972,309			206,878,423	
CO_2 (kg)	14,103,079			11,190,335	
Energy Saving Ratio (%)				12.3	
CO_2 Reduction Ratio (%)				20.7	

refrigerant flow rate, and temperatures [20]. This heat pump system has two operational modes. The first mode is a cooling mode that uses ice with water slurry, which is the typical mode during the summer. In this mode, the ice is produced using the heat pump connected with the latent heat-storage system. The second mode is the winter mode, in which the circulating water is heated by the heat pump connected with the heat exchanger system to collect heat from the seawater and the ambient air. Energy-saving effects and carbon dioxide-reducing effects of the heat pump system are estimated from the test results.

The objective of this study is to compare the actual operating characteristics and efficiency of a seawater-source heat pump using an ice-storage system to the predicted evaluation of the two assumed conventional systems. That is, an air-source heat pump without ice storage and an oil-fired absorption refrigerating system. The desired outcome would be to show that the seawater-source heat pump significantly uses less electricity than the air-source heat pump and the oil-fired system. Additionally, the CO_2 emissions for the seawater-source heat pump favorably compare as they might be less than those for the other conventional assumed systems described.

3.1. System Description

Shimane Aquarium (AQUAS) is located in an area facing the Japanese Sea (in the Shimane Prefecture, Japan). The building has two stories and a cellar with a total floor area of 10,293 m², and the volume of the fish tank is 3000 m³. The primary cooling loads at the aquarium are air conditioning for building, cooling of the ventilation air for the fish tank, and cooling and heating of the water in the fish tank. The system selected is one that combines two seawater-source heat pumps, WSHP001 and 002 (cw: 650 kW, hw: 732 kW) and a heat recovery type of air-source heat pump, AWSHP003 (cw: 510 kW, hw: 697 kW). Seawater-source heat pumps transfer heat to and from the seawater by means of circulated water and a heat exchanger. An air-source heat pump uses the outdoor air for heat absorption and rejection. This pump is more common because of its lower initial cost and ease of installation, but seawater-source heat pump is more energy efficient. The heat pump tested in this work transfers heat to the outdoor air and to the sea. It maintains some of the initial cost advantages of the air-source heat pump and some of the performance advantages of the seawater-source heat pump.

The systems provide water cooling using off-peak power. The primary heat source is the heat collected from the seawater and stored in the ice-storage tank, IS (ts: 4500 kWh × 2). The heat produced by the heat pump at night is stored in the ice-storage tank. **Figure 6** shows a diagram of the heat pump system in a summer mode (charging the ice storage). In general, the increased efficiency of the seawater-source heat pump is gained by two mechanisms. First, water is a much better heat transfer fluid than air, so heat is moved much more efficiently. Second, the seawater allows the heat pump to extract heat from water that is usually warmer than the outside air during the winter and cooler than the outside air during the summer. This allows a more efficient heat pump operation. The seawater-source heat pump usually provides warmer supply air temperatures during the winter and cooler supply air temperature during summer, which increases comfort levels.

Ice-storage technology has been shown to be effective in reducing the operating cost of cooling equipment during the summer time. By operating the refrigeration equipment during off-peak hours to recharge the ice storage and by discharging the storage during on-peak hours, a significant fraction of the on-peak electrical demand and energy consumption is shifted to off-peak periods. Cost savings are realized because utility rates favor leveled energy consumption patterns. The variable rates reflect the high cost of providing energy during relatively short on-peak periods. Therefore, they constitute an incentive to reduce or avoid operation of the cooling plant during peak periods by using cold storage. The large difference between on- and off-peak energy and peak consumption rates should make cold-storage systems economically feasible.

3.2. Loads of the Building throughout the Year

Figure 7 shows the ambient air temperature, the ambient air humidity, and the seawater temperature throughout the year. The maximum temperature in August was 35.9°C, and the mean temperature was 28.3°C. Relative humidity ranged from about 70% at night to 95% during the day in August. **Figure 8** shows the daily loads of the cooling air conditioning for the space above the fish tank, cooling water, and heating water for the fish tank, and the air conditioning for the building during the period from March 27, 2000 to March 10, 2001. The air-conditioning load for the building during the summer exceeded the predicted loads, but loads in the winter were about 70% of the predicted loads. The loads to provide cooling water

and cooling air conditioning for the fish tank have been stable and constant since May, 2001. The loads to provide cooling water for the fish tank could be generated only during the summer time, but the loads of heating water for the fish tank could be produced only during the winter.

The primary cooling loads at the aquarium cool the water in the fish tank and cool ventilation air in the building. **Figure 9** shows the typical building cooling requirements on a typical summer day between August 14 and August 20, 2000. The loads to provide cool air conditioning were nearly constant every day. The loads to provide cool ventilation air in the building and cool water in the fish tank were greatest in the afternoon because of hotter outside air. The outside air temperatures increased and, combined with solar gains, lighting, and a large audience energy gains, the cooling loads increased during the day. The relationship between cooling air-conditioning and cooling water loads is an important consideration regarding the type of heat pump system to install in the aquarium because the heat pump produces cooling water for the aquarium from seawater-source heat and air-source heat.

Winter days tend to be warm in Shimane Prefecture. For example, **Figure 10** shows the use of aquarium heating and airconditioning on a typical winter day between January 15 and January 21, 2001. These data are from operator records for these periods. On these days, the plant provided maximum heating using air to warm each individual building's system. By reducing the amount of outside air used in the building, the heat needed is also reduced, while the amount of heating (heat recovery) is increased. When minimum outside air is used, the major heat is utilized (after initial building warm-up) for air-conditioning. On an average winter day, large heating (heat recovery) loads could be generated from heating water for the fish tank.

3.3. Energy Usage

Figure 11 compares the daily electrical consumption of the three heat pumps from March 27, 2000 until February 26, 2001. During the first two months after the kilowatt-hour meters were installed, the AWSHP003 used about 22% of the electricity that the building used. Average energy usage for those months was 3900 kWh per day for the AWSHP003. This included both heating and cooling modes of operation with the transition from heating to cooling occurring in April. The electrical requirements of the AWSHP003 relatively remained constant throughout the year. AWSHP003 was running throughout the day. The load factor of this heat pump was 50% during the period of cooling and air conditioning in the summer time, and the load factor was about 60% during the period of heating in the winter.

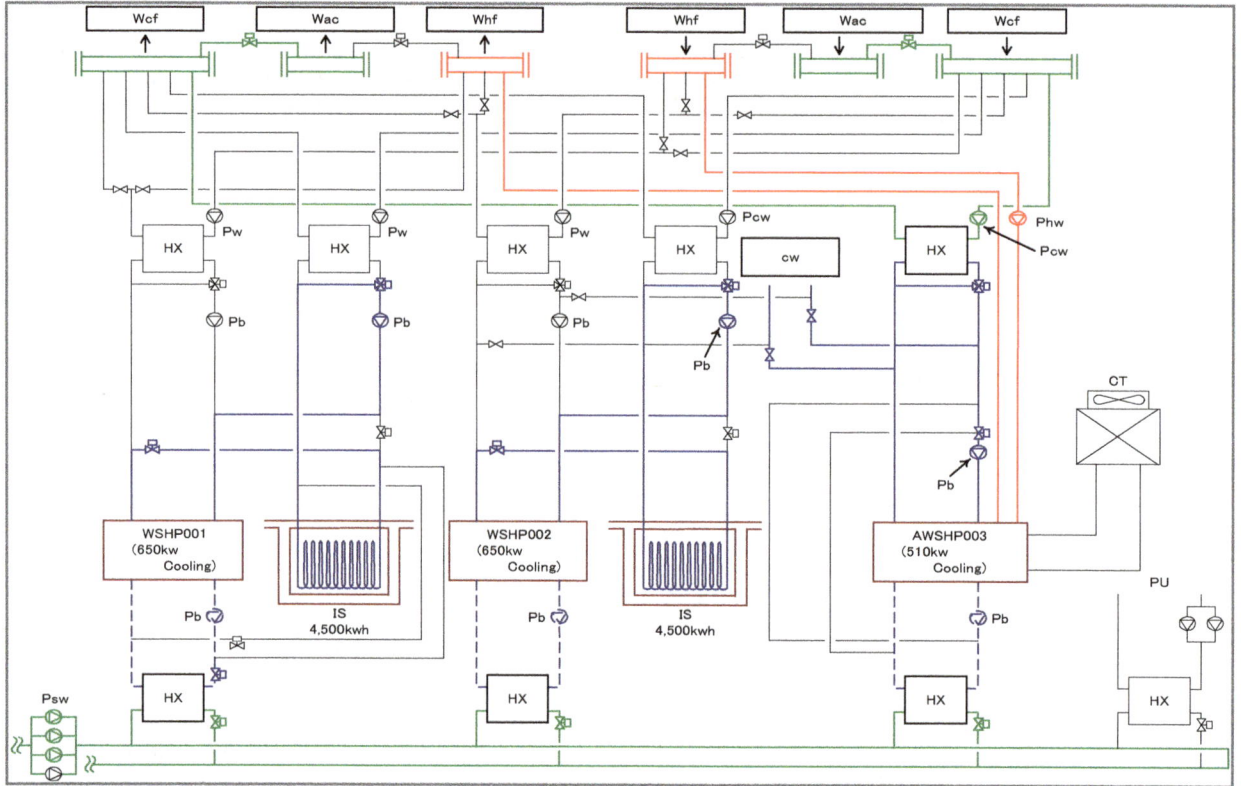

Figure 6. Diagram of the heat pump system in the aquarium (summer mode: charging the ice storage).

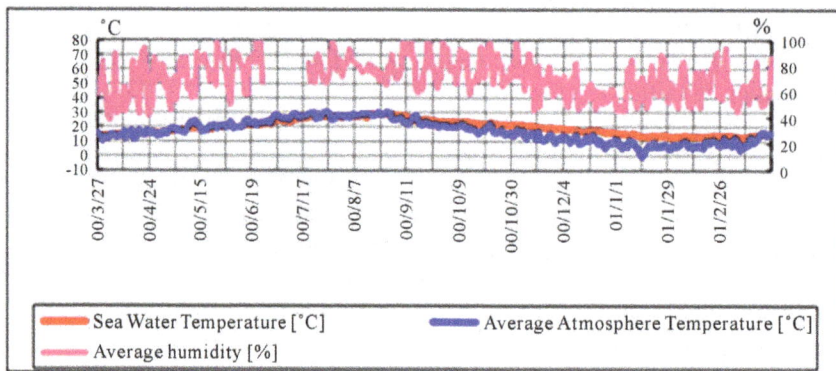

Figure 7. Seawater temperature, air temperature and humidity.

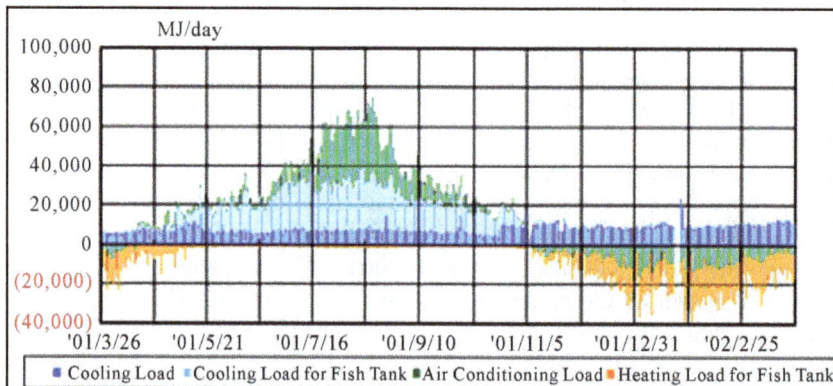

Figure 8. Daily loads of the building during the year.

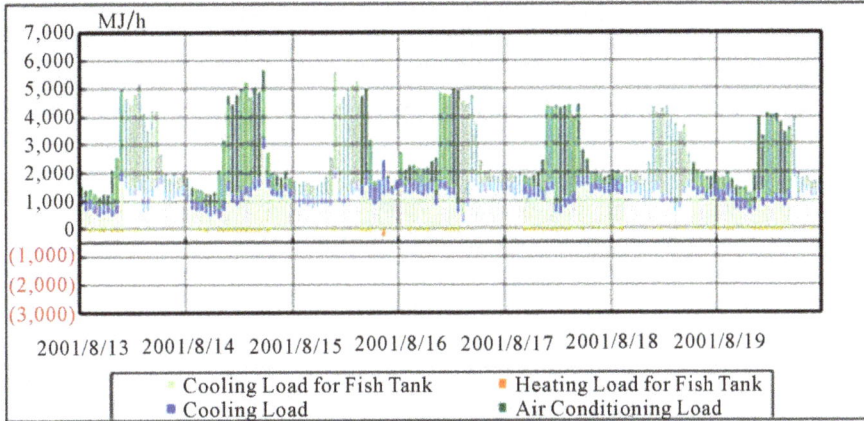

Figure 9. Hourly loads of the building during the summer.

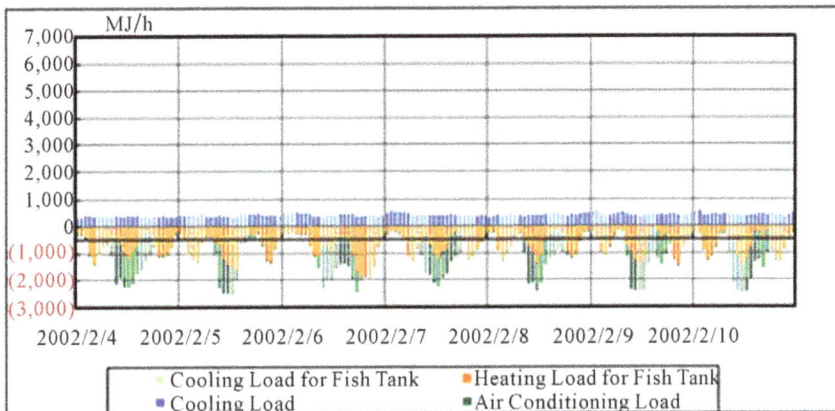

Figure 10. Hourly loads of the building during the winter.

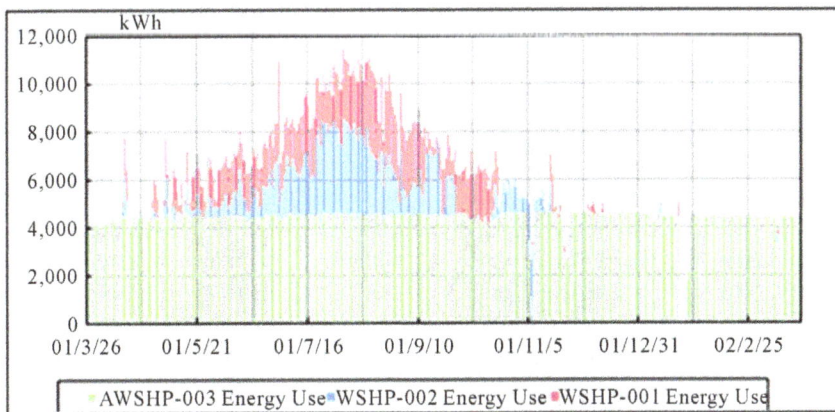

Figure 11. Daily electrical consumption of three heat pumps.

The WSHP001 and WSHP002 relatively used more energy during the summer than during the winter. Overall, during the period from April 15 to November 21, 2000, the WSHP001 and WSHP002 used about 80% of the energy that the AWSHP003 used. The electrical energy usage of both WSHP001 and WSHP002 peaked during the period from August 14 to August 20 because of the heavy air-conditioning load of the building. Both WSHP001 and WSHP002 supplied only the cooling energy for the air conditioning of the building and the cooling water for the fish tank during the summer time. With more data from the winter months, it was expected that the energy usage of the WSHP001 and 002 would approach about zero. WSHP001 and WSHP002 were not used for heating in the winter, except during the maintenance period of AWSHP003.

Figure 12 shows the typical energy consumption of the WSHP001 and WSHP002 for the period after the kilowatt-hour meters were installed in March 2000. Periodic readings began after August 14, and weekly readings were conducted until August 20, 2000. Both of the WSHP001 and WSHP002 fully ran after the initial building opening in the morning, stopped operations during the period from 1:00 p.m. to 4:00 p.m. in the afternoon, and ran again with the ice-storage tanks according to the loads from 4:00 p.m. Summer days tended to produce more energy from cooling than was needed, and the ice-storage tanks ran at night to produce ice and water slurry to charge the ice-storage tank. Cooling heat was picked up by the ice-storage system, resulting in warmer chilled water return. This warmer chilled water return was stored in the ice-storage tank. Viewed as a heat source for the building, this increase in ice and slurry water temperature was (potential) heat storage.

The heat pump operating at night removed energy from the seawater and produced cooling water, which was stored in the ice-storage tank. The ice and water slurry were used during the next day with the largest loads early in the afternoon. The ice-storage system provided a match in time between cooling availability and the time when heat was needed. By operating the refrigeration equipment during off-peak hours to recharge the ice storage and discharge the storage during on-peak hours, a significant fraction of the on-peak electrical demand and energy consumption were shifted to off-peak periods (see **Figure 13**). It also provided the means to recover and produce cooling at the lowest possible cost by using off-peak electrical energy.

3.4. Energy Efficiency

The summer mode was most efficient for two reasons. The owner received the full benefit of air conditioning the building and cooling water for the fish tank at the same time. In addition, the cold water from the ice-storage tanks maintained the condensing temperature, which in turn reduced the work performed by the heat pumps. When operating in the winter mode, the AWSHP003 obtained its heat from outdoor air and accrued air-conditioning benefits. However, the cold water benefit still occurred. We found from the monitoring data that the COP was relatively high, averaging 3.4 during the summer time. When operating in the summer mode with ice storage, the COP of WSHP001 was 2.6, and the COP of WSHP002 was 3.0 for the output for each unit of electrical energy used (see **Figure 14**). This was the best use of resources and was quite remarkable, considering that it operated in this mode 60% of the time on a year-round basis.

In winter, when the unit was mainly operating in the winter mode, the COP was quite low. The COP was even lower for AWSHP003 than WSHP001 and WSHP002, even though it was working throughout the year. By checking the daily results, we found that the load running was often performed for most of the running time to allow a very low COP. Another reason for the decreased COP of AWSHP003 was due to the heat-recovery running mode, even though the heating loads were almost zero.

The performance of the heat pumps with the ice-storage system was compared with those two other systems (assumed systems), which were installed in the aquarium where heating and cooling were supplied with combinations of the air-source heat pump system without ice storage and the oil-fired absorption refrigerating system. Energy consumption by the WSHP001 and WSHP002 was19% less than the consumption by the oil-fired absorption refrigerating system (see **Figure 15**).

The seawater-source heat pump system emitted 86 tons of CO_2 in a year. This favorably compared to the emissions of the other alternatives for heating and cooling in the two other systems (assumed results): the emissions of the air-source heat pump system without ice storage were 102 tons of CO_2, and the emissions of the oil-fired absorption refrigerating system were 176 tons of CO_2. The electric heat pump with ice storage emitted two times less CO_2 than did the oil-fired system for heating and cooling (see **Figure 16**). In **Figures 15** and **16**, the oil-fired absorption refrigeration system had a smaller percentage of the energy use in nighttime and daytime, because much of energy consumption depended on the energy use of heavy oil instead of electricity in the oil-fired absorption refrigeration system.

4. Conclusions

This paper described two case studies: a cogeneration system of a hospital and a heat pump system installed in an aquarium that used seawater as latent heat storage. First, this work discussed the technical viability and the pattern of operation suitable for a cogeneration system that supplied energy to the Shimane University Hospital. The analysis started with the energy consumption data available for the hospital; no technical obstacles were identified. The typical patterns for operational units of the cogeneration system were decided by the hourly energy demands during several seasons throughout the year. The average ratio between the electrical and thermal load in the hospital was suitable for the operation of a cogeneration system. A case study performed for a non-optimized cogeneration system predicted large potential for energy savings and CO_2 reduction. These will be precisely analyzed in future study.

The second case study examined a seawater-source heat pump system installed in a newly built aquarium in Shimane Prefecture, Japan, which provided simultaneous

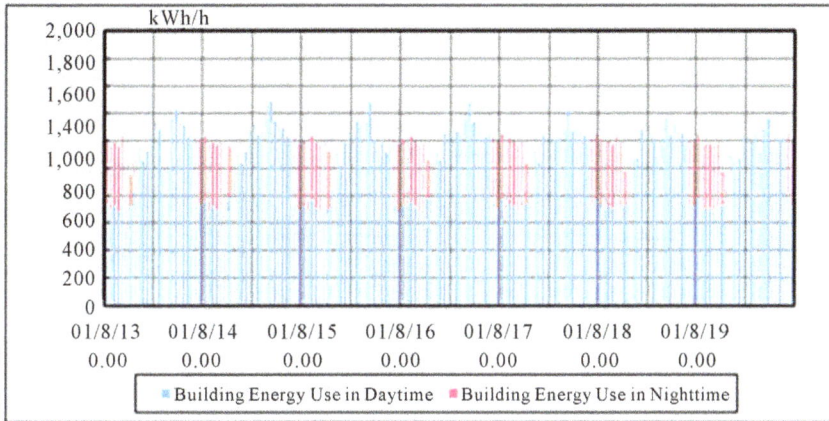

Figure 12. Hourly energy consumption of the building during the summer.

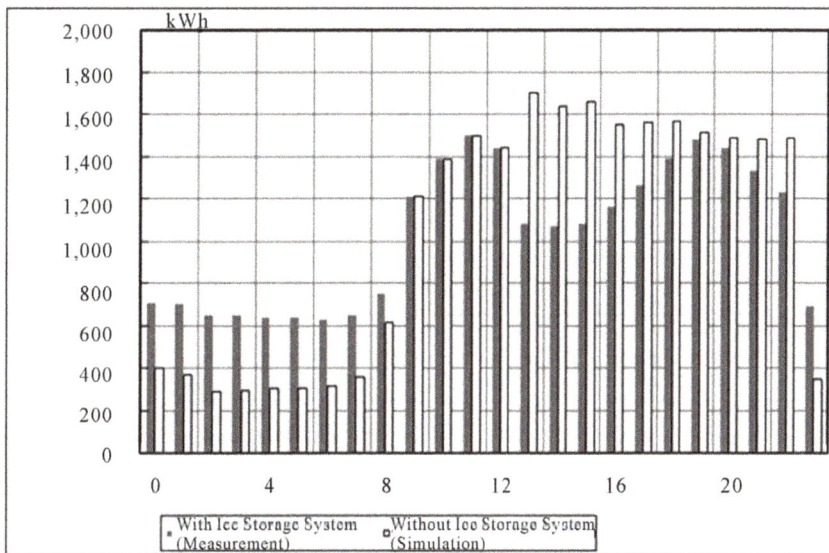

Figure 13. Shift of peak loads.

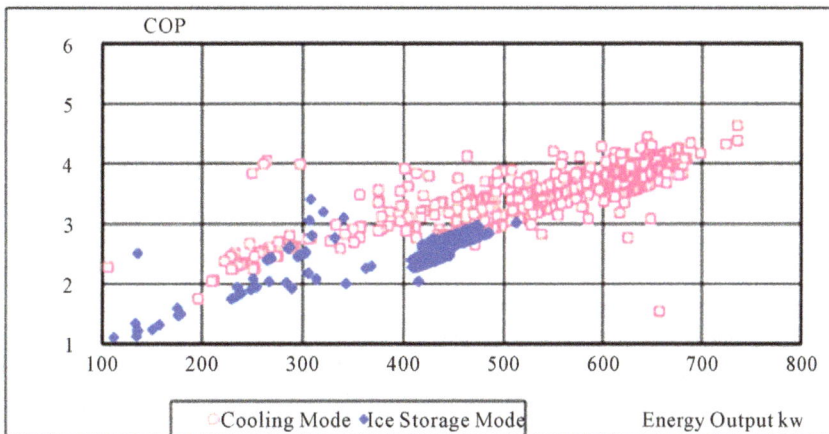

Figure 14. The COP of WSHP001.

heating and cooling. A COP of the WSHP001 system at running conditions was 3.4 for cooling and 2.8 for ice storage. By operating the refrigeration equipment during off-peak hours to recharge the ice storage and by discharging the storage during on-peak hours, a significant fraction of the on-peak electrical demand and energy

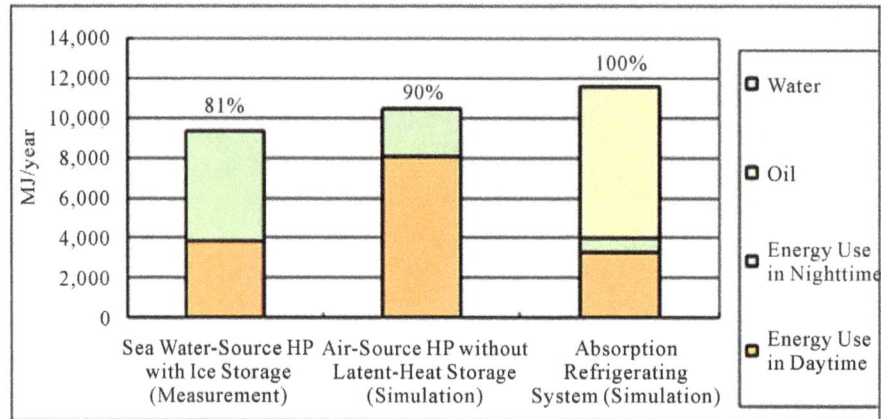

Figure 15. Reduction of energy consumption of WSHP001 and WSHP002.

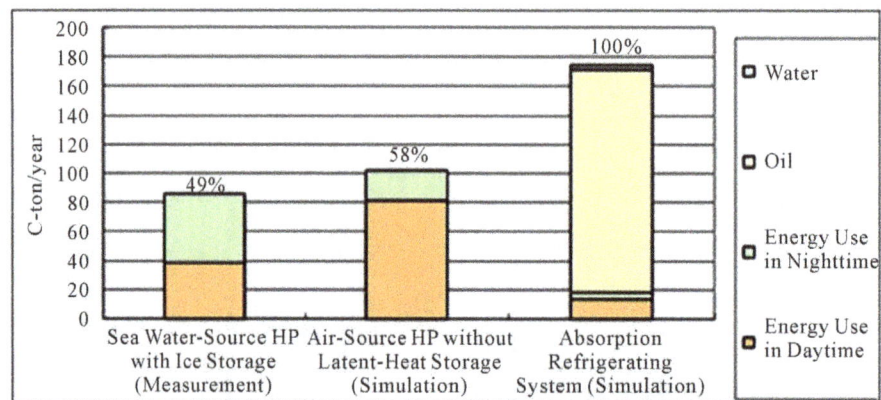

Figure 16. Reduction of CO$_2$ emissions of WSHP001 and WSHP002.

consumption were shifted to off-peak periods. Then, this case study was compared to two other assumed systems in which heating and cooling was supplied by a conventional air-source heat pump and a conventional oil-fired refrigerant. The energy consumption of the seawater-source heat pump for heating and cooling was 19% lower than the energy consumption of the oil-fired absorption refrigerating system. In addition, the CO$_2$ emissions for heating and cooling were favorably compared because the emissions of heat pump system were two times less than those for the oil-fired system.

5. Acknowledgements

The author gratefully acknowledges the contributions of the staff of the Shimane University Hospital. The concept of a specially designed seawater-source heat pump with ice-storage system was formerly introduced by The Chugoku Electric Power Co., Inc. I want to thank Mr. M. Koyama and Mr. T. Kuriyama of Nikken Sekkei Ltd. for their guidance in planning the present study. I also want to thank the staff of The Aquarium (AQUAS) in Shimane Prefecture for assisting with the project. The work was sponsored by The Chugoku Electric Power Co., Inc.

REFERENCES

[1]   G. Bizzarri and G. L. Morini, "Greenhouse Gas Reduction and Primary Energy Savings via Adoption of a Fuel Cell Hybrid Plant in a Hospital," *Applied Thermal Engineering*, Vol. 24, No. 2-3, 2004, pp. 383-400.

[2]   A. S. Szklo, J. B. Soares and M. T. Tolmasquim, "Energy Consumption Indicators and CHP Technical Potential in the Brazilian Hospital Sector," *Energy Conversion and Management*, Vol. 45, No. 13-14, 2004, pp. 2075-2091.

[3]   L. Barelli, G. Bidini and E. M. Pinchi, "Implementation of a Cogenerative District Heating: Optimization of a Simulation Model for the Thermal Power Demand," *Energy and Buildings*, Vol. 38, No. 3, 2006, pp. 1434-1442.

[4]   E. Bilgen, "Exergetic and Engineering Analyses of Gas Turbine Based Cogeneration Systems," *Energy*, Vol. 25, No. 12, 2000, pp. 1215-1229.

[5]   J. H. Santoyo and A. S. Cifuentes, "Trigeneration: An Alternative for Energy Savings," *Applied Energy*, Vol. 76, No. 1-3, 2003, pp. 219-227.

[6]   K. R. Voorspools and W. D. D'haeseleer, "Reinventing

Hot Water? Towards Optimal Sizing and Management of Cogeneration: A Case Study for Belgium," *Applied Thermal Engineering*, Vol. 26, No. 16, 2006, pp. 1972-1981.

[7] L. Gustavsson, A. Dodoo, N. L. Truong and I. Danielski, "Primary Energy Implications of End-Use Energy Efficiency Measures in District Heated Buildings," *Energy and Buildings*, Vol. 43, No. 1, 2011, pp. 38-48.

[8] S. Okamoto, "Saving Energy in a Hospital Utilizing CCHP Technology," *International Journal of Energy and Environmental Engineering*, Vol. 2, No. 2, 2011, pp. 45-55.

[9] P. E. Phetteplace and W. Sullivan, "Performance of a Hybrid Ground-Coupled Heat Pump System," *ASHRAE Transactions*: *Symposia*, Vol. 104, No. 1, 1998, pp. 763-770.

[10] S. P. Kavanaugh and X. Lan, "Energy Use of Ventilation Air Conditioning Options for Ground-Source Heat pump System," *ASHRAE Transactions*: *Symposia*, MN-00-5-2, 2000, pp. 543-550.

[11] J. R. Brodrick and D. Westphalen, "Uncovering Auxiliary Energy Use," *ASHRAE Journal*, Vol. 43, No. 2, 2001, pp. 58-61.

[12] J. Ni and H. Liu, "Experimental Research on Refrigeration Characteristics of a Metal Hydride Heat Pump in Auto Air-Conditioning," *International Journal of Hydrogen Energy*, Vol. 32, 2007, pp. 2567-2572.

[13] A. Satheesh, P. Muthukumar and A. Dewan, "Computational Study of Metal Hydride Cooling System," *International Journal of Hydrogen Energy*, Vol. 34, No. 7, 2009, pp. 3164-3172.

[14] S. Okamoto, "A Heat Pump System with a Latent Heat Storage Utilizing Seawater Installed in an Aquarium," *Energy & Buildings*, Vol. 38, No. 2, 2006, pp. 121-128.

[15] T. Savola and I. Keppo, "Off-Design Simulation and Mathematical Modeling of Small-Scale CHP Plants at Part Loads," *Applied Thermal Engineering*, Vol. 25, No. 8-9, 2005, pp. 1219-1232.

[16] D. Henning, S. Amiri and K. Holmgren, "Modelling and Optimisation of Electricity, Steam and District Heating Production for a Local Swedish Utility," *European Journal of Operational Research*, Vol. 175, 2006, pp. 1224-1247.

[17] Å. Marbeand S. Harvey, "Opportunities for Integration of Biofuel Gasifiers in Natural-Gascombined Heat-and-Power Plants in District-Heating Systems," *Applied Energy*, Vol. 83, No. 7, 2006, pp. 723-748.

[18] K. Difs, M. Danestig and L. Trygg, "Increased Use of District Heating in Industrial Processes: Impacts on Heat Load Duration," *Applied Energy*, Vol. 86, No. 11, 2009, pp. 2327-2334.

[19] S. Okamoto, "Energy Saving by ESCO (Energy Service Company) Project in Japanese Hospital," *Proceedings of IMECE2009*, Lake Buena Vista, 13-19 November 2009.

[20] A. Satheesh and P. Muthukumar, "Performance Investigations of a Single-Stage Metal Hydride Heat Pump," *International Journal of Hydrogen Energy*, Vol. 35, 2010, pp. 6950-6958.

Nomenclature

AWSHP003: seawater- and air-source heat pump 003
C: generated electricity, heating, and cooling energy by a cogeneration system
COP: coefficient of performance
CT: cooling tower
h, hr: hour
HX: heat exchanger
IS: ice-storage tank
P: pump
PU: power unit
RT: ton of refrigeration
t: ton
T: total energy consumption of conventional system
WSHP001: seawater-source heat pump 001

WSHP002: seawater-source heat pump 002
X: total amount of CO_2 from a conventional system
Y: amount of CO_2 from a cogeneration system

Subscripts

ac: air conditioning
b: brine
cf: cooling for fish tank
cw: cooling water
hf: heating for fish tank
hw: heating water
sw: seawater
ts: thermal energy storage
w: cooling and heating water

Appendix

The results in **Table 1** are as follows. The electricity provided by the public utility is the primary energy, and it is included in the analysis to accommodate its large consumption. It will be analyzed separately in future study.

Dimensions:
1 kWh = 9.83 MJ
Heating value of heavy oil = 39.1 MJ/L
Heating value of natural gas = 46.1 MJ/m^3
Electricity: 0.555 CO_2-kg/kWh
Heavy oil: 2.71 CO_2-kg/L
Natural gas: 2.365 CO_2-kg/m^3
Efficiency of cogeneration system generator = 0.41

$$T = \text{Electricity} + \text{Heavy Oil} + \text{Natural Gas}$$
$$= 17,274,380\,(\text{kWh}) + 1,592,684\,(\text{L}) + 84,408\,(\text{m}^3)$$
$$= 169,807,155\,(\text{MJ}) + 62,273,944\,(\text{MJ}) + 3,891,209\,(\text{MJ})$$
$$= 235,972,309\,(\text{MJ})$$
$$C = \text{Electricity} + \text{Natural Gas}$$
$$= 11,382,420\,(\text{kWh}) + 2,060,504\,(\text{m}^3)$$
$$= 111,889,189\,(\text{MJ}) + 94,989,234\,(\text{MJ})$$
$$= 206,878,423\,(\text{MJ})$$
$$\therefore \frac{T-C}{T} \times 100\,(\%) = \frac{235,972,309 - 206,878,423}{235,972,309} \times 100\,(\%)$$
$$= 12.3\,(\%)$$

18

To Improve Energy Efficiency via Car Driving Deduction by Land Use Planning

Mohammad Malekizadeh[1], M. F. M. Zain[1], Amiruddin Ismail[2], Ahmad Hami[3]
[1]Department of Architecture, Universiti of Kebangsaan Malaysia (UKM), Bangi, Malaysia
[2]Sustainable Urban Transport Research Centre (SUTRA), Faculty of Engineering and Built Environment,
Universiti Kebangsaan Malaysia (UKM), Bangi, Malaysia
[3]Department of Landscape Architecture, Universiti of Tabriz, Tabriz, Iran

ABSTRACT

Urban area consumes about main percentage of used energy. Cities need basic review in land management, structure and form to minimize the use of energy which creating environmental pollution. Urban planners and designers are looking for a solution and essential agreement in urban planning and designing principles that can decrease the pollution from rapid urbanization. Travelling is essential for daily needs of most people in urban area. Issues arise when one considers the amount of necessary fossil fuels used in the majority of daily commuting for accessibility to services. It is necessary to design a city to minimize the use of energy which creating environmental pollution. Research conducted in Subang Jaya in Malaysia in 2012 finds a variable which influences on use of car, propose of use of car and commuting distance by car. However it tried to find effect of train station and density on use of car for accessibility to this services and facilities. Findings illustrate neighbourhood distance from train station influences distance to facilities and services in neighbourhoods. However it illustrates derived distance by car was affected by residential lots distance from restaurant, work place, school, park, house area per person, and car ownership.

Keywords: Town & City Planning; Railway Systems; Urban Design; Transport Planning; Pollution; Energy Efficiency

1. Introduction

New form of urbanism took place on the demand for building. Consequently planning project started to accommodate the workers and their family near industrial plants after the industrial revolution. Low-income class community increases and upper-class community moved to suburbs, resulting in an urban sprawl of horizontal magnitude to about every community throughout the word [1]. In developing country this process was different; the developing countries mostly have high economic growths which increases growth of employment centres. Population move from other area for working to these centres. Capital of country and a few cities attracted these employment centres and population. The cities in developing countries have been facing with rapid urbanization and land fragmentation which are always faster than urban planning and designing. In this manner residential area takes place without considering facilities in accessible distance.

1.1. Back Ground Literature

Khattak and Rodriguez 2005 found that auto trip, travel distance, travel time, regional trip, external trip distance and trip duration of trip in neo-traditional neighbourhood residence is less than the conventional neighbourhood significantly [2]. Inner city zone that has integrated streets, parks and green spaces structure has more accessibility to facilities and services [3]. Amount of trip of inner city's resident is higher with shorter duration, and residents of this area spend least time in travel; in this area people walk and bike more and use less car travel. Residents of inter commuter belt spend the most time in travel with longest travel time [4]. Key elements of neighbourhood walkability are proxy and connectivity. First one come from mixed-land uses and the second one is related to street pattern. Handy and Cao *et al.* 2005 in their study indicated putting resident's proximity to destinations with alternatives ways to auto usage results to less driving; this means decrease in driving can be reach by increase in accessibility. Increasing in accessibility in existing areas can be possible by policies that include revitalization traditional neighbourhood design by main street program design, and filling undeveloped and redeveloped shopping centre [5]. Maleki, Zain, & Ismail, 2012 illustrated Street density, house density, house diversity, and non-residentia land use, positively and significantly influence accessibility to facilities and services. Distance to shops, distance to high school, distance to health centre, distance to train station, land diversity, block length and average lot size negatively and significantly influence accessibility to facilities and services [6]. Iacono and Krizek *et al.* 2009 say money, and other cost play as impedance role in travel, this impedance in recent none motorize travel study is distance. He used of distance and time both for calculating accessibility to recreation, restaurant, shopping, and work by walking and cycling [7].

1.2. Back Ground of Case Study

Malaysia has experienced continuous economic growth since 1985; this trend resulted in rapid urbanization especially in Kuala Lumpur and Klang Valley [8]. This rapid urbanization increases the need for more housing and urban development in a larger area. However, it increased economic growth as well as a desire for greater accessibility and mobility for all. Increases in travel needs that are not managed efficiently and effectively result in excess congestion in private motorize travelling mode. Haphazard growths make residential area develop without proper planning for acceptable distance to employment centres and facilities. In poor accessibility condition, residents have to commute long distances to get their daily needs to their workplace. Whenever facilities are located further than the walkable distance, residents use necessarily motorized travel. They prefer to use cars if public transport is not available. The public transportation master plan of Kuala Lumpur/Klang Valley is illustrated as such: The net result of increased car usage has been a rise in congestion across the region. However door to door travel times for private vehicles remain competitive against the use of public transport. Travel times are typically much higher by public transport resulting in poorer accessibility to jobs and facilities [9].

Malaysia as a developing country in sensitive natural environment needs to conduct rapid urbanisation growth in the way that minimise impact of urban development on local environmental resources. Kuala Lumpur, the capital of the country and its township like Subang Jaya are the centres of this growth and thus more attention needs to be given for environmental protection. The government desires to improve the urban environment and quality of life, and to ease the pressure on the infrastruc-

ture in general by balancing the development in Kuala Lumpur and the Klang Valley [10].

1.3. New Approaches in Land Use Management

Sustainability has multidimensional affects on the environment and many aspects of human life. An integrated systematic view of events can leads us to a better understanding interaction between urban land use and environmental pollution. Most cities grow faster than land use planning and land use managing. Rational use of recourses is to consider fairness between current generation and future generation for use of these recourses [11]. "Efficiency" versus "solidarity" is considered as key uncertainies in the Sustainability outlook. Efficiency is doing more with use of fewer materials and recourses. In land use management for population reduction this means residents participation in activities with use of less fossil foul and pollution reduction. The trend towards more "efficiency" means that decision making is increasingly based on economic rationality and market forces. This strategy mainly directed to facilitate market processes with the limited government intervention. The trend towards more "solidarity" involves decision making which is determined by values on social equity and solidarity, cultural identity and sustainability. Government coordination is important in solidarity and not restrained.

Supplying services by private investor sector in local centre shows they made it well, because capital tends to increase in the best manner. At the same time they increase welfare and level of accessibility of surrounding residents. They don't have claim on land value rising resulting from greater accessibility but they share the benefit that give rise to capitalization [12].

The compact city model is supported for a number of reasons which relate to sustainable urban development and include: conservation of the countryside, less need to travel by car, thus reduced fuel emissions supports for public transport and walking and cycling, more efficient utility and infrastructure provision Burton 2003 and revitalisation and regeneration of inner urban areas [13].

"New urbanism" is a broadly defined movement that seeks to end the cancerous sprawl and replace this type of expansion of growth with redesigned cultural residents that encourage genuine commitment to civic life. It tends to create new ways of guaranteeing the design of pedestrian, public and semi-private spaces, as well as vehicular movement. The importance of humanly scaled design in the metropolitan landscape and the city are the stirring call for New Urbanism. The New Urbanism pays attention to local condition and adjusted concepts principles, which will transcend local differences into a unique urban pattern. In the view of this trend, the physical environment is an integrated product of culture, religion, climate, socio-economic values and technology. These concepts had incorporated the following objectives since 1992:

- To minimize energy consumption through proper orientation and arrangement of the plan layout and building block forms.
- To optimize layout condition with respect to security, safety and comfort of the walker rather than the motorized travel. Both must find appropriate definitions in the layout of the scheme.
- To articulate physical solutions for religious, social and recreational necessities and unite them within an integrated neighbourhood concept.
- To decrease the initial and running cost of infrastructure by minimizing the roads and service lines of water, sewage, electricity and telephone [14].

2. Material and Method

2.1. Theoretical Frame Work

The key consideration in cost and energy efficiency is location and intensity of land usage areas [15]. Urban forms provide possibility for energy consumption in transport sector. One of the best ways to reduce vehicle travel is to build places where people can do more with less driving. Today's trend is looking for residential site which has pedestrian-friendly, mixed-use, high-density communities, with short and fast accessibility to public transportation instead of driving. Customer is looking for smaller housing units and mixed-land used communities projected in compact living residential sites.

Use of public transportation is the solution for energy consumption and pollution emission in urban areas. Compact dense green neighbourhood with energy efficiency and environmental friendly material completes this process. It is recommended that urbanism and policy makers should start to make land use which has acceptably lower impact on environment. They must focus on human scale in urban land use development. This can support the promoting of transportation that uses energy resources other than fossil fuel. Integration of connectivity between land use and transportation is necessary to improve non-motorized mode of transportation as an important factor in flexibility and adaptability of urban forms.

Households have wide range of income and dimensions. Sustainable community is the community answer to this multiplication by providing mixed housing types for a wide range of incomes and household structures in greater density with more opportunity for human contact. The mentioned characteristics with solution for human reinforcement on controlling use of automobile shaped about main part of new urbanism and neo traditional residential design [16].

2.2. Method

This research was conducted on 30 neighbourhoods in Subang Jaya, Kuala Lumpur in Malaysia. Neighbourhoods were selected random from the neighbourhoods which are near train station and other are far from train station. Method of this research is survey via questioner to residents of restrict. A sample of population was 750 persons return rate was 60%. Survey was conducted since May until July 2011. Researchers meet residents and asked them to answer to the questions. After data was prepared, first, Authors explain statistical descriptive on distance and travel mode; and in the next stage data was analysed to find relation between variable and influence of independent variable on commuted distance by car as indicator of fossil foul consumptions.

3. Results and Discussion

Sustainable accessibility in urban built environment is the interaction between land use and transportation modes in such a way that a person or group can participate in activities within a minimal distance and time while consuming less non renewable energy and prefer using renewable energy [17].

3.1. Travelling by Mode of Public Sustainable Transport

Travel by train is 34% of total travelling in area. The proposes for travelling by train are work 16%, school 4%, college/university 12%, city centre 34%, shopping 18%, entertainment 8%, visit family and friend 4%, and others 4% **Figure 1**.

3.2. Travelling by Modes and Purposes

Residents were asked to determined which mode they use for their purposes. Modes of travelling are car, walking, motorcycle, train, bus, and bicycle. Purposes categorized into 10 category including: post, bank, restaurant, health, mosque, park, high school, industry, primary school, shopping, work, and train station. Results illustrated that people who are going to train station use car 33%, walking 14%, bus 11% and motorcycle 4% as

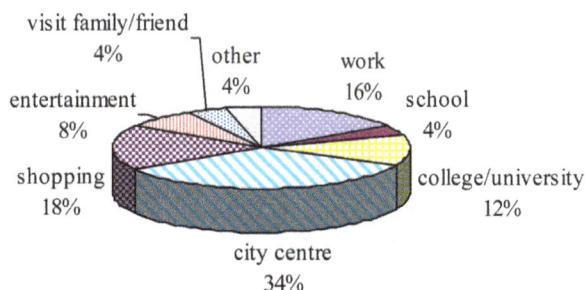

Figure 1. Proposes of travelling by train in Subang Jaya.

mode of travelling. They use of car 42%, walking 9%, train 7%, bus 5%, motorcycle 5%, and bicycle 2% for travelling to work place. This percents for travelling to shops are car 38%, walking 16%, bus 5% motorcycle 5% train 3% and bicycle 2%. Residents mostly walk for access to park and green space 34% travelling by car with 18% is second travelling mode for access to park, motorcycle 7%, bus 5% and bicycle 2%. **Figure 2** shows percents of trip by mode of travelling for each purpose.

3.3. Purposes of Travelling by Car and Car Ownership

Residents were asked to answer how many car they have, based on their answer, result of this survey shows 69.7% of residents have at least one car; 58.4% of people have 1 car, 37.7% of people have 2 cars, and 3.7% of people have 3 cars. Travel to work and shopping are main propose of use of cars in area. **Figure 3** illustrates proposes of travelling by car for each purpose in Subang Jaya. Population size and car ownership support railway significantly. Railway station which has neighbourhood with higher average car ownership household has higher positive patronage level compare with those neighbourhoods with lower average car ownership. Car ownership

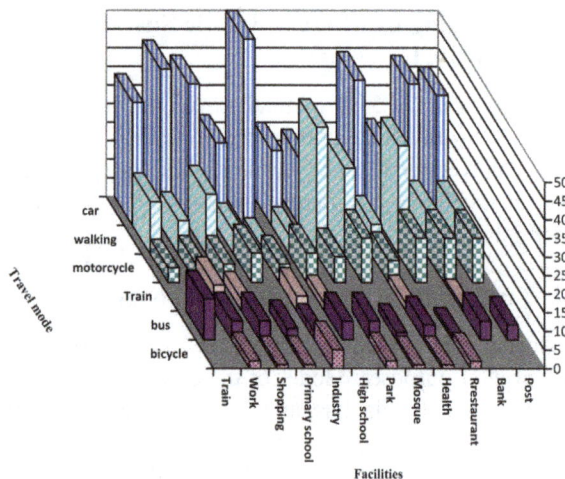

Figure 2. Travel mode for access to local facilities.

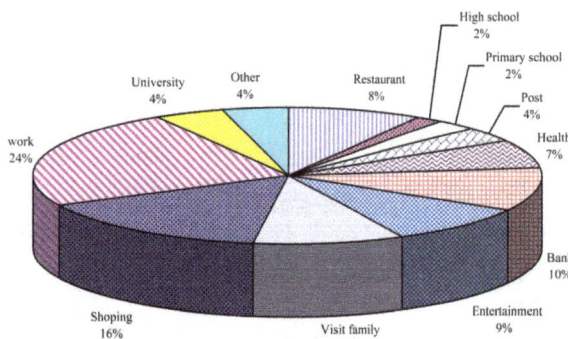

Figure 3. Traveling by care for accessibility to facilities.

associates with park and ride and pick-ups and drop-off to transit station for longer trip by transit system [18].

3.4. Impact of Train Station on Distance to Facilities

Travelling by train as sustainable public transportation influences fossil foul consumption. Accessibility to train station makes people use more this travel mode directly, but decrease distances to facilities near station decrease use of motorized travel indirectly [19]. Researchers looked to find interaction between nearness to train station and variables of distances to facilities and employment. Correlation coefficient was used in this analysis to interpret whether a relationship existed between distance to train station and distance to facilities and employment. Correlation was also used to test whether the relationship between distance to train station and distance to facilities are positive or negative.

The Pearson correlation coefficient, commonly abbreviated as "r" measures the degree to which a linear relationship exists between two variables. A perfect, positive linear relationship between two variables has a value of 1.00, while a perfect, negative linear relationship has a value of −1.00. If there is no relationship at all it is recorded as 0.00. Correlation coefficients test illustrated distance to train station and distance to facilities and employments have correlation significantly **Table 1**.

3.5. Multiple Regressions

Residents of communities with high-density, mixed-use, pedestrian-friendly, ideally with ready access to public transportation, drive a third fewer miles than those who are living in convection outskirts of a city [20]. Generally residents select the closest facility to their location. This selection may come from the fact that most services like post office, bank and so on are equal or based on hierarchy service provider which covers special area. But people may travel for other facilities which have differences in quality or are symbolic like cinemas, special shops and recreational facilities from their place to other places [21].

In the final analysis researchers used linear regressions model to test the contribution of independent variables that explain the variation in a dependent phenomenon, explain the nature (positive or negative) and slope of the relationship, and provide a means to control for intervening factors. In the last stage result of the previous stage will be used to develop a model of efficient residential site and neighbourhood via measuring influence of distance to restaurant, distance to school, distance to park, house area per person, cars ownership, and pedestrian quality as independent variable on derived distance by car ownership weekly **Table 2**.

Table 1. Correlation between distance from residents to train station and distance to facilities.

	Work	Shop	Primary school	Industry	High school	Park	Mosque	Health	Restaurant	Bank	Post
Distance to train station	0.608**	0.766**	0.554**	0.761**	0.404*	0.110	−0.032	0.660**	0.388**	0.757**	0.742**

** Significant in level P value 0.01 *significant in level P value 0.05.

Table 2. The result of linear regression of distance to facilities on driving by car.

Model	Unstandardized Coefficients		Standardized Coefficients	t	Sig.	Collinearity Statistics	
	B	Std. Error	Beta			Tolerance	VIF
	−120.472	62.369		−1.932	0.066		
	0.065	0.015	0.542	4.346	0.000	0.708	1.412
(Constant)	−0.002	0.001	−0.350	−2.776	0.011	0.694	1.440
Distance restaurant	0.003	0.001	0.338	2.723	0.012	0.712	1.404
Distance work Distance school House area/person Distance park	2.673	0.804	0.376	3.323	0.003	0.859	1.164
Cars ownership Pedestrian quality	−0.007	0.003	−0.294	−2.467	0.022	0.773	1.294
	104.659	16.418	0.701	6.375	0.000	0.910	1.099
	−2.594	11.832	−0.026	−0.219	0.829	0.813	1.231

a. Dependent Variable: Derived by cars by km.

4. Conclusions

There is a general agreement that the local shops are important to the neighbourhood and its stability. The walkable nature of the neighbourhood also helps the commercial trip to increase as most residents enjoy walking up to facilities and daily necessary services of neighbourhood [22].

Findings of this research illustrate that neighbourhood distance from train station influences distance to facilities and services including shops, primary school, high school, health centre, restaurant, bank, post office, industry and work. These findings show if distance to train station decrease distance to these facilities and services decrease and vice versa. However, weekly derived distance by car was affected by distance to restaurant, work place, school, park, house area/person, and car ownership.

REFERENCES

[1] A. Kirby, "Current Research on Cities and its contribution to Urban Studies," *Cities*, Vol. 29, Suppl. 1, 2012, pp. S3-S8.

[2] A. J. Khattak and D. Rodriguez, "Travel Behavior in Neo-Traditional Neighborhood Developments: A Case Study in USA," *Transportation Research Part A: Policy and Practice*, Vol. 39, No. 6, 2005, pp. 481-500.

[3] A. M. L. Stahle and A. Karlström, "Place Syntax Geographic Accessibility with Axial Lines in GIS," *5th Space Syntax Symposium Proceedings*, Delft, 1 June 2005, pp. 131-144.

[4] H. Millward and J. Spinney, "Time Use, Travel Behavior, and the Rural-Urban Continuum: Results from the Halifax STAR Project," *Journal of Transport Geography*, Vol. 19, No. 1, 2011, pp. 51-58.

[5] S. Handy, X. Cao and P. Mokhtarian, "Correlation or Causality between the Built Environment and Travel Behavior? Evidence from Northern California," *Transportation Research Part D: Transport and Environment*, Vol. 10, No. 6, 2005, pp. 427-444.

[6] M. Z. Maleki, M. F. M. Zain and A. Ismail, "Variables Communalities and Dependence to Factors of Street System, Density, and Mixed Land Use in Sustainable Site Design," *Sustainable Cities and Society*, Vol. 3, 2012, pp. 46-53.

[7] M. Iacono, K. J. Krizek and A. El-Geneidy, "Measuring Non-Motorized Accessibility: Issues, Alternatives, and Execution," *Journal of Transport Geography*, Vol. 18, No. 1, 2010, pp. 133-140.

[8] G. Knaap and E. Talen, "New Urbanism and Smart Growth: A Few Words from the Academy," *International Regional Science Review*, Vol. 28, No. 2, 2005, pp. 107-118.

[9] Suruhanjaya Pengangkutan Awam Darat, "Greater Kuala Lumpur/Klang Valley Public Transport Master Plan," 2011. http://spad.gov.my/images/stories/esgklkvlptmp.pdf

[10] I. J. D. Jebasingam, "Creating the Essence of Cities: The Planning & Development of Malaysia's New Federal Administrative Capital, Putrajaya," 2006.

[11] M. Neuman, "Notes on the Uses and Scope of City Planning Theory," *Planning Theory*, Vol. 4, No. 2, 2005, pp. 123-145.

[12] C. Webster, "The Donald Robertson Memorial Prizewinner 2003 the Nature of the Neighbourhood," *Urban Studies*, Vol. 40, No. 13, 2003, pp. 2591-2612.

[13] P. Howley, M. Scott and D. Redmond, "An Examination of Residential Preferences for Less Sustainable Housing Exploring Future Mobility among Dublin Central City Residents," 2009. www.elsevier.com/locate/cities,

[14] M. A. E. Saleh, 2"Learning from Tradition: The Planning of Residential Neighborhoods in a Changing World," *Habitat International*, Vol. 28, No. 4, 2004, pp. 625-639.

[15] F. S. Chapin and E. J. Kaiser, "Urban Land Use Planning," University of Illinois, St. Champaign, 1979.

[16] Y. R. Jabareen, "Sustainable Urban Forms: Their Typologies, Models, and Concepts," *Journal of Planning Education and Research*, Vol. 26, No. 1, 2006, pp. 38-52.

[17] C. Curtis and J. Scheurer, "Planning for Sustainable Accessibility: Developing Tools to Aid Discussion and Decision-Making," *Progress in Planning*, Vol. 74, No. 2, 2010, pp. 53-106.

[18] B. P. Y. Loo, C. Chen and E. T. H. Chan, "Rail-baSed Transit-Oriented Development: Lessons from New York City and Hong Kong," *Landscape and Urban Planning*, Vol. 97, No. 3, 2010, pp. 202-212.

[19] M. Z. Maleki and M. F. M. Zain, "To Promote Future Sustainability with Integrated Design of Urban and Transportation System," *6th Malaysian Universities Transport Research Forum Conference*, School of Housing, Building & Planning University Sains, Pulau Pinang, 2011.

[20] R. Binsacca, "Builder," 2008. http://www.ecohomemagazine.com/news/compact-cure.aspx?printerfriendly=true 1-3.

[21] P. Naess, "Accessibility, Activity Participation and Location of Activities: Exploring the Links between Residential Location and Travel Behaviour," *Urban Studies*, Vol. 43, No. 3, 2006, pp. 627-652.

[22] J. Distasio, "Reacting to Actions: Exploring How Residents Evaluate Their Neighbourhoods," University of Manitoba (Canada), Winnipeg, 2005.

19

Simulation Using Sensitivity Analysis of a Product Production Rate Optimization Model of a Plastic Industry

Mala Abba-Aji[1], Vincent Ogwagwu[2], Bukar Umar Musa[3]
[1]Department of Mechanical Engineering, University of Maiduguri, Maiduguri, Nigeria
[2]Federal University of Technology, Minna, Nigeria
[3]Department of Electrical and Electronics Engineering, University of Maiduguri, Maiduguri, Nigeria

ABSTRACT

This study analyzes the sensitivity analysis using shadow price of plastic products. This is based on a research carried out to study optimization problem of BOPLAS, a plastic industry in Maiduguri, North eastern Nigeria. Simplex method of Linear programming is employed to formulate the equations which were solved by using costenbol software. Sensitivity analysis using shadow price reveals that the price of wash hand bowls is critical to the net benefit (profit) of the company.

Keywords: Sensitivity Analysis; Simplex Method; Linear Programming; Optimization

1. Introduction

Shadow prices: the simplex-method provides more useful information than just the optimal solution to a linear programming problem. From the optimal tableau, the value of each resource in terms of its contribution to profits and overheads is determined. For the sensitivity analysis, the net benefit (or cost) of adjusting the amount of resource can also be determined. The relative value of a resource with respect to the objective function in a linear programming problem is called its shadow price. It is the amount of change in the objective function per unit change in its right-hand side value.

2. Methodology

2.1. Injection Molding Process

Injection molding is one of the most important plastics molding processes. It is carried out usually on horizontal hydraulic press.

Granular thermoplastic materials are gravity fed from a hopper into a pressure chamber ahead of a plunger.

The moving plunger causes the granular plastic to be compressed and then forced through a heating cylinder to palletize it. A torpedo shaped object in the centre of the heating cylinder, assists uniform heating.

The palletized plastic is then injected through an injection nozzle at great pressure into the die cavity to form the required component. The die is water-cooled; making the injected plastic to freeze almost immediately the die cavity is filled.

The plunger returns, and the mould open to eject the formed material. The mold closes and the cycle is repeated.

In modern machines, as used in the company, the feed plunger is replaced with a motor driven screw plasticizer. It serves the function of both part-heating the plastic granules by internal sheer and feeding it to the mould. (Resistance heater bands are still used on the heating cylinder). The screw–plasticizer helps to ensure that the thermoplastic fed through the injector nozzle is maintained at a constant and uniform temperature and viscosity.

The process requires the use of expensive dies, usually called molds; thus its use has to be justified by large production runs. The process is easily automated, and cycle times of just a few seconds are common, making injection molding the most widely used process for producing plastic items. Also a wide range of shapes and plastic materials can be molded [1].

2.2. Simplex-Method Algorithm

High customer demand of kettle, water jug, wash-hand bowl, Big Bowl, medium bowel and small bowel was observed within the period of August to February of every year; but the company is uncertain of allocating the optimal proportion of the products.

Let A = kettle B = water jug C = wash hand bowel D = big bowel E = medium bowel F = small bowel G = parker H = Hanger

Let X_1, X_2, X_3, X_4, X_5, be the proportions of products to be produced. These are decision variables of the model, and h, H, Φ, d, e, be duration of injection, charge and cooling of the various products respectively as shown in **Table 1**. These durations; injection, cooling, and charge time were recorded from the injection molding machine

Capacity; is the maximum time assigned to the injection molding machine through the function setting of a mini computer attached to the machine. Only an experienced machine operator could do this.

Contribution to profit;

Let: a, b, c, d, e, f and g be contribution to profits of the products A, B, C, D, E, F and G respectively.

The contribution to profit and overhead per unit of each product is determined. The company was uncertain about how many of each product to produce in order to maximize their profit. The simplex-method provides information more than just the optimal solution to linear programming problem. The optimal tableau determines the value of each resource in terms of its contribution to profits and overhead. We can also determine the net benefit (or cost) of adjusting the amount of resources.

The simplex equations can be written as;

Maximize $- ax_1 + bx_2 + cx_3 + dx_4 + ex_5$

Subject to

Injection $h_I x_1 + H_I x_2 + \Phi_I x_3 + d_I x_4 + e_I x_5 \leq C_I$

Charge $h_c x_1 + H_c x_2 + \Phi_c x_3 + d_c x_4 + e_c x_5 \leq C_{II}$

Cooling $h_g x_1 + H_g x_2 + \Phi_g x_3 + d_g x_4 + e_g x_5 \leq C_{III}$ [3]

$$x_1, x_2, x_3, x_4, x_5 \geq 0$$

Using Gauss Jordan Complete elimination method, series of tableau will be obtained, procedures of elimination being repeated until there are no further negative

Table 1. Resource and maximum capacities of products [2].

(a)

Resource type	A	B	C	D	Capacity
Injection time	h_I	H_I	Φ_I	d_I	C_I
Charge Time	h_c	H_c	Φ_c	d_c	C_{II}
Cooling Time	h_g	H_g	Φ_g	d_g	C_{III}

(b)

Resource type	E	Capacity
Injection time	e_I	C_I
Charge Time	e_c	C_{II}
Cooling Time	e_g	C_{III}

values in the last row i.e. the objective function row.

	₦
Sales;	230003059.97
Less Variable cost;	
Materials;	7146900.85
Machine operator's wages;	532,000
Diesel;	1,250,000
Metered power supply;	65,550
Overtime;	84,000 9078450.85
Total contribution	13924609.12
Less fixed costs;	
Accountant salary;	46,800
Courier service;	2650
Communication facilities;	16,500
Transportation;	84,000
Lubricants;	115,000 686,150
Profit	13238459.12 [2]

The contribution at any given level of sales can be found by using the formula;

Contribution = sales × p/v ratio

where p = profit v = volume [3].

The proportion to be produced so as to maximize the contribution to profit of each product and the cost implication of adjusting constraints could be achieved by solving the linear programming model. From the analysis above, the equation can be written as;

Maximize $40.98x_1 + 25.62x_2 + 2.65x_3 + 2.61x_4 + 4.23x_5$

Subject to

Injection $9.5x_1 + 7.5x_2 + 8x_3 + 6x_4 + 8x_5 \leq 15$

Charge $11.3x_1 + 9x_2 + 10x_3 + 6x_4 + 8x_5 \leq 14$

Cooling $15x_1 + 5x_2 + 6.5x_3 + 6x_4 + 8x_5 \leq 15$

where $x_1, x_2, x_3, x_4, x_5 \geq 0$ (non-negativity constraint) [4,5]

3. Results and Discussions

A computer program, academic version software was used to solve the generated equations and after ten iterations obtained the following results;

Value of the objective function = 47.150, yield

$x_1 = 8280$, $x_2 = 0.5159$, $x_3 = 0.0$, $x_4 = 0.0$, $x_5 = 0.0$ [4,6,7].

Assuming an incremental value of ₦ 5 to each of the five products of interest for five different values, keeping other products constants, employing sensitivity analysis, 25 different simulations were carried out, which gave the following results in **Table 2**.

When a shadow unit prices are used, with an increments of ₦ 5.00 on the initial unit prices, significant changes in the maximized profits of water jugs and wash hand bowls were noticed.

The maximum contribution to profit will be obtained when a shadow price of ₦ 25 increments on the initial unit price of wash hand bowls is used, yielding the value of the objective function, p = 78.00.

Table 2. Computer programmed results for shadow prices.

S/No.	PRODUCTS	Initial value plus ₦5	Initial value plus ₦10	Initial value plus ₦15	Initial value plus ₦20	Initial value plus ₦25
1	Water jugs	51.29	55.43	59.57	63.71	67.85
2	Wash hand bowls	49.73	55.40	63.18	70.00	78.00
3	Big bowls	47.15	47.15	47.15	47.15	47.15
4	Parker	47.15	47.15	47.15	52.76	64.42
5	Hanger	47.15	47.15	47.150	47.150	51.153

Table 3: Results showing the benefit of adjusting the constraint for wash hand bowls.

Products Proportion	S/P per month B/model (₦)	S/P per month A/model (₦)	C/p	Simplex results (proportions) based on C/p	Maximzed profits (P)	Q/month B/model	Optimum Q/month A/model	Net profits (₦)
	1) 785695.57		45.98	0.828	51.290			
	2) 833343.57		50.98	0.828	55.431			
X_1	3) 877988.57		55.98	0.828	59.570	8929		A/model = 221507778.734
	4) 922633.57		60.98	0.828	63.711			
	5) 967278.57		65.98	0.828	67.851			
	1) 951537.6		30.62	0.516	49.730			
	2) 1045137.6		35.62	1.556	55.409			
X_2	3) 1138737.6	2145257.28	40.62	1.556	63.187	18720	30,287.4	B/model = 220688459.1
	4) 1232937.6		45.62	1.556	70.000			Profit margin = 819319.634
	5) 1352937.6		50.62	1.556	78.000			

The cycle time for wash hand bowls is 35.5 seconds. The company is using 8-hours per day, quantity produced in a month = 60 × 60 × 8/35.5 × 1.5556 × 24 = 30287.4 Units

Selling price per month = Quantity/month X Unit price
Selling price per month = 30287.4 × 70.83 = ₦ 2,145257.284

Substituting the selling price back into the profit statement for the month of February, 2005: when the unit volume for wash hand bowls V = 18,720, S/month = ₦ 1325937.6 yielding total sales of ₦ 230,453,059

	₦
Sales;	230453059.9
Less Variable cost;	9078450.85
Total contribution	221374,609.1
Less fixed costs;	686,150
Profit	2220688459.1

Using the optimum quantity or volume of wash hand bowls V = 30287.4 Units and selling price of ₦ 2,145257.284,

	₦
Sales:	231272379.584
Less variable cost:	9078450.85
Total contribution:	222193928.734
Less fixed costs:	686,150
Profit:	221507778.734

From the results obtained after simulating the equations of the × linear programming using simplex method, the value of the objective function, which was the profit foregone was 47.150445528799 and the optimal proportions of the products to be produced using injection molding machine, based on their contribution to profits are:

X_1, proportion of water jugs to be produced = 0.8280

X_2, proportion of wash hand bowls to be produced = 0.5159

X_3, proportion of big bowls to be produced = 0.0

X_4, proportion of packer to be produced = 0.0

X_5, proportion of hanger to be produced = 0.0

Summary of the results present the net profit for water jug and wash hand bowl are presented in **Table 3**.

4. Conclusions

Since the maximum time the company used was 8 hours per day, and the cycle time for water jugs and wash hand bowls are 36 units and 35.5 units, then the optimum number of the two products to be produced per day will now be, 966.15 units for water jugs and 1572.457 units for wash hand bowls. The profit margin obtained was ₦ 25062868.41 per month [2,8].

This is a clear justification why the company needs to emphasize the production of water jugs and wash hand bowls as regard to injection molding machine. Furthermore, sensitivity analysis, using shadow price, revealed that the price of wash hand bowls is critical to the net benefit (profit) of the company. When the proposed unit selling price ₦ 60.5 is used for wash hand bowls, optimum quantity of 30287 units will be produced yielding a maximum net benefit of ₦ 819, 319.634 per month. The company needs reconsider the price of wash hand bowls as regard to injection molding machine, and also concentrates on the other products being considered in this analysis in order to improve their selling prices, taking into cognizance, the quality of the products, customer requirements and customer affordability.

In this paper, sophisticated cost model that requires the use of design parameters to provide design alternatives can be carried out.

Apart from the time constraint considered in this work, temperature is another constraint that affects the production of plastics. Further research can then be carried out when temperature constraints from blow film molding and extrusion units of the industry were obtained.

The procurement of raw materials is a major challenge facing plastic industries in Nigeria. The government should encourage petrochemical industries producing plastic raw materials, like the one in Port Harcourt, to be in full production. That will reduce cost of importing raw materials from abroad. Also the foreign raw materials have a very low melting point compared to the one produced in Nigeria. This is not pleasant for molding process.

Finally, it is strongly recommend that Nigerian industries should adopt the modern operation research techniques so that they can obtain optimum results and make proper decisions.

REFERENCES

[1] A. Ibhadode, "Introduction to Manufacturing Technology," AMBIK Publishers, Lagos, 2001, pp. 92-95.

[2] A. Aji, *et al.*, "Development of a Product Production Rate Optimization Model: A Case Study of BOPLAS Plastic Industry," *Continental Journal of Engineering Sciences*, Vol. 4, 2009, pp. 26-31.

[3] R. Hussey, "Cost and Management Accounting," P, (app. $C3$-C_{22}), 1989.

[4] J. Krawjewski, "Operations Management; Strategy and Analysis," Addison-Wesley Publishing Company, Canada, 1987, pp. 118-125.

[5] R. Costenoble, "Simplex Method of Solving General Linear Programming Problems". www.Zweigmedia.com/third Edsite/index.htm

[6] K. Crow, "Archiving Target Cost/Design to Cost Objectives,"2001. www.ndp-solutions.com/dtc.htm

[7] K. Crow, "Design for Manufacturability". www.ndp-solutions.com/dfm.htm

[8] C. Frederic, "Cost and Optimization Engineering," McGraw Hill Publishing Company, New York, 1985.

20

Towards Energy Conservation in Qatar

Mohamed Darwish

Qatar Environment and Energy Research Institute, Doha, Qatar

ABSTRACT

Qatar energy consumptions are among the highest in the world, and can easily serve double the present population. Energy conservation is a must, as the energy resources are finite, and their consumptions are increasing at alarming rates. The country depends on desalted seawater, which consumes extensive amounts of energy, and is produced by using the least energy efficient desalting system. The desalination process is vulnerable to many factors, and strategic water storage needs to be built. The high energy consumptions are ruining the air and marine environments. Several suggestions are introduced to conserve energy in the Cogeneration Power Desalting Plants (CPDP), by moving to replace the Multi Stage Flash (MSF) desalting system by the energy efficient Seawater Reverse Osmosis System (SWRO); fully utilizing the installed power capacity to desalt water in winter, when electric power load is low, and during summer non-peak hours for strategic water storage; and modifying the simple Gas Turbines (GT) Power cycle plants to GT combined cycle to raise the Electric Power (EP) generation efficiency (to about 50%).

Keywords: Energy; Fuel; Electric Power; Conservation; Cogeneration Power Desalting Plants; Gas Turbine Combined Cycle; Multi Stage Flash Desalination; Reverse Osmosi

1. Introduction

Energy conservation (*i.e.* using less energy for a service) is essential for the entire chain of energy; from fuel extraction, fuel energy conversion to Electric Power (EP) including EP generation, transmission and distribution to usage in industry, transportation, buildings, and for consumers; and consumed EP and fuel. In the hot, humid, and sunny Gulf Cooperation Countries (GCC), renewable energy sources such as solar can play a key role by utilizing plentiful solar energy to generate EP and to desalt seawater. The EP is necessary to meet cooling demands and other industrial and human activities. Desalting seawater and treating waste water for reuse are processes that consume extensive amounts of energy, and needed in this water-scarceregion. While renewable energy is important for sustainable future, Fossil Fuels (FFs) are continued to be the mainly used energy source for the foreseeable future. It's therefore important to ensure that FFs are used as efficiently as possible while applying renewable energy application should be increasing.

Qatar is a small country but is ranking worldwide among the highest per capita (/ca) in: income, production and proven reserves of oil and Natural Gas (NG), energy(fuel and electricity) and water consumptions, CO_2

emission, water scarcity, and shortage in food production. In 2010, Qatar's economy grew by 19.40%: the fastest in the world. The rapid growth is due to ongoing increases in production and exports of Liquefied Natural Gas (LNG), oil, petrochemicals and related industries.

Qatar, the largest exporter of LNG, is a major force in global energy markets. Qatar's Electric Power (EP) has grown considerably in recent years with amazing doubled installed EP capacity in just two years, 2009-2011, and doubled EP and Desalted seawater (DW) generations in 6 years (2004-2010). These raise concerns about the unusual vulnerabilities (or insecurity) in water, food, and environment.

The economy is steadily growing: oil and gas account for roughly 85% of export earnings and 70% of government revenues; these made the highest income/ca, [1]. Qatar's successful 2022 world cup bid accelerates large-scale infrastructure projects such as Qatar's metro system, Qatar-Bahrain causeway, and solar Air Conditioning (AC) of nine stadiums. Qatar GDP (purchasing power parity) increased to $181.7 billion (estimated in 2011), compared to $153 billion and 131.2 billion estimated in 2010 and 2009 respectively [2].

The fast-growing economy, rapid urbanization, and

increase of population result in raising demands on scarce water resources and energy, especially EP. These affect, negatively, air and marine environment, and make CO_2 emission/ca the highest in the world. The energy forecast is expected to increase in Qatar about 6.6 times from 2000 to 2020, compared to 2.4 in Bahrain, 2.15 in Kuwait, 3.55 in Oman, 2.98 in Saudi Arabia, 2.83 in UAE, and 3.11 for all GCC, [3].

Current FF consumption patterns are not sustainable. Managing domestic demands for FF and EP remain key challenges. An energy-wasteful culture has grown up by subsidizing fuel and electricity prices. Qatar should adopt more energy-efficient practices in building design and transport infrastructure. The need for green buildings should be recognized, and environmental issues should be addressed through architecture to reduce the dependence of heavily air-conditioned buildings absorbing vast amounts of EP.

To conserve energy, consuming sectors are to be recognized, and measures to make them more efficient should be outlined. The main consuming sectors are the Cogeneration Power Desalting Plants (CPDP) producing both EP and DW; and building AC. The CPDPs account for more than 50% the fuel energy consumptions in most Gulf Co-operation Council (GCC). The AC load account for 70% of the EP peak summer load, and about 50% of annual EP consumption. Energy conservations in these two sectors, (CPDP and building AC), are discussed in this paper. The state of these two sectors (CPDP and building AC) is outlined first, and followed by suggested measure to lower their energy consumptions.

2. Qatar Prime Energy Production and Consumption

In 2011, total oil production was 1.638 Million barrel per day (Mbbl/d); including crude oil of 1.296, and the consumption was 0.16 Mbbl/day. The oil export was 1.478 Mbbl/day, [4].Oil production increased about 2.3 folds from 2001 to 2011; see **Figure 1**, [5].

Latest data on Natural Gas (NG) production, consumption, and export are given in the next table in terms of Billion Cubic Meter (BCM), and Billion Cubic Feet (BCF), [5].

year	Production, BCM (BCF)	Consumption BCM (BCF)	year	Export BCM (BCF)
2010	96.33 (3402)	21.8 (770)	2011	107 (3779)
2011	116.96 (4121)	19.54(690)	2012	114.23 (4105)

The NG production increased about 5.4 folds from 27 BCM/y in 2001 to 146.8 BCM/y in 2011. Exported NG was 14.04 BCM/y in 2000 and became 113.7 BMC/y in 2011, almost 8 fold increase. In 2010 and 2011, Qatar's

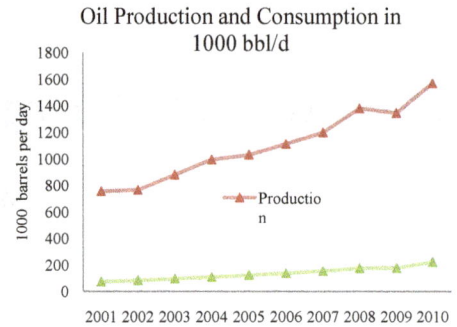

Figure 1. Oil production and consumptions in Qatar from 2001-2010 in 1000 bbl/d, [5].

NG exports by pipelines was 19.2 BCM/y, but as LNG export jumped from 76.1 BCM in 2010 to 102.6 BCM in 2011, or 35% in only one year, see **Figure 2**. The consumed primary energy (oil and NG) in Million ton oil equivalent (Mtoe) is given in **Figure 3**, [5].

The NG production increases are accompanied by continuously rising consumptions of total energy, oil and NG, **Figure 2**, [5]. NG consumption increased from 11 BCM in 2001 to 23.8 BCM in 2011, more than 2.16 folds in 10 years, and 16.4% annual increase from 2010 to 2011. Oil consumptions increased from 73,000 bbl/d in 2001 to 238,000 in 2011, 3.26 folds in 10 years; and 8.3% annual increase from 2010 to 2011. The total primary energy consumption in Million tons of oil equivalent (Mtoe) increases from 12.3 in 2001 to 29.4 in 2011; 2.38 folds in 10 years, and 14.1% increase from 2010 to 2011, **Figure 3**. In 2011, the 29.4 Mtoe includes 8 Mtoe for oil and 21.4 Mtoe for NG, *i.e.* 72.8% for NG and 27.2% for oil.

Table 1 summarizes the last 10 years history of fuel energy production and consumption in Qatar, [5].

The NG is a valuable resource that should be used efficiently. Its prices (in \$/Million Btu as given in **Figure 4**), is continuously increasing worldwide, except in the US from 2009 to 2011. This special case in US is due to the higher NG production rate compared to consumption rate. The production increased from 511.1 BCM in 2005 to 651.3 in 2011, 27.4% in 6 years; or annual 4.2% increase; while the consumption increases from 623.4 to 690.1 BCM in the same period, or 10.6% in 6 years, or annual 1% increase. From 2010 to 2011, the production increases 7.7%, while the consumption increases only 2.4%.

3. Cogeneration Power Desalting Plants (CPDP) in Qatar

3.1. Electric Power

EP and Desalted seawater (DW) are necessities for living in Qatar. DW represents 99% of municipal water supply. AC is expensive summer necessity to deal with the hot and humid (average high 42°C, although 50°C was reported), **Figure 5**. AC equipment consumes 50% of EP productions, and 70% of peak summer load, [6].

Natural Gas Production & Consumption

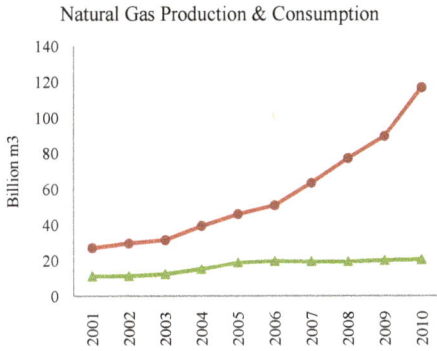

Figure 2. Natural gas production and consumptions in Qatar from 2001-2010 in Billion Cubic Meters (BCM), [5].

Prime Energy Consumption

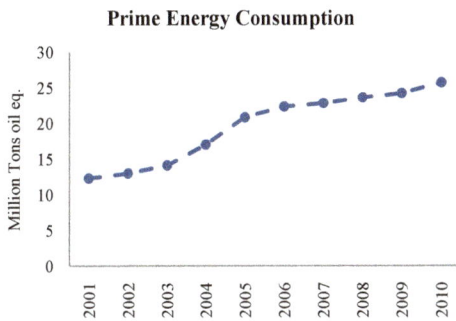

Figure 3. Primary energy consumption in Qatar from 2001-2010 in M tons oil equivalent (Mtoe), [5].

The EP power consumption in Qatar in kWh is given in **Figure 6**.

The installed capacity of the EP power plants increased from 1333 MW in 1995 to almost 9000 MW in 2011, as shown in **Figure 7** and next table, and as reported by KAHRAMA, [8]:

Year	1995	2000	2005	2009	2010	2012
Capacity in MW	1333	2248	2829	4893	7830	9000

Presently the EP has excess generating capacity (since September 2009) and supplying the surplus to Bahrain and Kuwait through GCC's Electric Grid. Qatar's power generating capacity has reached 8800 MW, up from 5300 MW in the first half of 2009. This is going to expand to 11,500 MW in late 2012 and expected to be 13,500 MW in 2013.

The peak EP load usually occurs in summer due to AC high EP consumption, and increased as given in the next table.

Year	1995	2000	2004	2005	2006	2007	2008	2009
MW	1244	1990	2520	2735	3230	3550	3990	4535

The required PP capacities are determined by the expected peak load; almost doubled in 6 years.

The EP generated was more than doubled from 2004 to 2010, **Figure 8**, and is given in the next table:

Year	1970	1980	1990	1995	2000	2005	2006	2007	2008	2009	2010
GWh	277	2416	4818	5415	9735	13,238	17,071	19,462	21,616	24,158	28,144

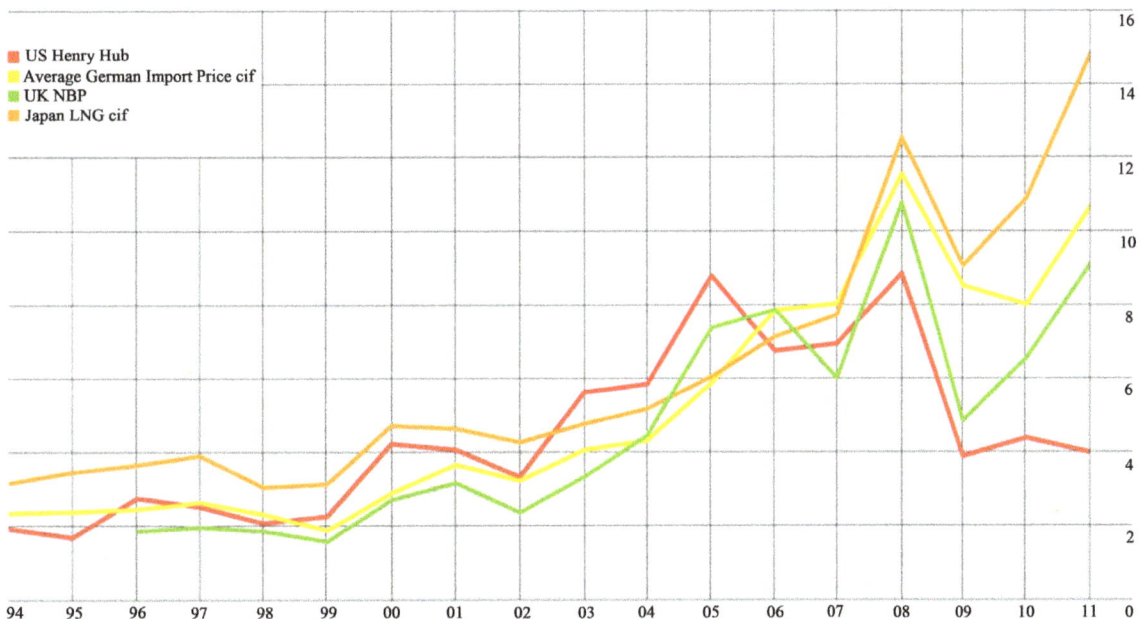

Figure 4. World prices of natural gas in US dollars per Million British Thermal Units (MBTU), [5].

Average High/Low Temperature for Doha, Qatar

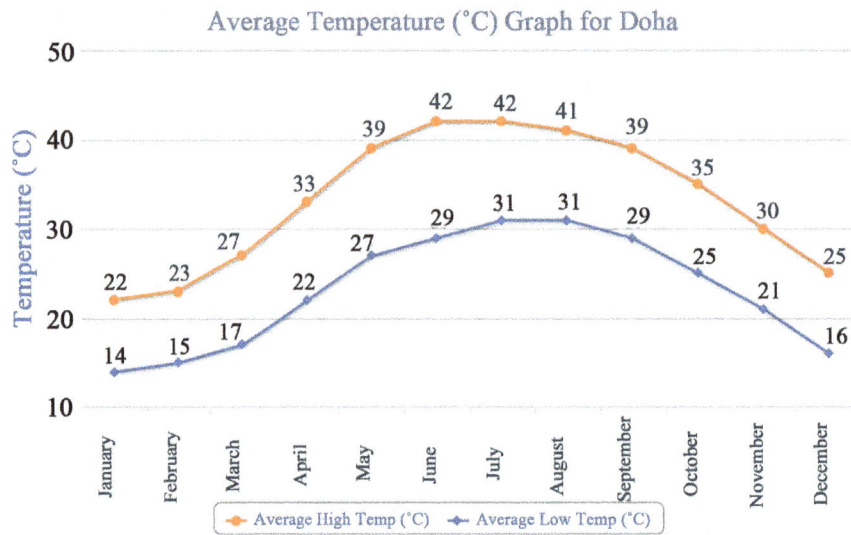

Figure 5. Average maximum and lowest temperatures in Doha, Qatar, [7].

Table 1. Fuel energy production and consumption in Qatar, [5].

Year	2001	2002	2003	2004	2005	2006	2007	2008	2009	2010
Oil Production (bbl/d)	754	764	879	992	1028	1110	1197	1378	1345	1569
Oil Consumption (bbl/d)	73	84	95	107	122	136	153	174	176	220
Natural Gas Production (Billion·m^3)	27	29.5	31.4	39.2	45.8	50.7	63.2	77	89.3	116.7
Natural Gas Consumption (Billion·m^3)	11	11.1	12.2	15	18.7	19.6	19.3	19.3	20	20.4
Million tons oil eq.	12.3	13	14.1	17	20.8	22.3	22.8	23.6	24.2	25.7

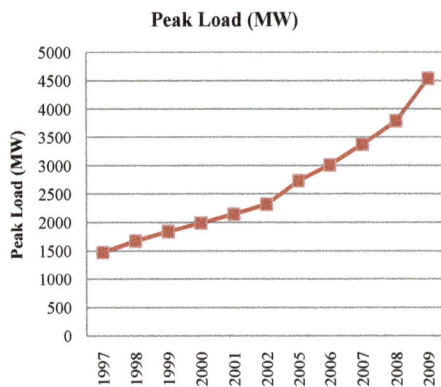

Figure 6. Peak EP load increased from 1997 to 2009, and 5090 MW in 2010, [9].

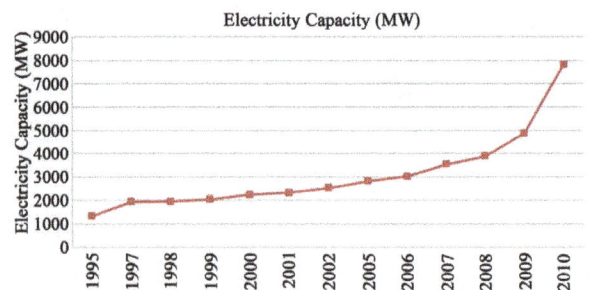

Figure 7. Qatar's EP installed capacity in the last 10 years, [9].

Figure 9 gives the increase in the EP consumption since 1980 to 2010.

In 2009, the EP consumption was shared as: 25% billed residential buildings, 19% unbilled or non-revenue electricity (Qatari residential and unaccounted losses), 12% government and street lighting, 15% commercial sector (including light industries), 29% industrial sector (including large industries which do not generate its own EP need such as the iron industries), and 9.9% other sectors, see **Figure 10**, [10].

The power sector and water desalination plants depend on natural gas. Their combined gas consumption is estimated by 550 MCF/day (15.58 MCM/d). According to QP, this would reach more than 850 MCF/d (27.08 MCM/d) in 2015. Normally, electricity demand has been

Annual Electricity Generation(GWh)

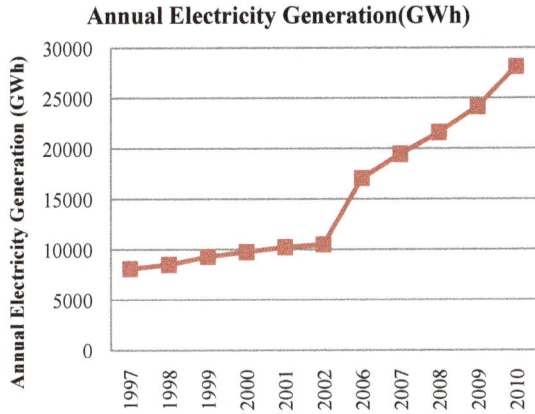

Figure 8. The EP generated in the last 30 years, [data from 9].

Electricity net consumption (Billion KWhrs)

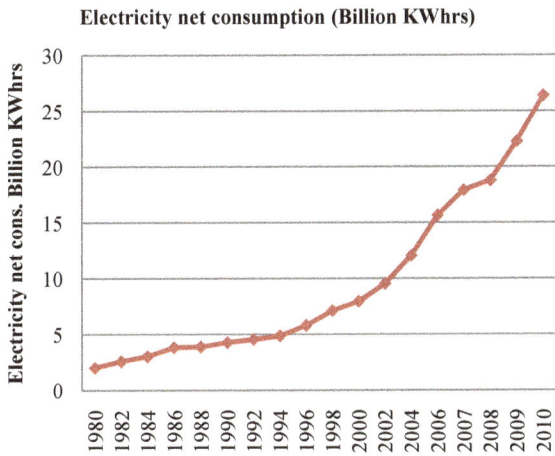

Figure 9. The EP consumption in the last 30 years, [data from Ref. 9].

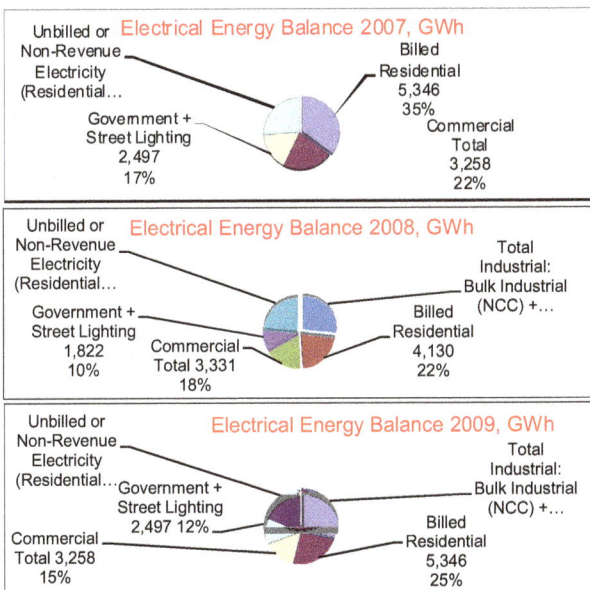

Figure 10. Electrical energy consumptions in different sectors, [10].

growing by about 7% per year , but since 2006 this has risen by 17% a year. The NG daily consumption in the power sector of 15.58 MCM/d (5.7 BCM/y) in 2010 represent 28% of the total consumed NG (20.4 BCM/y) in that year, [11].

Qatari citizens, who account for 26% of consumers and 42% of consumption, continue to receive water and power free of charge, one main reason why demand for water and power has been rising rapidly, [11].

The 4893 MW capacity in 2009 consists of 3230 MW simple Gas Turbine (GT), **Figure 11**, and 1650 MW GT Combined Cycle (CC) Power Plants (PP), **Figure 12**.

In 2010, the EP was produced in 6 main power plants (PP) of 7830 MW total capacity operated by NG. Some of these PP's belong to:

1. The state owned utility, Qatar General Electricity and Water Corporation (KAHRAMA), including Ras Abu Fontas (RAF) A, B, B1, and, B2, Sailiya, and Doha Super South. Others belong to private companies such as.

2. Ras Laffan (RL) Power Company (RL A, B, and C),

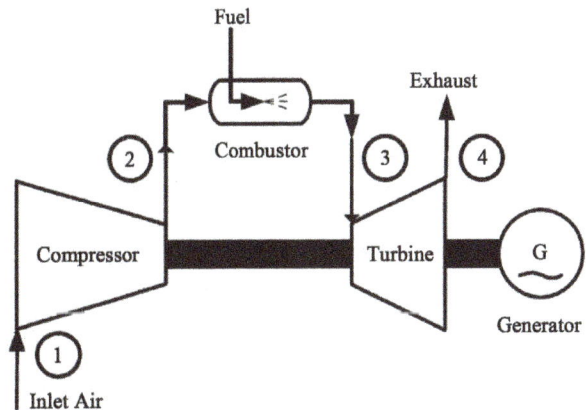

Figure 11. Schematic diagram of simple gas turbine (GT) cycle.

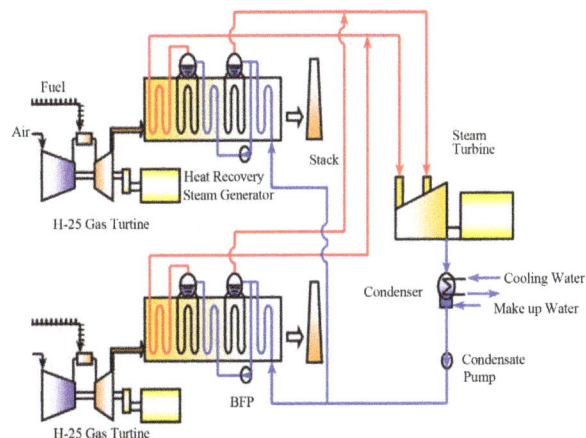

Figure 12. Combined gas turbine/steam turbine combined cycle (CC), [12].

and

3. Mesaieed Power Company.

Table 2 shows the EP shares of these companies, as 2009, [8].

3.2. Desalted Seawater Plant in Qatar

All large thermally operated Desalting Plants (DP) are combined with PP to satisfy the thermal energy needs of the DP. Desalting seawater is an energy intensive process. The Multi Stage Flash (MSF), **Figure 13**, is most predominantly used desalting seawater system in Qatar and all GCC. Thermal Vapor Compression-Multi Effect (TVC-ME), **Figure 14**, is another thermally operated distillation system recently introduced in the GCC. In Qatar, 80.6% of desalting capacity is by MSF system, and the balance (19.4%) is by TVC-ME system, [13]; while in Kuwait, all DP capacity is by MSF. Both consumes about 260-MJ thermal energy per each one m^3 distillate product, the pumping energy for the MSF and TVC-MED are about 4 kWh/m^3, and 2 kWh/m^3 respectively to move their streams. However the TVC-ME use higher pressure steam (of higher value) than that used for the MSF system. The thermal energy is supplied in the form steam at moderate pressure, say of 2 to 3 bar for the MSF and about 10 bar for the TVC system.

Steam supply to the MSF (or TVC-ME) units comes either from heat recovery steam generator (HRSG) utilizing the waste heat rejected from GT; **Figure 15**; or steam extracted (or discharged) from the Steam Turbine (ST) of the CC, **Figure 16**, [15]. Steam also can be supplied from Extraction-Condensing Steam Turbine (ECST), as shown in **Figure 17** in most Kuwait CPDP plants. In these plants, two DP can be combined with one or two steam turbines, **Figure 18**.

Most of power plants in Qatar are CPDP. Example is the RasLaffan CPDP producing 3550 MW of EP, and 163 MIGD of DW. The capacities of the five main de-

salting plants in Qatar are given in **Table 3**, [8]:

Ras Abu Aboud (RAA) had 11 MIGD, but was demolished in 2007.

3.3. CPDP in Qatar

The known PPs in Qatar are, [8,16-24]:

Table 2. Share of different Qatar power companies in the installed EP capacity, [8].

Independent power and water producers	Contracted capacity
Qatar Electricity & Water Company	
Ras Abu Fontas-A	497
Satellites:	
Al Sailiyah	121
Doha Super South	62
Ras Abu Fontas B	609
Ras Abu Fontas B1	417
Ras Abu Fontas B2	567
Ras Abu Fontas Sub-Total	2273
RasLaffan Power Company	
RasLaffan A	756
RasLaffan B	1025
RasLaffan C	1769
RasLaffan Sub-Total	3350
Mesaieed Power Company Limited	
MesaieedPowerstation	2007
Total Capacity	7830

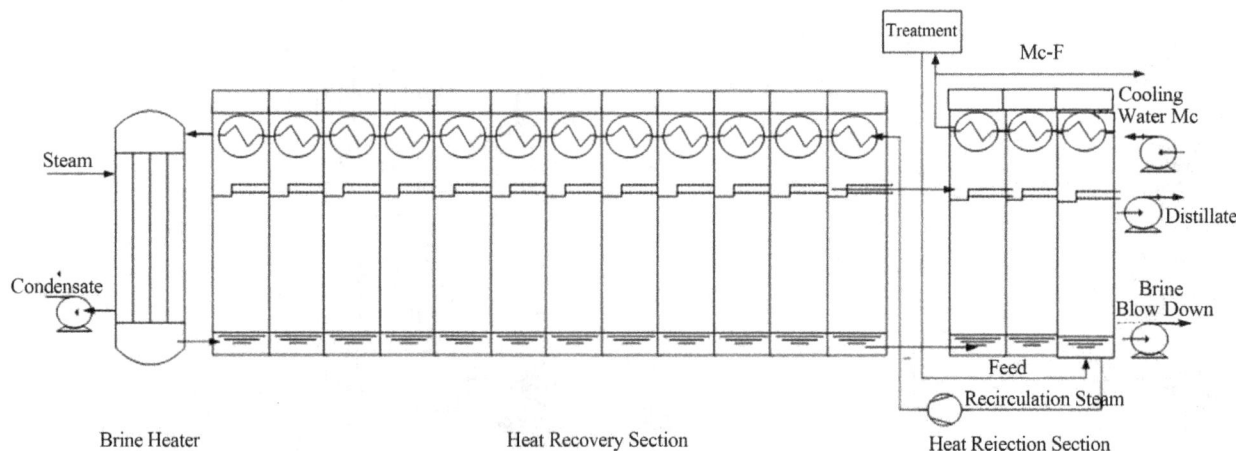

Figure 13. The multi stage flash desalination flowsheet.

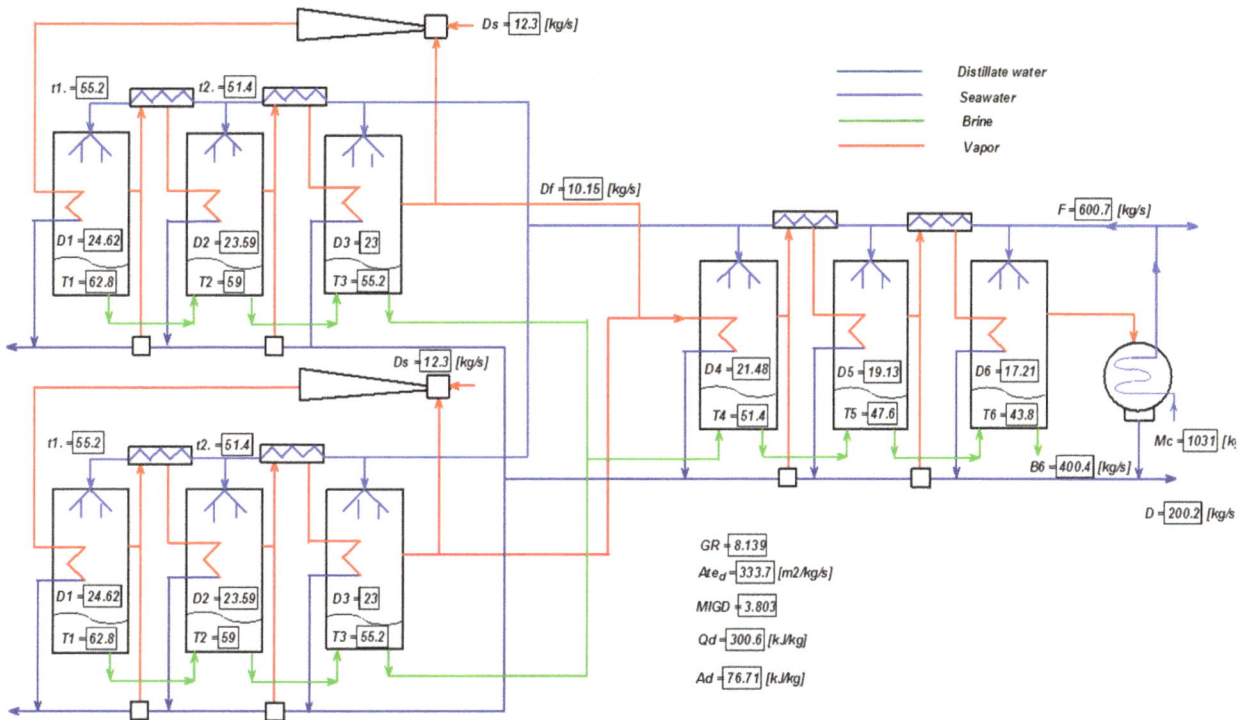

Figure 14. Thermal vapor compression-multi effect distillation system (TVC-MED) flow of Al Taweela plant, [14].

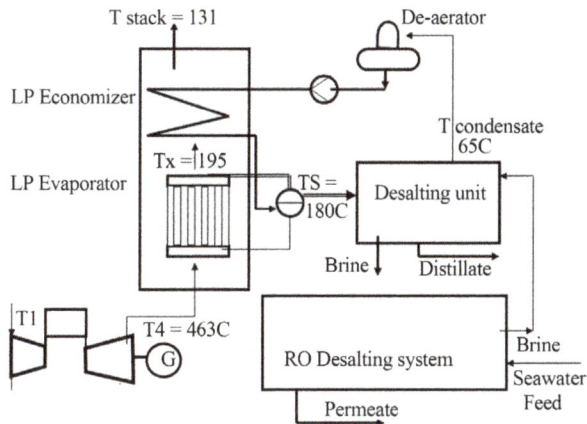

Figure 15. GT operating RO system with HRSG steam operating desalination plant.

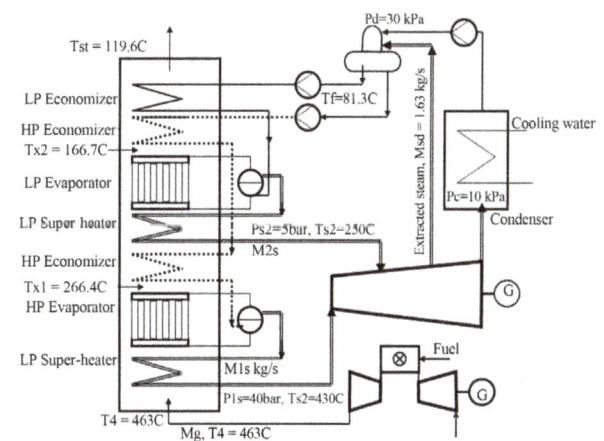

Figure 16. Combined cycle supplying steam to desalination plant.

Table 3. The capacities of the five main desalting plants in Qatar.

Desalting capacity	Plant name in MIGD
Ras Abu Fontas A (MSF)	55 (was 70)
Ras Abu Fontas B&B1 (MSF)	33
Ras Abu Fontas B2	29
RasLaffan A	40
RasLaffan B (MSF)	60
RasLaffan C (TVC-ME)	63

- RasAnuFontas (RAF) A: It has EP capacityof497 MW including 6 GT × 32 MW, 6 × 48 MW, and 2 × 9 MW. It has 10 × 7 MIGD capacity MSF desalting units. The plant was commissioned in 1980. The desalting capacity was reported as 55 MIGD in 2010.
- RAF B & B1: RAF B started with 5 CC × 121.8 MW each; operated in 1995-1996 with total EP capacity of 609 MW. RAF B1 was added in 2002 by 3 GT×125.5 MW each; total 387 MW from Alstom GT13E2 gas turbines. All eight turbines are installed in one air-conditioned gas turbine building. The plant has 5 × 6.6 = 33 MIGD capacity MSF desalting units. The

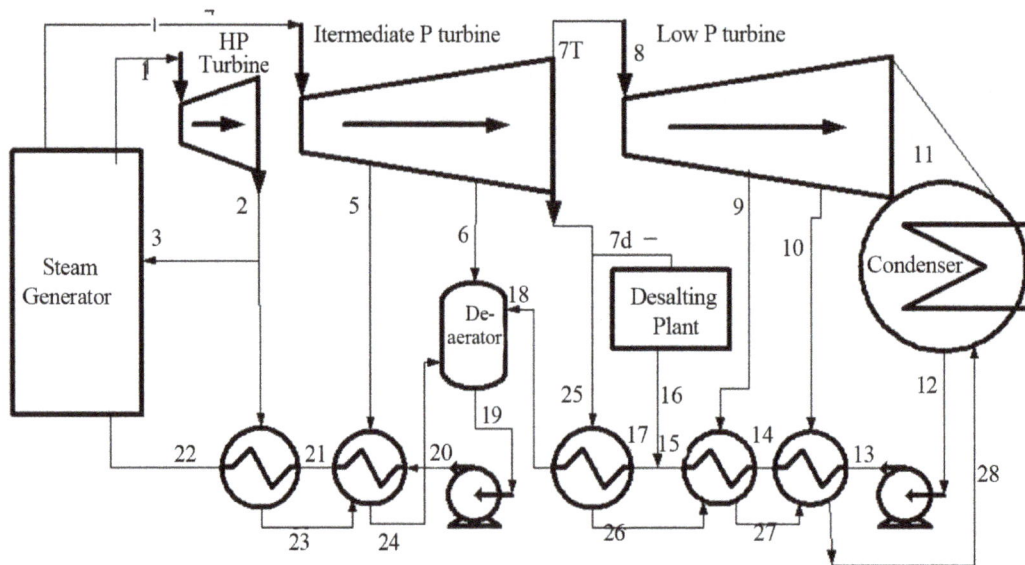

Figure 17. Steam supply to desalination system from steam extraction-condensing turbine.

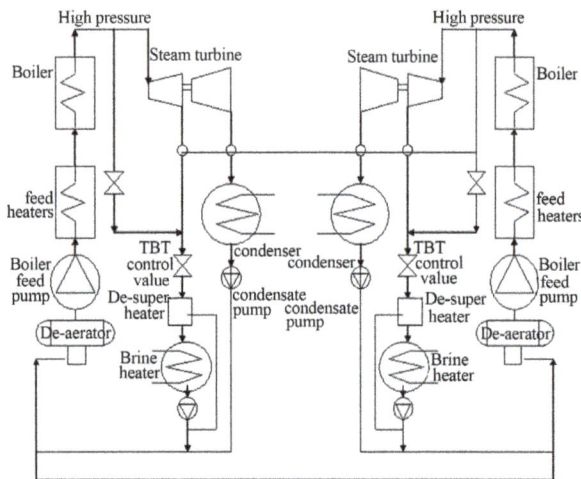

Figure 18. Steam supply to desalination system from two steam turbines.

Figure 19. Ras abu fontas power and desalting plant, [20].

MSF plant designed capacity is 52.7 MIGD. The Qatar Electricity and Water Company (QEWC) and the Qatari General Electricity and Water Corporation (KAHRAMAA) signed a contract on January 7, 2012 to build 2 MSF units of 18 MIGD each MSF type at total cost of $M500 in Ras Abu Fontas; and to be completed in the first half of 2015,

RAF B2 has 608 MW capacities consisting of 5 GT with 5 HRSG supplying steam to five MSF distillers, providing a desalination output of 33MIGD (150,000 m³/d).

- Also RAF A1 desalting plant of has 45 MIGD desalting capacity. It uses the waste heat from RasAbou-Fontas B1, and was commissioned in 2010, **Figure 19**.
- Saliyah of 2 × 67 MW 9001B gas turbines.

- Doha Super South has 62 MW EP capacity.

So, the total EP capacity belongs to the Qatar Electricity and Water Company is 2273 MW, and its desalting water capacity is 162 MIGD.

- RasLaffan (RL)-A has two CC of 378-MW each. Each CC has 2 GT (known as SGT5-4000F) ×100 MW plus Steam Turbine (ST) of 178 W. Each CC is combined with 2 MSF desalting unit of 10 MIGD each. So, RA has 756 MW total EP capacity and 40 MIGD desalting capacity. RL A plant started operation in May 2004, and cost of $720 m.
- RL B of electric power capacity equal 1025 MW using CC; and 4 MSF × 15 MIGD desalting plants; and cost $M900 and commissioned in 2006. The plant has three largest Siemens SGT5-4000F gas turbines, two 220MW steam turbines supplying steam to the four MSF units.

- RL C (known also as RasGirtas) is the largest power and water plant in Qatar with EP capacity of 2730 MW using CC, and was opened in May 2011; and cost US3.9 billion. It has 10 TVC-ME desalination units built by Sedimof 28,640 m^3/d (6.3 MIGD) each, [25]. Besides satisfying Qatar with 30% of EP need and 20% of potable water need; it will also provide off-peak electricity to neighboring GCC countries via a planned regional power grid, [26],

- Mesaieed Power plant: Its EP capacity is 2000 MW power plant. It was commissioned in 2009. The plant has three sections, and each section contains: 2 × 233 MW GT + 2HRSG + 1 × 233 MW steam turbine. All GT and Steam turbines are manufactured by GE. The plant was built by the Spanish firm IBERDROLA Ingeniería at cost $1.63 billion dollars, [27]. A5760 m^3/d TVC-MED desalination plant was added by Tri-Tech Holding Inc., [28].

The desalted seawater annual production, in Million cubic meters per year (Mm^3/y) increased from 225.1 Mm^3/y in 2006 to 373.6 Mm^3/y in 2010, see **Figure 20**, [8].

Water consumption/ca is among the highest in the world. Water production in 2010 was 373.6 Mm^3/y, [8] and population was 1,699,435, see **Table 4**. This gives annual cubic meters per capita (m^3/y.ca) consumption of 220 m^3/(y.ca); or daily 602 liters per person (l/d.ca), [8]. A total of 1,903,447 people were inside of Qatar on Jan. 31. That's an all-time high, surpassing the 1.85 million record set in November 2012, [29].

4. Water and Energy Relation in Desalted Seawater (DW) Production

4.1. Energy Consumed by Multi Stage Flash (MSF) Desalting System

Energy and water are closely linked. In this section, the extensive amounts of energy used in desalting seawater are calculated.

In calculating the energy consumed in the commonly used Multi Stage Flash (MSF) system, a practical example of 6 Million Imperial Gallons per Day (MIGD) MSF unit operating in Kuwait is given here.

Each desalting plant requires mechanical energy to: extract seawater from the sea, desalting process, brine discharge, and DW distribution. Thermally operated desalting units need also heat in the form of steam supply at moderate pressure (2 - 3 bar), besides the pumping energy. For the considered MSF of 6 MIGD, it has Gain ratio (Distillate product D/Supplied steam S) equal to D/S = 8, consumes 280 kJ of thermal energy/kg of distillate. The flow rates of the distillate output D is 316 kg/s (6 MIGD), the consumed steam S is 316/8 = 39.5 kg/s, the required cooling water is about 7.7D = 2433 kg/s, and the re-circulating stream is 12D. The nominal capacity and heads of used pumps are given in the following **Table 5**, [30].

The nominal pumping energy used to produce 316 kg/s distillate is 5250 kW, or 16.6 kJ/kg, but the actual consumed pumping energy is 14.4 kWh/m^3.

If the steam supplied to the desalting unit is at high temperature T_h= 127C (400 K), was supplied to Carnot cycle operating between this steam temperature and low temperature T_L = 47C (320 K), the average condenser temperature in the Gulf area, the Carnot cycle would have an efficiency of

$$h\left(Carnot\right) = \left(1 - T_L/T_h\right) = \left(1 - 320/400\right) = 0.2$$

For an actual cycle, its efficiency would be lower than that of ideal Carnot cycle, say 15% less.

So the real work equivalent to the 280 kJ/kg thermal energy supplied the MFS is:

280 × 0.85 × 0.2 = 47.6 kJ/kg (or 13.22 kWh/m^3), or the equivalent of work to the 280 kJ/kg of heat supplied to the MSF unit per kg of distillate. In practice this number reaches 16 kWh/m^3.

So, the total equivalent work (for the heat supplied and pumping energy) is about 17.22 kWh/m^3.

By knowing that the electrical energy cost is the US ranges from cheap cost $0.08/kWh in the state of Idaho,

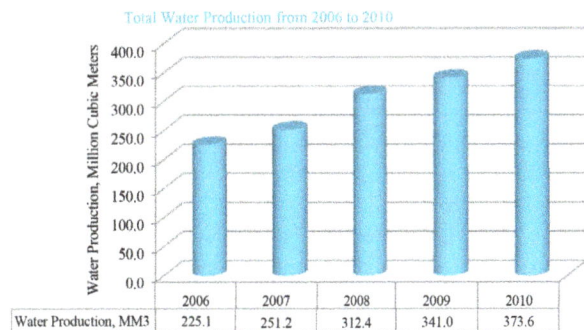

Figure 20. Desalted seawater production in Qatar; representing 99% of municipal water, [8].

Table 4. Historical Qatar population, [29].

Year	1960's	1986	2000	2002	2004	2006	2007	2008	2009	2010
Population in 1000	70	369	744.55	793.34	840.29	1041.73	1226.21	1553.73	16317.3	1699.44
% annual increase						17.30	17.70	26.70	5	4.10

Table 5. Characteristics of several pumps in a MSF unit of 6 MIGD operating in Kuwait, [30].

Pump duty	Flow rate in l/s	Head in m	Motor rating, kW
Re-circulating pump	4200	73	4100
Seawater cooling	2675	25	890
Distillate pump	348	57	170
Condensate pump	45	114	90
Total Power			5250

where EP is generated by hydropower, to very expensive cost of 0.332/kWh in Hawaii, where EP is produced by burning oil, the average cost for most of other states can be considered equal to $0.12/kWh, and will be considered here in calculating the consumed energy of the desalting processes, [31].

If the average US EP cost is considered here, the energy cost in $/m^3 of distilled water is $2.1/m^3, and if this represents 60% of distilled water cost, the actual DW cost would be $3.5/m^3.

The commonly used technology of MSF and newly used Thermal Vapor Compression-Multi Effect Distillation (TVC-MED) are very energy extensive, and should be stopped, and substituted by the more energy efficient Seawater Reverse Osmosis (SWRO) desalting system.

4.2. Energy Consumed by Seawater Reverse Osmosis (SWRO) Desalting System

In seawater Reverse Osmosis (SWRO) desalting system, **Figure 21**, membranes are used to separate fresh water from saline feed-water. Feed-water is pumped, after being pre-treated to selective semi-permeable membranes which allow water, but not salt to pass through the membranes. The pressure of the pumped feed seawater to the membranes should be much higher than the osmotic pressure for the freshwater to pass through the semi-permeable membranes, leaving the solid salt behind. Examples of applied pressures at different salinities are given in **Table 6**, [32].

The RO plants are very sensitive to the feed-water quality (salinity, turbidity, temperature), while other distillation technologies are not so demanding in this respect. High-salinity seawater requires high feed pressure to the membranes and thus requires more energy. High-turbidity feed-water can cause fouling where membrane pores are clogged with suspended solids, and thus the SWRO needs extensive feed water pretreatment.

The energy consumed by the SWRO is calculated here for a typical example.

For one m^3/second (s) of permeate (P), and recovery ratio R is:

R = P/F = 1/3, the feed (F) would be 3 m^3/s.

Table 6. Required Feed water (F) applied pressure for several F's salinity, [32].

Water source	salinity, mg/l	pressure range, bar
Brackish water	500 - 3500	3.4 - 10.3
Brackish to saline	3500 - 18,000	10.3 - 44.8
Seawater	18,000 - 45,000	44.8 - 82.7

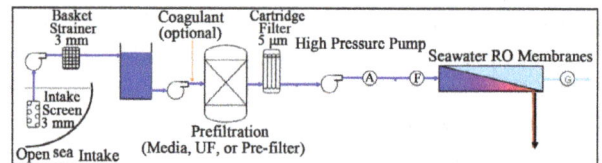

Figure 21. Simplified reverse osmosis scheme without energy recovery system.

For 70 bar applied feed pressure (P$_f$), typical feed pressure for high salinity water in GCC), 0.8 pump efficiency (η_p), the consumed feed pump energy is:

$$W(\text{feed pump}) = (F \text{ in } m^3/s)(DP \text{ in kPa})/h(\text{pump})$$

$$= 3 \times 7000/0.8 = 26250 \text{ kW}/(m^3/s)$$

$$= 26250 \text{ kJ/m}^3 = 7.29 \text{ kWh/m}^3$$

Energy recovered: the brine (B) flow rate becomes 2 m^3/s, and leaves the membranes at pressure little less than the feed pressure, say 67 bar, and enter an energy recovery turbine of 0.9 turbine efficiency (η_t) would give work output

$$W(\text{turbine}) = (B \text{ in } m^3/s)(DP \text{ in kPa}) \times h(\text{turbine})$$

$$= 2 \times 6700 \times 0.9 = 12060 \text{ kW} \left(3.35 \text{ kWh/m}^3\right)$$

So, the net feed water pump power = 7.29 − 3.35 = 3.94 kWh/m^3.

The SWRO consumed energy is almost 1.2 the energy consumed by the feed water pump, or 4.728 kWh/m^3.

This power consumption depends on feed water temperature, see **Figure 22** and the seawater salinity, which varies with time.

The most energy efficient desalting Seawater Reverse Osmosis (SWRO) should be the only used method, where the pumping energy is only used, in the range of 4 - 6 kWh/m^3, and the desalting cost is in the range of $1 - $1.5/m^3.

For 5 kWh/m^3 energy consumption by SWRO method, the energy cost for $0.12/kWh/m^3, the energy cost is $0.6/m^3. If the energy cost represents 50%, the DW cost is in the range of $1.2/m^3.

The high cost of producing DW by the thermal technology is the reason behind its decreasing share from about 55% in 2003 to 34.8 in 2012 as shown in **Figure 23(a)**, [33] and **Figure 23(b)**, [34].

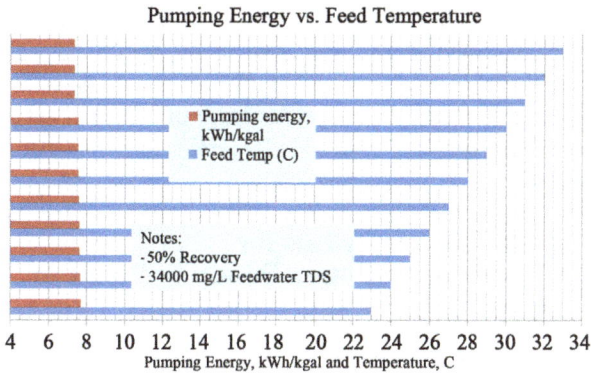

Figure 22. Dependence of pumping energy (in kWh/kgal = 3.75 kWh/m³) in SWRO on seawater temperature, [32].

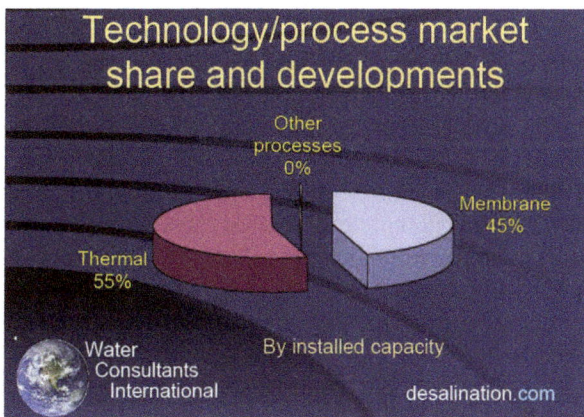

(a)

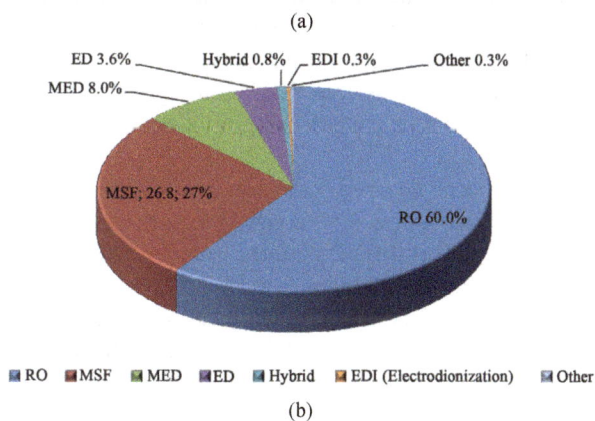

(b)

Figure 23. (a) The share of different dealing technologies in 2003, [33]; (b) Desalination Technology Market (IDA in Koschikowski, 2011, [34].

This and the previous cost of the desalting from the MSF do not include the heavy cost encountered of marine environment due to disposing very high salinity brine loaded with polluted chemical back to the sea. These do not include the cost of air pollution caused by burning fuel to produce the energy required to the desalting process.

It may be noticed here at the end of COP 18 held in Doha, Qatar in Dec. 2012, two major announcements were given:

The first that HH the Amir pledged at least 2% of EP generated in Qatar should be by solar energy by 2020.

The second is by Fahad bin Mohammed Al Attiya, Chairman of the Organizing Sub-Committee for COP18/CMP8. It stated that final design stage of putting together the solar plant which will generate 1800 MW power from 2014 onwards and would potentially produce 80 percent of our water needs through harnessing solar power", [35].

5. Impact of CPDP Consumption on the Environment

The use of fossil fuel (NG and oil) combustion in PPs is accompanied by the CO_2 emission.

In 2011, the estimated NG consumption was 21.4 M tons of oil equivalent (toe) out of the total NG production of 132.2 M toe. This means that 16.2% of the production in that year was locally consumed. Meanwhile the consumed crude oil was 8 M ton, and represents 11.25% of the oil production of out of 71.1 M ton.

The NG combustion causes the CO_2 emission of ($21.4 \times 0.75 \times 44/12$) = 58.85 Mt/y of CO_2.

The oil combustion of 8 M ton causes the emission of ($8 \times 0.85 \times 44/12$) = 24.93 Mt/y of CO_2.

So, the total emission of CO_2 due to fuel combustion in Qatar in 2011 is calculated as 83.78 Mt/y of CO_2. For population of 1.85 at the end of 2011, the per capita CO_2 emission is 45.3 t/y.ca. This is compared with 44 t/ca reported in 2009, [36].

Desalting seawater has negative effects on the environment. A percentage of 20% - 24% of the fuel consumed in the CPDP is for DW generation; and thus the same ratio of CO_2 emission from CPDP is due to seawater desalting. More than CO_2 emission, the desalting of 373.6 Mm³ in 2010, means that on the average 6 times of this amount is withdrawn from the sea, and after partially desalted, 5 times of this amount is (or 1868 Mm³ of salty water rejected back to sea). The rejected water is at temperature of about 11°C higher than that of seawater, or dumping thermal energy of 82.192 Million (M) GJ in the sea. Also, part of the rejected water (brine), about twice of the desalted water is rejected at higher salinity than that of seawater, say at 70,000 part per million (ppm), if the seawater salinity is 45,000 ppm. These affect badly the marine environment by fish impingement in the intakes, high salinity and chemical contents of the rejected brine.

6. Need of Enhancing the Efficiency of the CPDP

Qatar's PPs are large fuel energy consumer. The consumed fuel energy to produce EP and DW can be approximately calculated. In 2010, the EP production was given as 28,144 GWh; and the DW production were given as 373.6 Mm^3 (225.2 MIGD or 1.0236 Mm^3/d). When the equivalent mechanical energy consumed to desalt one m^3 of desalted water is assumed as 20 kWh/m^3, [16], the daily equivalent EP consumed for desalting seawater is 20.47 GWh/d, or annual consumption of 7472 GWh/y. This gives total equivalent EP (for both electricity and DW productions) as 35,616 GWh/y, and the share of DW units in fuel consumption is 21%. The installed capacity of the PP in 2010 was 7830 MW, and thus the capacity factor can be estimated by 0.519. The PPs in Qatar in 2010 were simple GT of maximum efficiency of 30% and CC with maximum efficiency of 44%. When 36% average efficiency is assumed for the whole PP at all average loads, the consumed fuel energy in 2010 is 356.16 million GJ. This is equivalent to 337.57 BCF (9.563 BCM) of NG. The 2010 NG production and consumption were reported as 4121 BCF, and 772 BCF respectively. So, the NG consumed by power plants represents 43.7% of the total NG consumption and 8.15% of the total NG production in 2010. It was noticed that the NG production in Qatar increased from 3154 BCF in 2009 to 4121 BCF in 2010, amazing 30.6% in one year. The annual consumed EP/ca as shown in **Table 7** is among the highest/ca in the world.

The high rate of consumed EP in reference to other countries necessitates its reduction, raising EP generation efficiency, and having demand side management (DSM) programs. The reduction of the peak load would decrease the required capacity of new power plants, and the GHG emissions. The DSM should include building capacity in the design, supervision, and assessing the DSM measures. The power plants should be checked (audit) to use fuel efficiently, and lower its energy losses. Buildings consume the largest portion of EP during the peak load (about 70%) for air conditioning equipment. So, building construction energy code should be for prepared and strictly

Table 7. Electric power consumption kWh per capita for selected countries, [37].

Country/year	2008	2009	2010
Kuwait	17,751	17,610	18,320
Qatar	16,217	15,340	14,388
United Arab Emirates	11,841	11,121	11,044
Bahrain	9687	9214	9814
Egypt, Arab Rep.	1484	1549	1608
Oman	5110	5599	5933
Saudi Arabia	7127	7427	7967
United States	13,663	12,914	13,394

applied. Existing building should be audited to decrease their consumed energy. Better standards for efficient air conditioning equipment, and appliances should be established. It is clear that the performance of CPDP in Qatar can be drastically improved by several suggestions mentioned here.

7. Better Arrangements for Cogeneration-Power Desalting Water in Qatar

The performance of CPDP in Qatar can be improved by several suggestions mentioned below.

7.1. Suggestion 1: Use of GT to Operate SWRO in Winter, and Off-Peak Hours in Summer

Qatar's electric load varies along the day, and differs daily along the year. It is generally low in winter and high in summer. However, even during the hottest day in summer with the highest peak load, there is spare capacity that can be utilized during off-peak hours. In year 2010, the maximum load reached 5090 MW in July, while the reported minimum load was 1570 MW in February. Even in the hottest days of July when the peak load of 5090 MW, the load was less than 4400 MW for about 10 hours, see **Figure 24**.

This means that the power plants are grossly underutilized in winter with low load factor, about 30%, as well as in summer but with less extent. The load factor can be raised by operating the unused capacity to operate seawater reverse-osmosis (SWRO) desalting plant, mainly in winter, or chilling water for district cooling in summer.

The produced desalted water can be used later, stored, or injected in aquifers. Every one MW power output can produce up to 0.88 MIGD desalted water by SWRO if its specific energy consumption is 6 kWh/m^3. Moreover the GT efficiency η_{gt} as well as their power output are higher in winter than in summer; and the efficiency increases by its operation close to nominal load.

The difference between the maximum loads between the summer and winter months (say $5100 - 2300 = 1800$ MW) can be fully utilized to run Seawater reverse osmosis (SWRO) desalting plant of up to 160 MIGD. The demand for water is usually higher in summer than in winter, but the difference is not as large as the case of electricity. In 2010, it varies between 0.879 Mm^3/d in February to 1.133 Mm^3 in July; see **Figure 25**. Desalting and storing desalted seawater at low demands time can decrease the required desalting capacity to the average demand, and not the maximum water demand. It also can provide water storage capacity which is critically needed for the country water security. The country plans to build five "mega" reservoirs on the outskirts of Doha by 2016.

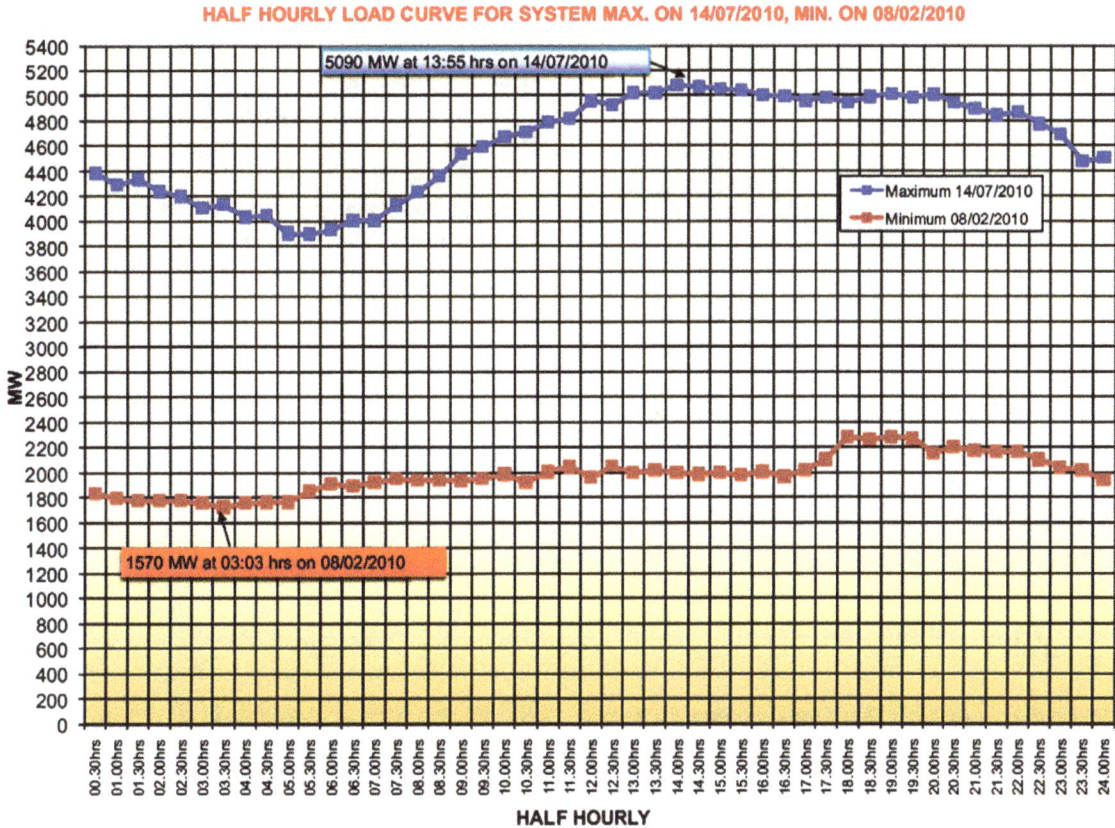

HALF HOURLY LOAD CURVE FOR SYSTEM MAX. ON 14/07/2010, MIN. ON 08/02/2010

5090 MW at 13:55 hrs on 14/07/2010

1570 MW at 03:03 hrs on 08/02/2010

Maximum 14/07/2010
Minimum 08/02/2010

Figure 24. Qatar EP hourly load variation during the days of maximum load and minimum load in year 2010, [8].

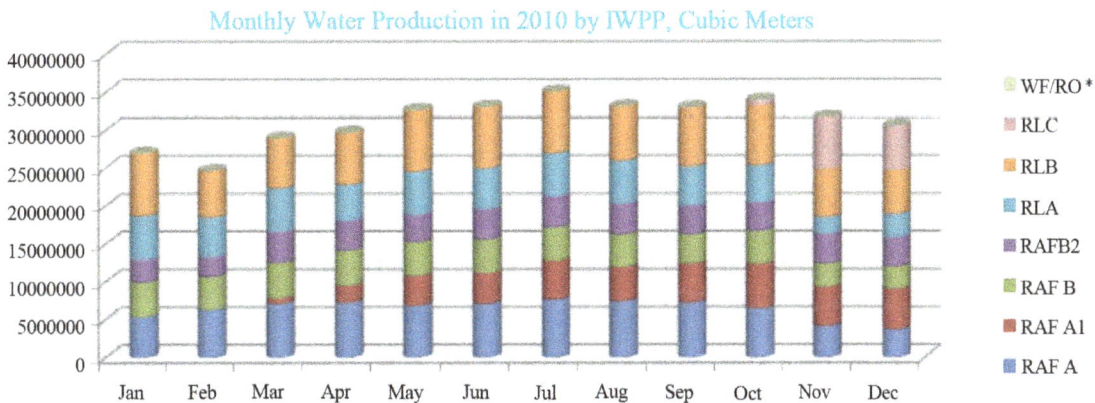

Monthly Water Production in 2010 by IWPP, Cubic Meters

WF/RO *
RLC
RLB
RLA
RAFB2
RAF B
RAF A1
RAF A

Figure 25. Monthly water production by desalting seawater and brackish (less than 1%) plants, [7].

Each reservoir will be around 200 m in diameter, 12m deep and store 73 million imperial gallons, which is around twice the size of the largest tanks ever built. Now, the storage capacity is sufficient for only two days, [38]. The planned increase in the storage fresh water capacity is shown if **Figure 26**, [39].

The main advantage of this suggestion is saving the energy consumed for desalting. In the present practice of using MSF desalting system, the average consumed thermal energy is 270 MJ/m^3 plus pumping energy of 14.4 MJ/m^3 (4 kWh/m^3). The total MSF thermal and

pumping energies can be expressed by equivalent mechanical energy of 20 kWh/m^3. Meanwhile, the SWRO with energy recovery devices (ERD) from the rejected brine consumes only about 5 kWh/m^3 pumping energy, or about one fourth of the MSF system. The average daily DW production is more than one Mm^3/d; and this can be easily doubled in 7 years, if the annual increase in DW is 7% prevails. Desalting one Mm^3 daily by SWRO can save 15 GWh/d, or 5475 GWh/y. This saving can be expressed by saving of fuel energy equal to 65.7 MGJ (62.415 BCF of NG). This is more than 15% of all con-

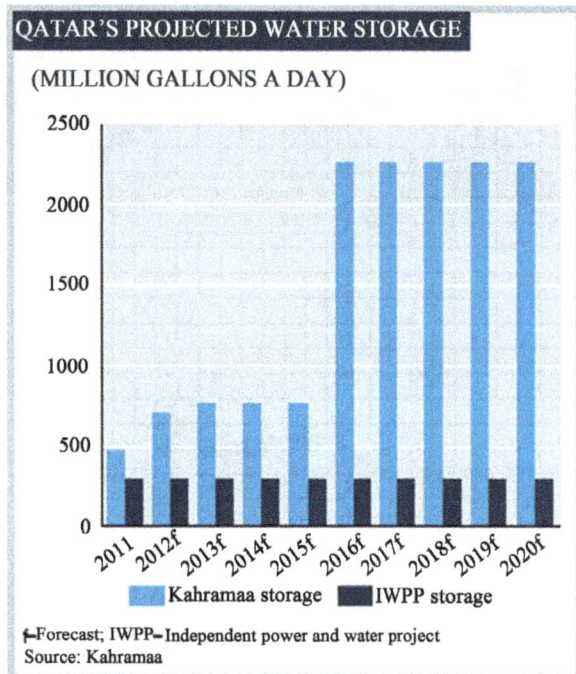

QATAR'S PROJECTED WATER STORAGE

(MILLION GALLONS A DAY)

f—Forecast; IWPP—Independent power and water project
Source: Kahramaa

Figure 26. Planned increase in fresh water storage capacity, [39].

sumed fuel energy in 2010, and the corresponding CO_2 emission. This also satisfies the needed increase of installed desalted water capacity to face its increasing demand.

7.2. Suggestion 2: Transferring the Simple Gas Turbine to Combined Cycle

This is common and known practice to increase the output of the simple gas turbine (GT) by adding heat recovery steam generator (HRSG) utilizing the heat content of the hot gases leaving the GT. Then steam turbine is added to utilize the steam generated from the HRSG and forming with the GT and the HRSG a combined gas/steam cycle (GTCC. The added steam turbine produces about 50% of the original GT output, without consumption of additional fuel, and thus can raise their efficiencies from that of simple GT, in the range of 30% - 35%, to that of the GTCC in the range of 45% - 50%.

The objective is to convert the simple GT cycle to a combined cycle plant in order to address the future electricity needs of Qatar, provide additional electrical capacity, utilize existing simple GT infrastructure to reduce environmental impacts and costs, and enhance the reliability of the state's electrical system by providing power generation near the centers of electrical demand.

The project would include the addition of two heat recovery steam generators, a steam turbine generator, an auxiliary boiler, an air-cooled dry condenser unit, and a 115-kilovolt (kV) electrical switchyard. The proposed

project would use existing infrastructure, including the existing natural gas pipeline, water supply pipeline and electric transmission line.

Some major components and features of the proposed project include:

• Addition of two new heat recovery steam generators (HRSG), each receiving the exhaust from one of the existing General Electric Frame 7EA combustion turbine generators (CTGs), and equipped with 324 MMBtu/hr, HHV capacity, natural gas-fired duct burners.

• Addition of a new 85 MMBtu/hr capacity natural gas-fired auxiliary boiler equipped with ultra low NOx burner(s) and 50-foot-tall, 48-inch-diameter stack.

• Addition of new nominal 145 MW (net output) condensing steam turbine generator (STG).

8. Conclusions

Giving the high oil and NG productions in Qatar, energy conservation may look to be unnecessary, and why should the Qatar worry about conserving energy resources? The answer should be that conserve energy is a must. The oil and NG resources are finite, and conservation makes financial sense since their local usage means having less available for export. The UAE turns from NG exporter to importer in 30 years. All the GCC, except Qatar for the time being, are in short of NG used in generating EP due to the EP staggering increase.

Though Qatar's petroleum production has grown steadily since 2002, Qatar's fields are maturing, and output at Dukhan—formerly the largest producing field—is in decline. In 2009, Qatar consumed approximately 147,000 bbl/d of petroleum. Although still relatively small compared to total production levels, consumption has more than tripled since 2000. The Electric power generation increased from 19.08 Billion kWh in 2009 to 22.28 Billion kWh in 2010. The CO_2 emission increased from 63.62 Mt/y in 2009 to 64.68 Mt/y in 2010.

This paper presents the status of energy (both fuel and EP), and DW productions and consumptions. It shows that energy should be conserved to sustain life in Qatar, and to save its environment from continuous deterioration.

REFERENCES

[1]   "Initial National Communication to the United Nations Framework Convention on Climate Change," Ministry of Environment, State of Qatar 2010.

[2]   Cia World Factbook and Other Sources, "Qatar Economy," 2012.
http://www.theodora.com/wfbcurrent/qatar/qatar_economy.html

[3]   "The GCC in 2020: Resources for the Future," A Report from the Economist Intelligence Unit Sponsored by the

Qatar Financial Centre Authority.
http://graphics.eiu.com/upload/eb/GCC_in_2020_Resour
ces_WEB.pdf

[4] Qatar, Overview/Data.
http://www.eia.gov/countries/country-data.cfm?fips=QA

[5] BP Statistical Review of World Energy, 2012, bp.com/
Statistical Review.
http://www.bp.com/assets/bp_internet/globalbp/globalbp_
uk_english/reports_and_publications/statistical_energy_re
view_2011/STAGING/local_assets/pdf/statistical_review
_of_world_energy_full_report_2012.pdf

[6] G. Sarraf, W. Fayad, T. El Sayed and S.-P. Monette,
"Unlocking the Potential of District Cooling, The Need
for GCC Governments to Take Action," Booz & Company, New York.
http://www.booz.com/me/home/thought_leadership_strat
egy/40007409/40007869/50737873

[7] Qatar Weather Averages.
http://www.worldweatheronline.com/Doha-weather-avera
ges/Ad-Dawhah/QA.aspx

[8] Qatar General Electricity and Water Corp, (KAH-
RAMAA) Statistical Yearbooks 2007, 2008, and 2010.

[9] Annual Statistical Report 2010, Organization of Arab
Petroleum Exporting Countries (OAPEC).

[10] Personal Communication with Dr. Youssef El Gendy
from KAHRAMAA.

[11] QATAR—The Power Sector.
http://www.thefreelibrary.com/QATAR+-+The+Power+S
ector.-a0265495758

[12] Hitachi Combustion Turbines.
http://www.hitachipowersystems.us/products/combustion
_turbines/index.html

[13] A. S. Al Malki, "Business Opportunities in Water Industry in Qatar".
http://www.siww.com.sg/pdf/Biz_Opps_in_Water_Indust
ry_in_Qatar.pdf

[14] A. O. B. Amer, "Development and Optimization of ME-
TVC Desalination System," *Desalination*, Vol. 249, No.
3, 2009, pp. 1315-1331.

[15] A. A. Alsairafi, I. H. Al-Shehaima and M. Darwish, "Efficiency Improvement and Exergy Destruction Reduction
by Combining a Power and a Multi-Effect Boiling Desalination Plant," *Journal of Engineering Research*, Vol.
1, No. 1, 2013, pp. 289-315.

[16] Ras Abo Fontas B1.
http://globalenergyobservatory.org/geoid/5020

[17] Ras Abu Fontas "B" Power and Desalination Plant.
http://midmac.net/projects_details.asp?project_id=12&cat
_id=9

[18] Ras Abu Fontas B-2.
http://globalenergyobservatory.org/geoid/5021

[19] Ras Abu Fontas A1.
http://www.impregilo.it/public/impregilo/en/engineering_
key_current_projects.php?s=&d=&p=367

[20] Ras Abu Fontas Power Station.
http://www.panoramio.com/photo/1922664

[21] RasLaffan A.
http://globalenergyobservatory.org/geoid/5022

[22] RasLaffan B.
http://globalenergyobservatory.org/geoid/5023

[23] RasLaffan C.
http://globalenergyobservatory.org/geoid/42860

[24] Mesaieed CCGT Power Plant.
http://globalenergyobservatory.org/geoid/42859

[25] MED/TVC Desalination Plant Inaugurated in Qatar.
http://www.desalination.biz/news/news_story.asp?id=593
9&title=MED/TVC+desalination+plant+inaugurated+in+
Qatar

[26] RasQatras Energy Plant.
http://en.wikipedia.org/wiki/Ras_Qartas_Energy_Plant

[27] Mesaieed Power Plant. MPCL-IBERDOLA INGENI-
ERIA & CONSTRUCCION (Mesaieed).
http://wikimapia.org/13712833/Mesaieed-Power-Plant-M
PCL-IBERDOLA-INGENIERIA-CONSTRUCCION

[28] Tri-Tech Infrastructure LLC, the Chinese company's US
Subsidiary, Was Awarded a US$8.3 Million Contract for
the Seawater Desalination Unit for the Utility Plant of
Qatar Petrochemical Co. Ltd. (QAPCO) at Mesaieed Industrial City in Doha.
http://www.desalination.biz/news/news_story.asp?id=625
4&title=Qatar+MED%2FTC+desalination+plant+goes+to
+Tri-Tech

[29] Qatar Population Surpassed 1.9 Million.
http://dohanews.co/post/42188848936/qatars-population-s
urpasses-1-9-million-three-years

[30] M. A. Darwish, *et al.*, "Technical and Economical Com-
parison between Large Capacity Multi Stage Flash and
Reverse Osmosis Desalting Plants," *Desalination*, Vol. 72,
No. 3, 1989, pp. 367-379.

[31] J. Jiang, "The Price of Electricity in your State," Planet
Money.
http://www.npr.org/blogs/money/2011/10/27/141766341/
the-price-of-electricity-in-your-state

[32] "Seawater Desalination Power Consumption," White
Paper, Water Reuse Association, 2011.
http://www.watereuse.org/sites/default/files/u8/Power_co
nsumption_white_paper.pdf

[33] J. Tonner, "Desalination Trends," Water Consultants
International.
Siteresources.worldbank.org
EXTWAT/Resources/4602122-1213366294492/5106220
-1213366309673/6.2JohnTonner_PPT_Desalination_Tren
ds.pdf

[34] Water Desalination Using Renewable Energy, Technology Brief, International Renewable Energy Agency, IRENA.
http://search.fastaddressbar.com/web.php?s=Water+Desalina
tion+Using+Renewable+Energy%2C+Technology+Brief&fi
d=65017

[35] Qatar to Get Solar-Powered Desalination Plant, 2012.
http://thepeninsulaqatar.com/qatar/216388-qatar-to-get-so
lar-powered-desalination-plant.htmlsiteresources.world-
bank.org/EXTWAT/Resources/4602122-1213366294492/
5106220-1213366309673/6.2JohnTonner_PPT_Desalinat

ion_Trends.pdf

[36] "CO$_2$ Emissions (Metric Tons per Capita)," The World Bank.
http://data.worldbank.org/indicator/EN.ATM.CO2E.PC

[37] "Electric Power Consumption (kWh per Capita)," The World Bank Data.
http://data.worldbank.org/indicator/EG.USE.ELEC.KH.P C

[38] C. McElroy, "Quenching Qatar's Thirst," Construction Week on line.com, by Cathal McElroy on April 7, 2012.
http://www.constructionweekonline.com/article-16307-qu

enching-qatars-thirst/1/

[39] Verity Ratcliffe Doha Shifts Focus to Water Security, Issue 2, 13-19 January 2012.
http://www.agentschapnl.nl/sites/default/files/bijlagen/M EED%20-%20Doha%20shifts%20focus%20to%20water %20security.pdf

21

A Homeowner-Based Methodology for Economic Analysis of Energy-Efficiency Measures in Residences

Nelson Fumo, Roy Crawford
The University of Texas at Tyler, Tyler, USA

ABSTRACT

Residential energy-efficiency measures, besides energy savings, provide opportunities for improvement of thermal comfort, air quality, lighting quality, and operation. However, all these benefits sometimes are not enough to convince a homeowner to pay the incremental cost associated with the energy-efficiency measure. The objective of this work is to develop a methodology for the economic evaluation of residential energy-efficiency measures that can simplify the economic analysis for the homeowner while taking into consideration all factors associated with the purchase, owner-ship, and selling of the house with the energy-efficiency measure. The methodology accounts for direct and indirect economic parameters associated to an energy-efficiency measure; direct parameters such as the mortgage interest and fuel price escalation rate, and indirect parameters such as savings account interest and marginal income tax rate. The methodology also considers different cases based on the service life of the energy-efficiency measure and loss of efficiency through a derating factor. To estimate the market value, the methodology uses the future energy cost savings instead of the cost of the EEM. Results from the methodology offer to homeowner annual net savings and net assets. The annual net savings gives the homeowner a measure of the annual positive cash flow that can be obtained from an energy-efficiency project; but more important, the net assets offer a measure of the added net wealth. To simplify and increase the use of the methodology by homeowners, the methodology has been implemented in an Excel tool that can be downloaded from the TxAIRE's website.

Keywords: Methodology for Economic Analysis; Energy-Efficiency Measures for Residences; Energy Savings in Residences

1. Introduction

According to the U.S. Energy Information Administration, the residential sector, with 23% of the total energy consumption in the United States, compares significantly with the energy consumption of the other sectors: commercial (19%), industrial (31%), and transportation (28%) [1]. Therefore, efforts at all levels are constantly supported by the government in order to promote energy efficient homes.

Energy used in buildings is a multi-variable phenomenon. The variables can be grouped into four categories: users, equipment, construction material, and weather. The study of how the variables in each of the categories affect the energy consumption offers the opportunity to decrease energy use without affecting the activities of the occupants and their thermal comfort. Users/occupants will be willing to adjust their energy use patterns when the benefits are tangible. Although technical matters can be improved, the economic aspect is always a major factor for decision making. Therefore, residential energy efficiency projects should be treated as a financial investment and the attractiveness of a project depends upon the return expected by the owner or investor. Thus, an appropriate economic analysis is a key factor to show the actual potential benefits of the investment which can be highlighted when environmental benefits are also illustrated.

Approaches from different authors to perform economic analysis on residential energy efficiency projects show their similarities. Martinaitis *et al.* [2] mention calculations of payback time, net present value, internal rate of return, and the cost of conserved energy (CCE) as approaches relatively easy to use for appraisal of renovation in residential buildings. In their proposal of separating investments into those related to energy efficiency improvements, and those related to building renovation, they use the CCE for the appraisal of energy efficiency investments. The CCE takes into consideration both the lifespan of measures and the cost of borrowing money. Sadineni *et al.* [3] develop cost benefit data to be used by a local electric utility in defining a rebate program to encourage energy efficient construction in the Desert Southwest region of the USA. Benefit/cost analysis was performed and payback periods were calculated. Payback period is commonly used because is easy to understand by people of any background. However, although the payback period is a valid economic metric, it leaves out the analysis other economic parameters considered in the proposed methodology. Gorgolewski [4] uses the "savings-to-investment ratio" (SIR) to rank predicted savings from retrofit investments. The SIR is defined as the ratio of the present value of the total life time energy savings and the investment cost, with the present value computed by discounting of all future savings to their equivalent present value. It can be noticed that the SIR is the inverse of the payback period. Ouyang *et al.* [5] performed life-cycle cost economic analysis to evaluate energy-savings effects on thermal simulations. They accounted for initial and maintenance costs, as well as the electricity rates. However, they did not account for added costs due to the interest rate on the loan for the initial and maintenance investment alleging that those costs can be counteracted or exceeded by the increased value of the property due to rapid economic development in China. Lekov *et al.* [6] presented the method used to conduct the life-cycle cost (LCC) and payback period analysis for gas and electric storage water heaters. The LCC accounted for consumer expenses during the life of an appliance, including equipment, installation, and operating costs (expenses for energy use, maintenance, and repair). To compute LCCs, they discounted future operating costs using a rate that reflects rates in various debt or asset classes that might be used to purchase the appliance. For new construction installations, the discount rate reflects after-tax real mortgage rates. The payback period is calculated using the change in purchase cost (normally higher) at a higher efficiency level, divided by the change in annual operating cost (normally lower). Ouyang *et al.* [7] used a standard life-cycle cost analysis to conduct economic analysis of upgrading residential buildings in China. The net present value and simple payback time were used to evaluate the investigated retrofits. Fixed loan interest rate, inflation rate, and an increase of electricity price were used for the economic analysis.

It is the opinion of the authors that most of the economic analysis does not account for all the factors involved in the life-cycle cost analysis from a homeowner point of view. Therefore, this paper introduces the methodology being developed at the Texas Allergy, Indoor Environment and Energy (TxAIRE) Institute [8] Research and Demonstration Houses to perform economic analysis of energy-efficiency measures (EEM's). This methodology has been developed with the purpose of accounting for all relevant economic factors and criteria in order to offer homeowners with a real and understandable economic analysis for decision making regarding residential EEM's. To simplify and promote the use of the methodology, it has been implemented in an Excel tool that is available from the TxAIRE's website. Throughout this paper, although some terminology are given in terms that a prospective owner can use to evaluate an EEM through the economic analysis of the entire house, the methodology is more oriented to be used for the incremental evaluation of a single EEM. Analysis of a single EEM for new houses or retrofits can be done by knowing the required information for the EEM and the reference case that the project will be compared with. This means that the project can be evaluated through incremental costs and benefits (savings) when the EEM is compared with a reference house without the EEM.

Using the methodology of Taylor *et al.* [9] as a reference, the authors consider that the following characteristics of the methodology introduce value to a homeowner's economic analysis: 1) beyond cash flows the net savings are more truthful to illustrate the economic benefits of the EEM; 2) the net assets illustrate the added net wealth of the homeowner if the EEM is implemented; 3) the methodology considers the market value based on future energy cost savings instead of the cost of the EEM; 4) the methodology proposes analysis of EEM's that have useful service life equal or lower than the house service life; 5) the methodology allows considering other parameters such as reduction of energy costs due to an energy efficiency derating factor, finance of the initial cost through mortgage and/or savings, and impact of savings account interest, and the methodology also allows to illustrate the cumulative benefits in case the homeowners want to consider selling the house in a year that is lower than the period of analysis.

2. TxAIRE Research and Demonstration Houses

The Texas Allergy, Indoor Environment and Energy (TxAIRE) Institute [8] was created to be a catalyst for the identification, development, demonstration, evaluation and promotion of technology products that improve the energy efficiency and indoor environmental quality of buildings. The TxAIRE Research and Demonstration Houses have been designed to serve as realistic test facilities for developing and demonstrating new technologies related to energy efficiency, indoor air quality, and sustainable construction materials and methods. The TxAIRE Houses are fully instrumented testbeds, making possible full testing and analyses of roof, wall, window, and slab building envelope components. All mechanical systems are also fully instrumented, and include multiple systems to facilitate comparison of performance. The potential of TxAIRE houses to demonstrate and promote technology for energy use reduction in the residential sector suggest the need for a consistent methodology for the economic evaluation of the tested technologies.

3. Residential Energy Efficiency Projects

The concept of Net-Zero Energy Homes can be considered as the ultimate goal of energy efficiency projects. However, the high cost associated with this kind of projects is a drawback for its implementation beyond few cases such as homes located in remote areas or owners really committed to the environment. If energy efficiency projects are treated as a financial investment, the attractiveness of residential energy efficiency projects depends upon the return expected by the prospective owner or investor. The estimated return from the investment is computed using economic parameters and the energy used to satisfy the occupants activities and thermal comfort. Energy efficiency projects can be evaluated individually since the cost-benefit varies from one to another alternative. However, several energy efficiency projects can be evaluated in conjunction in order to consider the overall cost-benefit from the synergetic effect on energy use.

When accounting for energy use, it is important to be aware that the efficiency of equipment may decrease over time due to normal wear, and replacement of technology often implies an increase in efficiency. Due to derating factor of installed technology or increment of efficiency when technology is replaced, the annual energy consumption varies even with constant weather conditions and house operation (occupancy patterns or schedules). Since simulations of energy use usually are given for one-year period, it is assumed that the given annual energy consumption is the average energy consumption for the years considered as period of analysis. As example, BEopt [10] defines the average energy use as the average of the annual energy use over the period of analysis. This is important for the consideration of uncertainty. For the economic analysis presented in this paper, the energy use is considered through the estimated utility bills. The utility bills should account for the average price of fuels (electricity, natural gas, etc.) for the first year of analysis and the energy consumption obtained in the same way and conditions for the projects to be compared.

4. Previous Definitions

The methodology proposed in this paper is intended to cover all significant scenarios of economic outflows and inflows associated with buying and owning of a house during the defined period of analysis, as well as considering the option of selling the house at any year during the period of analysis. Therefore, for better understanding of the use of some parameters in the proposed methodology, some definitions are given.

4.1. Period of Analysis

Period of analysis refers to the number of years the economic analysis will be performed. Although the useful life of a house can be more than a hundred years, shorter periods of analysis decrease the uncertainty on the actual performance of components, advances in technology on replacements, and variation of economic parameters with time. For residential projects a reference period of analysis can be 30 years since it is a common period for mortgage loan and allows accounting for equipment replacement. Since service life of equipment may be shorter than the period of analysis of an energy efficiency residential project, **Table 1** shows some examples of service life estimates that can be used as a comparison reference when considering a thirty year period of analysis.

4.2. Inflation

Inflation is a familiar concept of money value change over time. It is well understood that most things that are bought today or most services received today will cost more in the future because of the inflation, which also means that money loses purchasing power. LCC analysis can be done by considering cash flows in constant dollars (purchasing power does not change over time) or in current or nominal dollars (actual purchasing power for the year the cash flows are expected to occur in the future). The NIST Handbook 135 [13] recommends analysis with constant dollars using two methods to arrive at constant dollars amounts in an LCCA:

Method 1: "Estimate future costs and savings in constant dollars and discount with a 'real' discount rate, *i.e.*, a discount rate that exclude the rate of inflation."

Method 2: "Estimate future costs and savings in current dollars and discount with a 'nominal' discount rate,

Table 1. Service life estimates of residential equipment.

Equipment	Median Service Life (years)
Residential single or split air conditioning [11]	15
Residential air-to-air heat pump [11]	15
Hot-water heater [11]	20
Solar PV modules [12]	30

i.e., a discount rate includes the rate of inflation."

Polly *et al.* [14] use Equation (1) to model cash flows others than loan payments such as annual utility bill costs, replacement costs in the future, and residual values. The cash flows are inflated according with the estimated inflation (i_i) based on the year (j) the cash flow will occur,

$$F_j = PV \cdot (1+i_i)^j \qquad (1)$$

where F_j is the cash flow at current dollars at the end of year j and P is the cash flow at current dollars at the beginning of the period of analysis.

4.3. Discount Rate

The discount rate is an interest rate used to discount future cash flows to the present value. In general, the discount rate considers the possible growth of available money because of earnings, the risk or uncertainty of the anticipated future cash flows, and the variation in purchasing power due to inflation. Roberts [15], point out that for homeowners the appropriate discount rate depends on their particular financial circumstances giving as reference "The discount rate should be the APR (Annual Percentage Rate) of the highest risk-adjusted rate of return that you can obtain by investing your money, or the lowest rate at which you can borrow money, whichever is higher". In the same order of ideas, Lawrence [16] suggests two discount rates as references for residential real estate analysis. The first one is the Treasury Inflation-Protected Securities (TIPS) that can be used by the buyer as an alternative investment that, although has a very low interest rate, it is a risk-free investment guaranteed to grow with the rate of inflation. The second one is to use the interest rate on the loan used to acquire the property. Eliminating interest expense provides a return on investment, as money not spent, equal to the interest rate on the loan.

Since market interest rates consider the general inflation, they are normally used as nominal discount rates. To compute the real discount rate (i_{dr}), Equation (2) can be used when the nominal discount rate (i_d) and inflation (i_i) are known.

$$i_{dr} = \frac{1+i_d}{1+i_i} - 1 \qquad (2)$$

4.4. Escalation Rate

The escalation rate allows estimating the annual change in the price levels of the goods and services to occur in the future. Since price change for home-related items other than fuels have a zero relative price change ([13], pg. 3-13), the price escalation rate for all non-energy-related items is equal to the general inflation and there-

fore the real escalation rate is zero. The increase of goods or services with time, due to the escalation rate, it can be computed as a compound interest rate as shown in Equation (3)

$$F_j = PV \cdot (1+i_e)^j \qquad (3)$$

When the escalation rate of the item in analysis is different from the general inflation, a real escalation rate can be computed using Equation (4) as

$$i_{er} = \frac{1+i_e}{1+i_i} - 1 \qquad (4)$$

Since technologic development can make a future cost lower than the present cost, such as have happened with computers, a real price escalation rate can be negative. In this case, the future cash flow can be lower than the present cash flow even in presence of inflation.

4.5. Replacement Costs

An energy efficiency project may have a useful life lower than the period of analysis, which is particularly true for equipment (e.g. heating and cooling equipment) when compared with the useful life of a house. Therefore, cash flows associated with equipment replacement must be adjusted based on the assumed inflation between the beginning of the analysis period and the time of replacement using Equation (1).

In the process of obtaining the energy use (e.g. simulations), it should be taken into consideration that equipment is replaced with minimum standard efficiency equipment or the same equipment, whichever is more efficient [14].

4.6. Market Value

In real state, several definitions may exist for market value. For example,

1) Market value is generally defined as the price a willing buyer would pay a willing seller for a property in its present condition with neither buyer nor seller under pressure to act (such as career relocation, death of a family member, divorce, etc.) [17].

A number of factors may affect a residential property's market value, including:

- External characteristics—"curb appeal", home condition, lot size, popularity of an architectural style of property, water/sewage systems, sidewalk, paved road, etc.
- Internal characteristics—size and number of rooms, construction quality, appliance condition, demonstrated "pride of ownership", heating type, energy efficiency, etc.
- Supply and demand—the number of homes for sale versus the number of buyers; how quickly the homes

in your area sell, and
- Location—desirability for a particular school district, neighborhood, etc.

2) Market value is the price at which a particular house, in its current condition, will sell within 30 to 90 days [18].

This definition contains three elements:
- Particular house
- Current condition
- 30 to 90 days

In the same order of ideas, Braun [19] identifies four components to understand a real property market: demand, relationship between supply and demand, competition, and marketability. Since this study focuses on how properties with energy-efficiency characteristics are compared to others for the same market conditions, only the marketability is of interest. *Marketability* is defined as "the relative desirability of a property in comparison with similar or competing properties in the area," with the *value* defined as the desire expressed in an economical concept on a monetary basis [19]. For this study, as illustrated later, a reference market value for the energy-efficiency measure is estimated based on the future energy savings discounted with an effective mortgage rate.

5. Economic Analysis Methodology

The proposed methodology is based on the incremental cost and benefits of a house with an EEM when compared with the same house without the EEM. The approach that the houses are the same means that the EEM does not change the appearance of the house.

With similar opinion from Ouyang *et al.* [5], the authors consider that normally the house price is related to what a prospective owner sees or perceives from a house when compared with other houses available in the market for the present economic conditions. This suggests that details on aspects such as efficiency of components and remaining life expectancy of components are not properly considered or understood. The authors also consider that market value generally accounts only for an economic analysis that is mainly based on the initial investment. Since a better economic analysis should account for cash flows (positive and negative) associated with purchasing, owning, and selling a house over the period of analysis, the economic analysis of an EEM should be made based on the net assets at any year during the period of analysis.

5.1. Previous Considerations

- Although an EEM may not increase parameters normally affecting the computations of property taxes (square footage, heating and cooling area, etc.), it may increase the relative cost when compared with

another house without the EEM, thus a property tax may still apply.
- The annual initial estimated utility bill is increased for all years by the nominal fuel price escalation rate. It is recalled that the nominal price escalation rate accounts for general inflation. The escalation rate is assumed to be constant over the period of analysis.
- The annual initial estimated incremental maintenance cost associate with the EEM is adjusted annually by the general inflation rate.
- The residual value of the EEM is assumed to be zero at the end of its service life.
- A derating factor is used to decrease the energy cost savings in order to account for any decrease of efficiency of the technology over time. When the service life of the EEM compares with the service life of the house (for example walls insulation), the derating factor is neglected.
- The EEM initial cost should account for all expenses associated with the implementation of the project. For example, if a mortgage is being requested only for the cost of the project (not for the house itself) or a fraction of the closing costs can be associated to the project, these closing costs should be added to the actual cost of the project.

5.2. Methodology

The developed methodology allows the estimation of the homeowner's assets, at any year for the period of analysis, as consequence of the implementation of an EEM. Therefore, the computations performed through the methodology are based on the incremental value of cash flows as result of the implementation of the EEM. In other words, all computations are performed as the difference in cash flows when a house with an EEM is compared with a reference house that does not have the EEM.

As justified in Section 4, the methodology uses 30 years as a reference period of analysis; which is related to the common maximum mortgage term. However, the service life of energy-efficiency measures forces the consideration of two cases. One case relates to EEMs with service lives that compare with the service life of the house; for example, wall insulation and windows. The other case relates to EEMs with service lives that are shorter than the service life of the house. This case involves mainly equipment; for example, heating and cooling systems. For practical purposes on the implementation of the methodology as illustrated in Section 5.3, the maximum service lives of EEMs with service lives lower than the service life of the house is limited to 30 years.

For cases when the service life of the project is shorter than the service life of the house, results must be interpreted according to the technology being analyzed. This implies two options of analysis based on the replacement of the technology. The replacement option should be considered for technologies that must be replaced; for example, heating and cooling systems. The no-replacement option should be considered for technologies that do not necessarily need to be replaced; for example, solar photovoltaic systems. For the option that the technology must be replaced, the period of analysis is reduced to the service life of the project. This is because a new analysis should be done at the replacement year to account for variation in equipment efficiency and variation of technology cost other than inflation. For the option that the technology need not be replaced, although the energy savings and project market value become zero at the end of the service life, the analysis must be completed for the mortgage term if this is greater than the service life. This is done to account for the impact of mortgage payments on the net assets associated to the project.

To estimate the homeowner's assets for the implementation of an EEM, the following computations are needed. As reference, these computations are presented in the order they were implemented in the Excel Tool developed for the implementation of the proposed methodology. The Excel Tool is described in Section 5.3 and it can be downloaded free of charge from the TxAIRE's website.

5.2.1. Mortgage
The down payment (DP) is defined based on a down payment rate (r_{dp}) over the initial cost of the EEM (C) as shown in Equation (5).

$$DP = C \cdot r_{dp} \qquad (5)$$

Thus, the initial mortgage (M) is computed as

$$M = C \cdot (1 - r_{dp}) = C - DP \qquad (6)$$

As for standard analysis of mortgages, the fix annual mortgage payment (A) can be estimated for each year (j) as

$$A_j = M \cdot \frac{i_m \cdot (1 + i_m)^{n_m}}{(1 + i_m)^{n_m} - 1} \qquad (7)$$

where i_m is the annual mortgage rate and (n_m) is the mortgage term in years. However, since mortgage payments are equal monthly payments, in order to increase the accuracy of the computations from Equation (7), the equation needs to be modified to account for the monthly payments as

$$A = M \cdot \frac{i_m \cdot (1 + i_m/12)^{12j}}{(1 + i_m/12)^{12j} - 1} \qquad (8)$$

The mortgage interest (MI) for each year of analysis

are calculated based on the unpaid mortgage principal (P) of the previous year as shown in Equation (9).

$$MI_j = P_{j-1} \cdot i_m \qquad (9)$$

where the unpaid mortgage principal is computed as

$$P_j = P_{j-1} - (A_j - MI_j) \qquad (10)$$

Equations (9) and (10) indicate that the mortgage principal decreases faster as payments are made. This is important because the assets during the period of analysis are affected by the mortgage monthly payment which is an aspect considered by the methodology.

5.2.2. Energy Cost Savings
The energy cost savings (S) are computed as the product of the estimated energy savings and the fuel price. It changes over the years (j) because the energy savings can be reduced if a derating factor (r_d) is considered and because the price of the fuel changes with the fuel price escalation rate (i_f). Equation (4) can be used to obtain the nominal fuel escalation rate from a real fuel escalation rate (i_{fr}). If the estimated first year energy cost savings (S_1) is computed using the estimated energy savings from energy simulations and the estimated average cost of fuels (electricity and/or natural gas) for the first year, S for any year can be calculated using Equation (11). In Equation (11), the energy savings are assumed to decrease linearly during the EEM service life (n_l).

$$S_j = S_1 \cdot \left[1 - (j-1) \frac{r_d}{n_l - 1} \right] \cdot (1 + i_f)^{(j-1)} \qquad (11)$$

In Equation (11), $n_l - 1$ is used since the first year energy savings is estimating at the nominal efficiency.

5.2.3. Maintenance Cost
Annual maintenance cost (MC) associated to the EEM is increased by inflation (i_i) during the years (j) of service life as

$$MC_j = MC_1 \cdot (1 + i_i)^{j-1} \qquad (12)$$

where MC_1 is the first year maintenance cost.

5.2.4. Mortgage Tax Savings
Since mortgage interest (MI) are tax deductible, mortgage tax savings (MTS) are achieved at the rate of the marginal income tax rate (r_t) as

$$MTS_j = MI_j \cdot r_t \qquad (13)$$

5.2.5. Property Tax
Although an EEM may not affect the parameters considered to estimate the value of a house for taxes purpose, local or state property taxes must be considered if the

incremental cost of the house is taxable. To account for the increase in cost of the EEM that may affect the property value for tax purposes, the EEM cost is increased over the years with the general inflation rate. Equation (14) is used to account for property taxes (PT) associated to the EEM according with a property tax rate (r_{pt}) and inflation.

$$PT_j = C \cdot (1 + i_i)^j \cdot r_{pt} \qquad (14)$$

5.2.6. Annual Net Savings

Annual net savings (NS) refers to the direct benefits expected from the implementation of the EEM. This net savings is cash that will be added or deducted (if it is the case) from the savings account; which affects the homeowner's net assets. As shown in Equation (15), the project net savings is computed based on the energy cost savings, mortgage tax savings, annual mortgage payment, maintenance cost, and the effective property tax. Effective property tax $(PT_j \cdot (1-r_t))$ is used since the property tax can be deducted from the income tax.

$$NS_j = S_j + MTS_j - A_j - MC_j - PT_j \cdot (1-r_t) \qquad (15)$$

Project net savings is one of the two parameters to be reported and considered by the homeowner for decision making on the implementation of the EEM. Therefore, in order that the homeowner can compare options on today dollars, net savings for each year along the useful life of the EEM are adjusted using the general inflation rate as the discount rate.

$$ANS_j = NS_j \cdot (1 + i_i)^{-j} \qquad (16)$$

5.2.7. Savings Account

Savings account balance (SA) takes into consideration the cumulative incremental net savings. As mentioned, the down payment is considered as a withdrawal from the saving account at the beginning of the analysis. Thus, the down payment will cause an initial negative incremental balance on the savings account.

The savings account is increased or decreased by the project net savings, but the balance on the previous year is also affected by the savings interest rate (i_s). On the other hand, since the earned interests are taxable, the marginal income tax rate must be applied to the earned interests. Equation (17) shows how the balance on the savings account is computed from the balance of previous year $(j-1)$.

$$SA_j = SA_{j-1} \cdot [1 + i_s (1-r_t)] + NS_j \qquad (17)$$

5.2.8. EEM Market Value

The EEM market value (MV) is computed as the future energy cost savings increased according to the fuel price escalation rate (electricity, natural gas, etc.) and discounted using an effective mortgage rate (i_{me}) (mortgage rate after tax). An effective mortgage rate, Equation (18), is used since it is the actual cost of borrowing money for the implementation of the EEM.

$$i_{me} = i_m \cdot (1-r_t) \qquad (18)$$

Since the EEM market value is estimated based on future energy cost savings, two analyses arise based on the EEM service life. When the EEM's service life is lower than the service life of the house, the market value for each year is estimated based on the remaining years of the service life as shown in Equation (19a). When the EEM's service life compares with the service life of the house, the market value for each year must be estimated based on a specific number of years the energy savings will be projected (n_s) and the market value can be estimated using Equation (19b). Although any reasonable number of years may be used, the mortgage period can be used as a reference for the analysis. The mortgage period is suggested as a reference because it will be the period of analysis a prospective owner may use if the house is put up for sale.

$$MV_j = S_{j+1} \cdot \frac{1+i_f}{i_{me} - i_f} \left[1 - \left(\frac{1+i_f}{1+i_{me}} \right)^{n_l - j} \right] \qquad (19a)$$

$$MV_j = S_{j+1} \cdot \frac{1+i_f}{i_{me} - i_f} \left[1 - \left(\frac{1+i_f}{1+i_{me}} \right)^{n_s} \right] \qquad (19b)$$

From Equation (19), it can be noticed that the market value at the end of year j is computed based on the estimated energy cost savings of the following year $(j+1)$. This is done since the house market value at the beginning of the year $j+1$ must reflect the savings that can be achieved during this year.

5.2.9. Net Assets

The EEM net asset (NA) value is a measure of the added net wealth of the homeowner if the EEM is implemented. The net assets are computed based on the amount of cash the owner will have in the savings account if the house is sold at the market value and the mortgage principal is paid off. Equation (20) shows how the net assets are computed.

$$NA_j = MV_j - P_j + SA_j \qquad (20)$$

In Equation (20), the incremental market value is based on the future energy cost savings of the EEM. However, it does not reflect anything about housing market variations, but this does not introduce an error on the computation of the incremental assets. This is justified since the EEM does not change the appearance of the house. In other words, the market value of two houses with the same appearance and construction characteristics will change similarly under the same housing market. However, the actual market value of the house with the EEM should reflect the benefits from the EEM. It must be understood that the net assets computed with Equation (20) is based on the expectation that the house can be sold at year *j* at the estimated market value. If the house is sold for less than the estimated market value, the net assets would be decreased proportionally.

Project net assets are the second parameter to be reported and considered by the homeowner for decision making on the implementation of the EEM. Therefore, in order that the homeowner can compare options on today dollars, net assets for each year along the useful life of the EEM are adjusted using the general inflation rate as the discount rate.

$$ANA_j = NA_j \cdot \left(1 + i_i\right)^{-j} \qquad (21)$$

6. Excel Tool

The methodology has been implemented in an Excel tool. The tool is available at the TxAIRE's website and it can be downloaded free of charge. As mentioned, the methodology proposes two parameters to be used by the homeowner for decision making on the implementation of the EEM. An example of the two parameters, net savings and net assets adjusted by inflation, are illustrated in **Figure 1** and **Figure 2**, respectively. **Figure 1** shows the net savings obtained at a specific year at dollars that reflect the current purchase power. It can be seen as the net cash flow that may happen at that year after accounting for costs and energy savings. **Figure 2** shows the incremental net assets of the homeowner. It indicates the actual incremental wealth, at dollars that reflect the current purchase power, of the homeowner if the house is sold at any of the years. It reflects the cumulative net assets through the years.

Results illustrated in **Figure 1** and **Figure 2** were obtained using the following parameters:

Energy-Efficiency Measures (EEM)

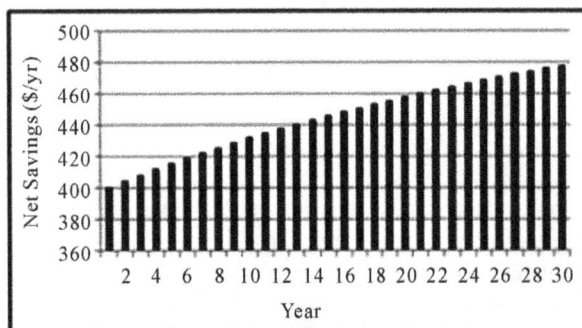

Figure 1. Net savings adjusted by inflation.

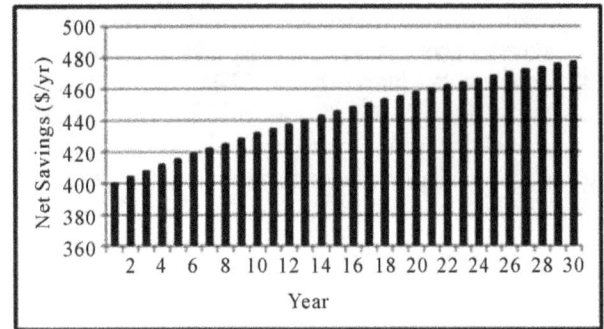

Figure 2. Net assets adjusted by inflation.

Initial cost ($): 4000
First year energy cost savings ($): 600
First year maintenance cost ($): 0
Efficiency derating factor (%): 0.00%
Service life equal to house life: TRUE
Time limit for energy savings (yr): 25
Fuel
    Fuel escalation rate (%): 2.50%
Mortgage
    Down payment rate (%): 0.0%
    Down payment ($): 0
    Mortgage ($): 4000
    Mortgage term (yr): 30
    Mortgage rate (%): 4.00%
    Annual mortgage payment ($): 229
    Effective mortgage rate (%): 3.00%
Other factors
    General inflation rate (%): 2.50%
    Marginal income tax rate (%): 25.00%
    Property tax rate (%): 0.00%
    Savings interest rate (%): 1.00%

7. Conclusion

Government agencies and professional organizations such as the American Society of Heating, Refrigerating and Air-Conditioning Engineers (ASHRAE) are working on different strategies to promote and develop technology and legislation (code, standards, etc.) to address energy use in residences. Nontraditional construction material and new technology, especially renewable energy technology, are well accepted as means for energy use reduction but their higher initial investment has been a drawback. Therefore, analysis involving all of the economic parameters as well as hidden benefits must be available for homeowners in the decision making process. The methodology presented in this paper has a homeowner point of view given as resulting annual net savings and cumulative annual incremental assets. The incremental assets obtained from the implementation of the energy-efficiency measure is a metric of the added net wealth of the homeowner if the house is sold at the estimated market value computed based on the future energy cost savings. To simplify and promote the use of the methodology, it has been implemented in an Excel tool free of charge.

REFERENCES

[1] The U.S. Department of Energy, "Annual Energy review 2010," Energy Information Administration. http://205.254.135.7/totalenergy/data/annual/pdf/aer.pdf

[2] V. Martinaitis, E. Kazakevicius and A. Vitkauskas, "A Two-Factor Method for Appraising Building Renovation and Energy Efficiency Improvement Projects," *Energy Policy*, Vol. 35, No. 1, 2007, pp. 192-201.

[3] S. B. Sadineni, T. M. France and R. F. Boehm, "Economic Feasibility of Energy Efficiency Measures in Residential Buildings," *Renewable Energy*, Vol. 36, No. 11, 2011, pp. 2925-2931.

[4] M. Gorgolewski, "Optimizing Renovation Strategies for Energy Conservation in Housing," *Building and Environment*, Vol. 30, No. 4, 1995, pp. 583-589.

[5] J. L. Ouyang, J. Ge and K. Hokao, "Economic Analysis of Energy-Saving Renovation Measures for Urban Existing Residential Buildings in China Based on Thermal Simulation and Site Investigation," *Energy Policy*, Vol. 37, No. 1, 2009, pp. 140-149.

[6] A. Lekov, V. Franco, S. Meyers, L. Thompson and V. Letschert, "Energy Efficiency Design Options for Residential Water Heaters: Economic Impacts on Consumers," *ASHRAE Transactions*, Las Vegas, January 2011, pp. 103-110.

[7] J. L. Ouyang, M. J. Lu, B. Li, C. Y. Wang and K. Hokao, "Economic Analysis of Upgrading Aging Residential Buildings in China Based on Dynamic Energy Consumption and Energy Price in a Market Economy," *Energy Policy*, Vol. 39, No. 9, 2011, pp. 4902-4910.

[8] The University of Texas at Tyler, "The Texas Allergy, Indoor Environment and Energy (TxAIRE)".

www.2.uttyler.edu/txaire/

[9] T. Taylor, N. Fernandez and R. Lucas, "Methodology for Evaluating Cost-Effectiveness of Residential Energy Code Changes," Pacific Northwest National Laboratory, 2012.

[10] National Renewable Energy Laboratory (NREL), "Building Energy Optimization Software (BEopt)," 2012. http://beopt.nrel.gov

[11] 2011 ASHRAE Handbook, HVAC Applications.

[12] A. W. Czanderna and G. J. Jorgensen, "Accelerated Life testing and Service Lifetime Prediction for PV Technologies in the Twenty-First Century," National Renewable Energy Laboratory, 1999. www.nrel.gov/docs/fy99osti/26710.pdf

[13] S. K. Fuller and S. R. Petersen, "Life-Cycle Costing Manual for the Federal Energy Management Program," NIST Handbook 135, 1995 Edition. www1.eere.energy.gov/femp/program/lifecycle.html

[14] B. Polly, M. Gestwick, M. Bianchi, R. Anderson, S. Horowitz, C. Christensen and R. Judkoff, "A Method for Determining Optimal Residential Energy Efficiency Retrofit Packages," U.S. Department of Energy, Energy Efficiency and Renewable Energy, Building Technology Program, 2011. http://www1.eere.energy.gov/library/default.aspx?page=2&spid=2

[15] J. Ritchter, "Financial Analysis of Residential PV and Solar Water Heating Systems," 2009. http://www.michigan.gov/documents/dleg/Thesisforweb_283277_7.pdf

[16] L. D. Roberts, "The Appropriate Discount Rate for Residential Real Estate Analysis," *Ezinearticles*, 2012.

[17] The New York State Department of Taxation and Finance, "How to Estimate the Market Value of Your Home." 2012. http://www.tax.ny.gov/pubs_and_bulls/orpts/mv_estimates.htm

[18] J. Vaughan, "What Is Market Value?" Creative Real Estate Online, 2012. http://www.creonline.com/what-is-market-value.html

[19] D. A. Braun. "An Introduction to Market Components and Interactions," *The Appraisal Journal*, Vol. LXXXI, No. 1, 2013, pp. 63-73.

Nomenclature

General:
EEM: Energy Efficiency Measure
F : Future cash flow
j : Index for year of economic analysis

Cost parameters for EEM's ($):
A : Mortgage payment (annual)
ANA : Adjusted net assets
ANS : Adjusted annual net savings
C : Initial cost
DP : Down payment
M : Initial mortgage amount
MC : Maintenance cost
MI : Mortgage interest
MV : Market value
MTS : Mortgage tax savings
NA : Nominal net assets
NS : Nominal annual net savings
P : Mortgage principal
PV : Present value
PT : Property tax
S : Energy cost savings
SA : Savings account balance

Annual economic rates (non-dimensional):
i_d : Nominal discount rate
i_{dr} : Real discount rate
i_e : Escalation rate
i_{er} : Real escalation rate
i_f : Nominal fuel price escalation rate
i_{fr} : Real fuel price escalation rate
i_i : General inflation rate
i_m : Nominal mortgage rate
i_{me} : Effective mortgage rate
i_s : Savings interest rate

Other factors (non-dimensional):
r_d : Efficiency derating factor
r_{dp} : Down payment rate
r_{pt} : Property tax rate
r_t : Marginal income tax rate

Time parameters (years):
n_m : Mortgage term
n_l : Service life of the EEM
n_s : Time limit for future energy savings

22

Indicators of Energy Efficiency in Buildings. Comparison with Standards in Force in Argentina

María Belén Salvetti[*]**, Jorge Czajkowski, Analía Fernanda Gómez**
Laboratorio de Arquitectura y Hábitat Sustentable (LAyHS),
Facultad de Arquitectura y Urbanismo, Universidad Nacional de La Plata,
La Plata, Argentina

ABSTRACT

In this work we make a comparative study of the energy behaviour in different building types. We analyze three cases of office buildings and three residential buildings, and compare them with a previous sample. We seek to find correlations or differences in behavior in terms of potential energy losses and gains, and UL values compared with Argentinian Standards to verify the degree of efficiency. For energy analysis we used a software which allows the analysis of thermal and energy building performance at steady state on a monthly basis. This software is called EnergoCAD and it also determines formal indicators based on IRAM standards. We conclude that the indicators used are clear to energetically "grade" buildings and to facilitate comparisons. In turn, smaller buildings are relatively less energy efficient than larger ones. At the same time it is noteworthy that the energy inefficiency has been growing rapidly over the years. Finally it is noted that none of the cases analyzed meets the National Standards.

Keywords: Energy Efficiency; Buildings; Standards; Indicators

1. Introduction

The following work was developed in LAyHS-FAU-UNLP. It is framed within one of the main research lines developed in the laboratory, aimed at energy efficiency in building for urban areas. It is part of the objectives of PICT 06#956 "Energy efficiency in building for metropolitan areas" and of the project accredited by UNLP "Energy Efficiency and Sustainability for the materialization of Buildings in the Context of Climate Change Adaptation."

The importance of this topic is related to two current issues: scarcity of resources and global warming. As far as is known, these two problems are responsible for the environmental deterioration that the world has been suffering, in which architecture and urban construction have a significant degree of influence [1].

For a long time we have relied on the development of appropriate technologies for the management of large-scale natural resources so that it is possible to meet the needs of the population. However, nowadays we can see this is not true because resources have been exhausted and the risk that this entails for the lives of millions of people and for the environment is growing [2].

The rapidly growing world energy use has already raised concerns over supply difficulties, exhaustion of energy resources and heavy environmental impacts (ozone layer depletion, global warming, climate change, etc.). The International Energy Agency has gathered frightening data on energy consumption trends. During the last two decades (1984-2004), primary energy has grown by 49% and CO2 emissions by 43%, with an average annual increase of 2% and 1.8% respectively. Current predictions show that the growing trend will continue [3].

The construction industry is one of the most important consumers of raw materials and non-renewable resources, and represents an important source of contamination during the different phases in the life cycle of a building. This implies a significant environmental impact not only during the process of extraction and processing of raw materials, but also during the construction and actual use of buildings, and also later when the building is demolished and recycled [4].

Fossil fuels are the main source of energy used in

buildings. In Argentina, for instance, 96% of power generation is through combined cycle while natural gas is intensively used for heating [5]. The scarcity and the potential risk posed by carbon emissions, generated by their use, make it necessary to discuss such intensive employment.

Building energy efficiency has come to the forefront of political debates due to high energy prices and climate change concerns. Improving energy efficiency in new commercial buildings is one of the easiest and lowest cost options to decrease a building's energy use, owner operating costs, and carbon footprint. Conventional energy efficiency technologies such as thermal insulation, low-emissivity windows, window overhangs, and day lighting controls can be used to decrease energy use in new commercial buildings by 20% - 30% on average and up to over 40% for some building types and locations [6].

In Europe there are numerous examples of sustainable construction which are aware of the current problems of scarcity of resources. Every day several print media, specializing in architecture, show more examples of corporate buildings that seek to improve their image by appealing to sustainable design. It is appropriate to take advantage of this growing wave of interest in the environment which is gaining ground in the field of construction, to develop buildings that are not only efficient in energy consumption, but also show respect for the environment.

The last seventy years of urban architecture history in Argentina show the emergence and development of buildings that grew on the constraints of urban sites. Building codes supported growth in height, increasing the profitability of soil [7]. However, during the period from 1900 to 1990, the thermal quality of buildings, especially the residential ones, went down despite growing technology offering. On the other hand, although the

standards of buildings" thermal quality that have existed since the late "70 s, private production has never ceased to lower quality standards to the point that in 1986 a revision of the Argentine standards was approved further lowering quality requirements" [8].

In recent years there has been progress in building energy efficiency indicators for tower blocks. In Argentina there already exist standards on energy saving in heating and cooling of residential buildings. They are IRAM 11659-2 [9] and IRAM 11604 [10]. However, there is the need for an antecedent for other uses such as office buildings, in public or private. There has not been a consensus in the country as to which is an appropriate efficiency indicator to hold energy demand and move towards "low energy" or "zero energy" building proposals.

2. Objective

The aim of this paper is to make a comparative analysis of the energy performance in different cases, both residential buildings and office ones, taking into account the many variables needed for its realization.

3. Materials and Methods

To carry out the work we took a sample of six buildings, three homes and three offices, with different structural characteristics. The choice of examples was randomly selected from various print media specializing in architecture. The intention was that the various examples chosen presented diverse forms of window and facade design, in terms of glazed surfaces and opaque walls, in solar protection systems, among others.

At the same time a previous work [11] was taken as part of the sample for a comparative analysis (see **Table 1**).

Table 1. Summary table previous sample.

Building	Envelope area (m^2)	Building area (m^2)	Volume (m^3)	Form Factor	CLF(W/°C)	$Q_{heating}$ (kWh/year)	$Q_{heating}$ (kWh/m^2/year)
Comega	19,045	13,357	40,071	0.48	4.00	909,212	68
Chacofi	16,975	10,736	32,208	0.52	4.23	972,647	91
Conurban	22,776	15,118	45,354	0.50	4.14	1,377,589	91
Esmeralda 116	9358	26,555	79,666	0.12	4.14	1,966,292	74
IBM	32,612	18,067	54,201	0.60	6.48	1,799,576	100
Libertad 565	3848	1323	3970	0.97	3.64	213,527	161
Madero	24,153	25,142	75,426	0.32	4.77	1,815,493	72
Malecom	8761	5045	15,136	0.58	4.16	532,326	106
Corrientes 1427	7451	4219	12,656	0.59	4.48	403,810	96
Libertad 731	7582	56,191	16,858	0.45	2.52	544,560	97
La Plata	14,479	12,681	38,043	0.38	4.20	1,018,820	80

Once the sample was taken, the characteristics of the envelope were determined from the graphic data and technical reports of the buildings. At the same time the formal design analysis was deepened in the different examples and their relation to the thermal quality of buildings, and the rational use of energy among other aspects.

For energy analysis we used the software *EnergoCAD* [12]. This software allows the analysis of thermal and energy building performance at steady state on a monthly basis. It also determines formal indicators based on IRAM standards. Then the results are exported to Excel for statistical analysis.

We proceeded with the completion of synthesis forms (**Figures 1-6**), which contain all the basic information of the case studies to its use with different graphics.

4. Results

As a result of the information processing we obtained the following results, which can be grouped as follows:

4.1. Office Building

We took three examples of office buildings (**Figures 1-3**)

11 de Septiembre Building	
Project:	Busnelli
Year:	2007

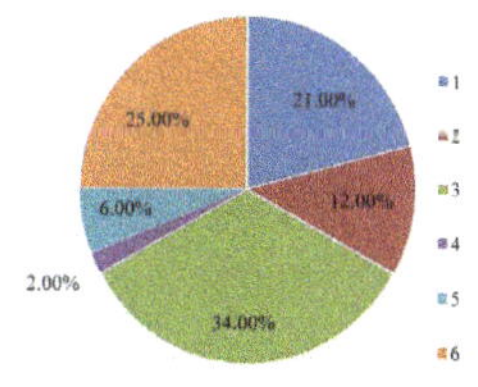

1. Walls 2. Roofs 3. Windows 4. Doors 5. Floors 6. Air Ren.

Dimensional Aspects:	
Building Area (m^2):	2014
Volume (m^3):	5138
Compactness (dimless):	0.71
Form Factor (dimless):	0.58
Thermal Aspects:	
Coef. UA (W/°C):	14,100
CFL (W/°C):	2.74
CFL adm.(W/°C):	1.16
Energy Aspects:	
Q (KWh/year)	417,496
Qs(KWh/m^2/year)	198

Figure 1. Basic description of 11 de Septiembre building (dimensional aspects, thermal aspects and energy aspects).

Fox LA Channel Building	
Project:	Alberto Varas
Year:	2005

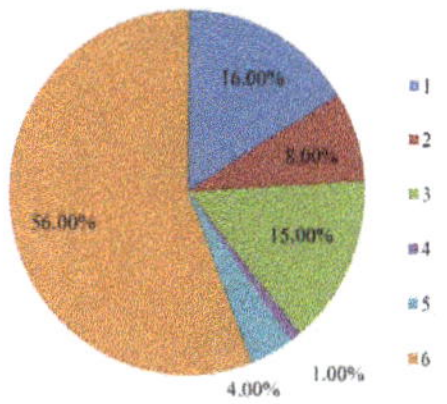

1. Walls 2. Roofs 3. Windows 4. Doors 5. Floors 6. Air Ren.

Dimensional Aspects:	
Building Area (m^2):	2690
Volume (m^3):	7262
Compactness (dimless):	2.19
Form Factor (dimless):	0.17
Thermal Aspects:	
Coef. UA (W/°C):	9150
CFL (W/°C):	1.26
CFL adm.(W/°C):	1.13
Energy Aspects:	
Q (KWh/year)	270,926
Qs(KWh/m^2/year)	101

Figure 2. Basic description of Fox LA Channel building (dimensional aspects, thermal aspects and energy aspects).

with different structural characteristics. They have different fenestrations, facade treatment, opaque and glazed surfaces, solar protection, etc.

4.2. Apartment Building

In this instance three cases of residential buildings were taken as examples (**Figures 4-6**), which also, as in previous cases, have different structural characteristics, have different fenestrations, opaque and glazed surfaces, sun protection, etc.

Figure 7 shows the relationship between the volume to be air conditioned and the heat load of the buildings analyzed in this study. There is a big difference in dimensions between the ANSES building (**Figure 3**) and the rest of the buildings discussed, as well as the heat load required for air conditioning it.

For its part, **Figure 8** shows the relationship between the volume to be heated and the heat load of the buildings analyzed, and it compares them with the buildings of a previous work. The trend lines for each of the works are similar.

$$Q_{heating} = 39.749 \times volume + 139849 \, (KWh/year) \quad (1)$$

ANSES Building	
Project:	M SG S S V
Year:	1974

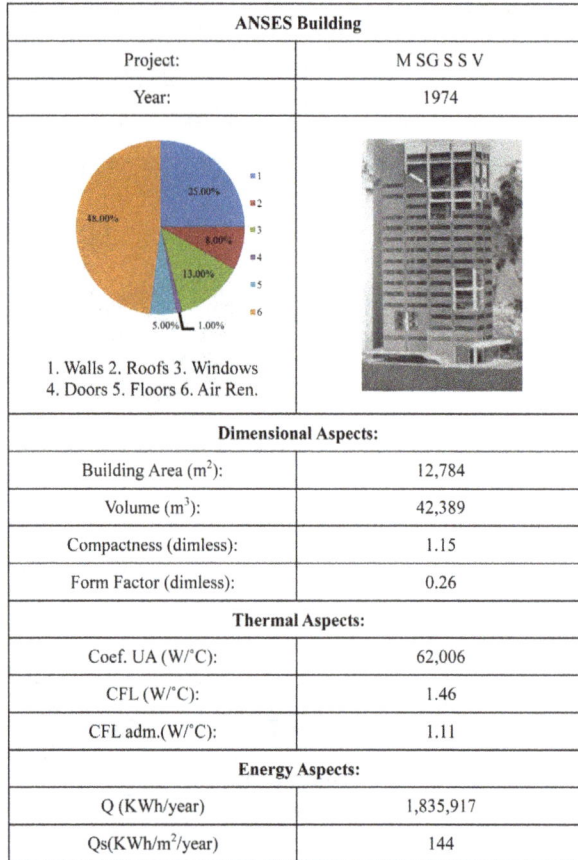

1. Walls 2. Roofs 3. Windows
4. Doors 5. Floors 6. Air Ren.

Dimensional Aspects:	
Building Area (m^2):	12,784
Volume (m^3):	42,389
Compactness (dimless):	1.15
Form Factor (dimless):	0.26
Thermal Aspects:	
Coef. UA (W/°C):	62,006
CFL (W/°C):	1.46
CFL adm.(W/°C):	1.11
Energy Aspects:	
Q (KWh/year)	1,835,917
Qs(KWh/m^2/year)	144

Figure 3. Basic description of ANSES building (dimensional aspects, thermal aspects and energy aspects).

$$Q'_{heating} = 23.788 \times volume + 155940 \left(KWh/year \right) \quad (2)$$

The steeper slope of the term (1), which corresponds to cases of recent construction, shows a growth in energy demand for heating, which adds to the energy inefficiency in recent years.

Among the cases analyzed in this work, two are prominent: the *LAChannel Fox* building (**Figure 3**) and the *ANSES* building (**Figure 4**). The first one is below the trend line (**Figure 8**), while the second one has values above the rest. Additionally, the latter can also be compared to the *IBM* building (**Table 1**). The volume to be heated in the *IBM* building is higher than in the *ANSES* one, however the heat loads are similar.

Figure 9 shows the heat load per square meter for the different cases analyzed. The sector "A" distinguishes those buildings to be climate-controlled with a volume less than three thousand cubic meters. These provide a wide variation in the heat load per square meter. Our hypothesis is that smaller buildings involve heating systems which do not impact significantly on the initial cost and show less concern for energy conservation. For their part, the buildings with a greater volume than three thousand cubic meters show a correlation with the nega-

Glamis Building	
Project:	Mardones-Viviani
Year:	2003

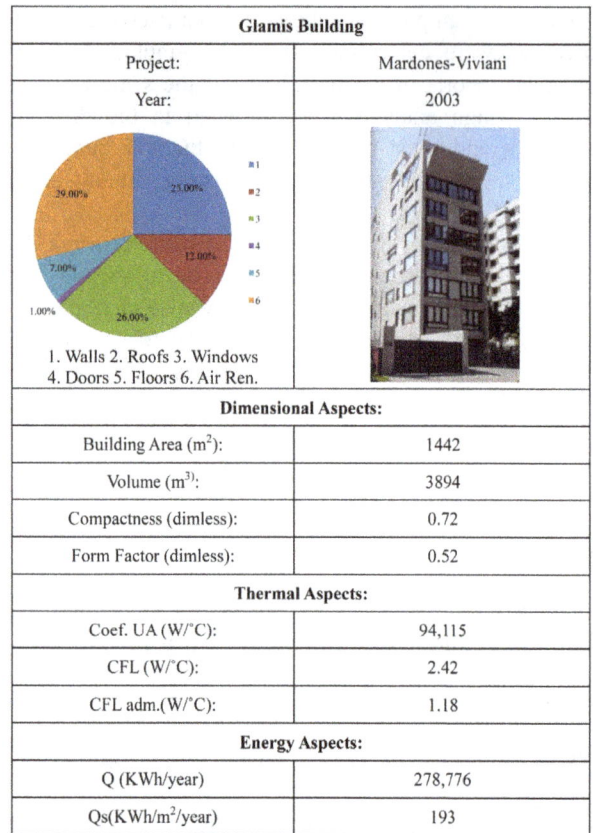

1. Walls 2. Roofs 3. Windows
4. Doors 5. Floors 6. Air Ren.

Dimensional Aspects:	
Building Area (m^2):	1442
Volume (m^3):	3894
Compactness (dimless):	0.72
Form Factor (dimless):	0.52
Thermal Aspects:	
Coef. UA (W/°C):	94,115
CFL (W/°C):	2.42
CFL adm.(W/°C):	1.18
Energy Aspects:	
Q (KWh/year)	278,776
Qs(KWh/m^2/year)	193

Figure 4. Basic description of Glamis building (dimensional aspects, thermal aspects and energy aspects).

tive slope, where the heat load per square metre increases as the size of the building decreases.

Figure 10 in turn relates the volume of different buildings with their Global Heat Loss Coefficient *(UL)* [W/m^3°C]. It can be seen as in all cases that the *UL* of the building exceeds the allowable *UL*, consequently there are no cases of present or previous building samples which comply with unbinding Argentina standard IRAM 11604. The correlation between the heated volume and the *UL* is low in the current sample (R2 = 0.352) and in the previous sample it is even lower (R2 = 0.195). This shows the lack of regulation of buildings' energy quality in Building Codes in Argentina. Moreover, in the current sample the relationship glazed/opaque is lower than in office towers of the previous sample, and this leads the UL to be significantly lower and closer to the allowable values of the IRAM standards.

4.3. Comparison among Deployed Buildings in Intermediate Cities in Temperate Climates

We performed an analysis which aims to show the energy performance of different buildings in some of the most important cities of Argentina (**Table 2**), with different weather characteristics, but all within the temper-

Table 2. Summary table of the cities analyzed.

City, Province	Latitude (°)	Size (population[*])	Heating Degree Days (°C[**])
Paraná, Entre Rios	−39.80	237,000	591
Córdoba, Córdoba	−31.40	1,316,000	608
La Plata, Buenos Aires	−35.00	563,000	992
Bahía Blanca, Buenos Aires	−38.70	318,000	1369
Mar del Plata; Buenos Aires	−38.10	542,000	1653

[*]INDEC 2001, [**]IRAM 11603 (18°C).

Gernika Building	
Project:	Miguel A. Roca
Year:	2004

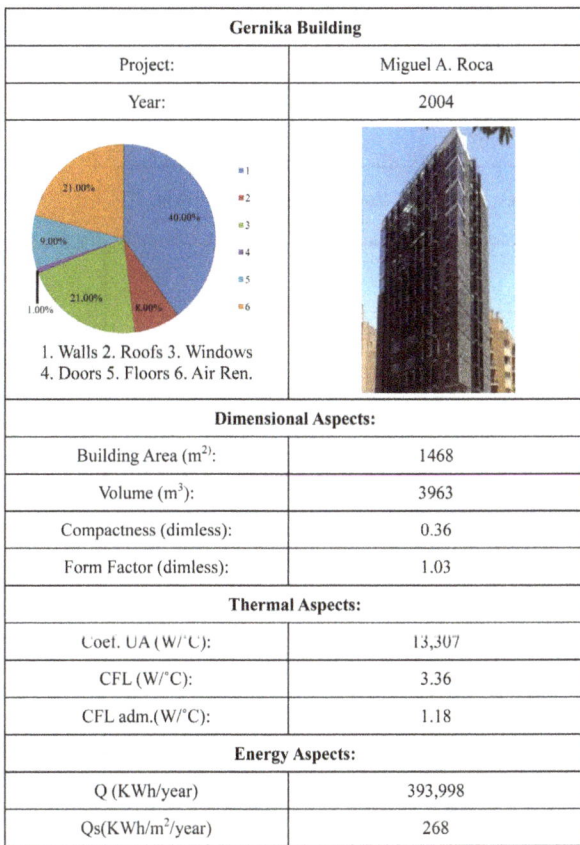

1. Walls 2. Roofs 3. Windows
4. Doors 5. Floors 6. Air Ren.

Dimensional Aspects:	
Building Area (m²):	1468
Volume (m³):	3963
Compactness (dimless):	0.36
Form Factor (dimless):	1.03
Thermal Aspects:	
Coef. UA (W/°C):	13,307
CFL (W/°C):	3.36
CFL adm.(W/°C):	1.18
Energy Aspects:	
Q (KWh/year)	393,998
Qs(KWh/m²/year)	268

Figure 5. Basic description of Gernika building (dimensional aspects, thermal aspects and energy aspects).

Terrazas Building	
Project:	Queixallos-Trull
Year:	2007

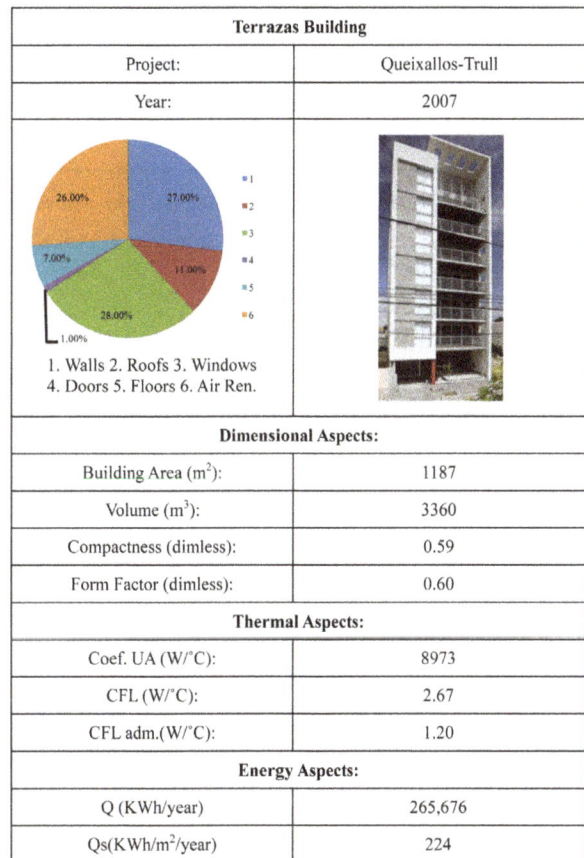

1. Walls 2. Roofs 3. Windows
4. Doors 5. Floors 6. Air Ren.

Dimensional Aspects:	
Building Area (m²):	1187
Volume (m³):	3360
Compactness (dimless):	0.59
Form Factor (dimless):	0.60
Thermal Aspects:	
Coef. UA (W/°C):	8973
CFL (W/°C):	2.67
CFL adm.(W/°C):	1.20
Energy Aspects:	
Q (KWh/year)	265,676
Qs(KWh/m²/year)	224

Figure 6. Basic description of Terrazas building (dimensional aspects, thermal aspects and energy aspects).

ate/mesothermal climate (Group C), [13].

Figure 11 shows the energy performance of the *Gernika* building (**Figure 6**), and its annual gains and losses according to different cities of Argentina.

It can be seen that in the case of Córdoba city (31°21'S, 64°05'W) the building shows the possibility of making gains of approximately 2,000,000 kWh/year, far higher than what could be obtained in other locations.

In most cases, the losses do not reach 500,000 kWh/year, marking an important difference between both values, while in other cities like La Plata (34°55'S, 57°57'W), Mar del Plata (38°00'S, 57°33'W) and Bahía Blanca

(38°44'S, 62°16'W) the gains and losses do not differ greatly. However, this is not the case with Paraná (31°44'S, 60°32'W); for this location, the percentage of annual heat loss is lower than in the other cities examined, and the potential gains are not as important as in the case of Córdoba (31°21'S, 64°05'W).

For its part, **Figure 12** shows the energy performance of the ANSES building (**Figure 4**). When located in La Plata city (34°55'S, 57°57'W) the building heat losses are approximately 1,800,000 kWh/year while heat gains fall short of 1,000,000 kWh/year. Contrary to what happens in Córdoba (31°21'S, 64°05'W) and Paraná (31°44'S,

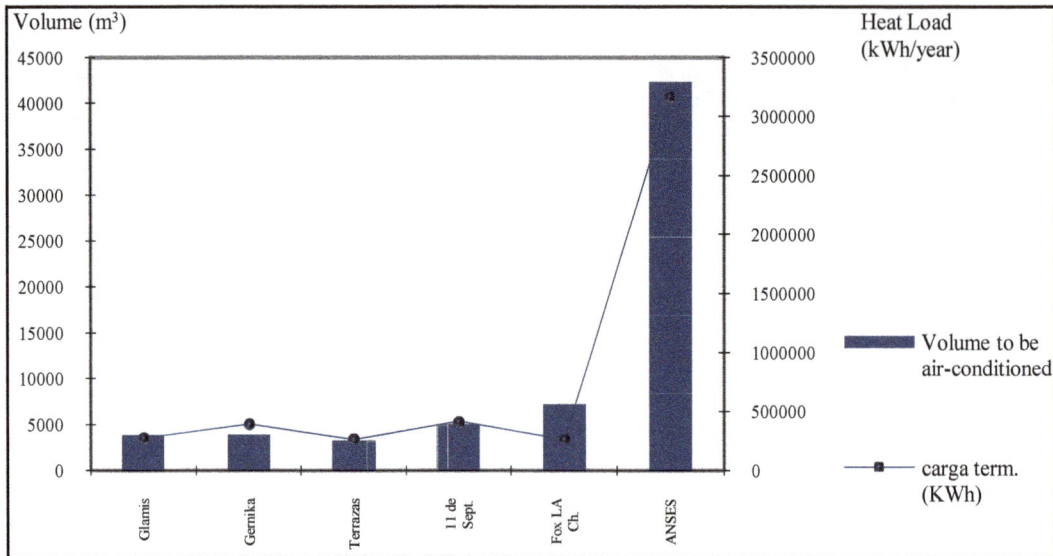

Figure 7. Relationship between the volume to be heated up/cooled and the heat load. Synthesis.

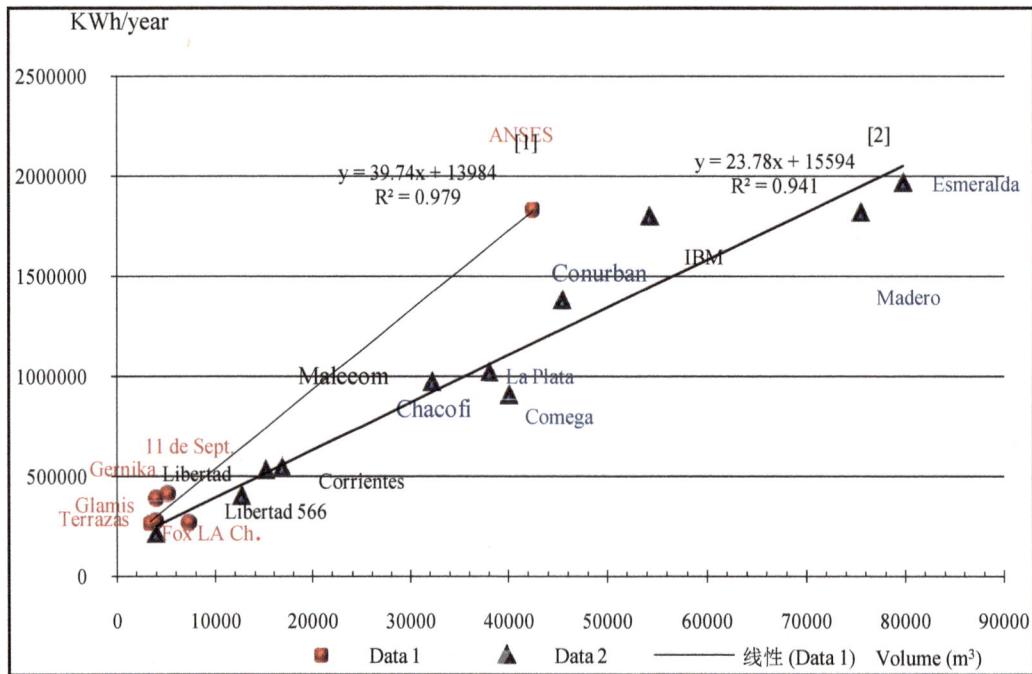

Figure 8. Relationship between the heated volume and the annual heat load. Comparison with previous work.

60°32'W) where the building has a balance of heat losses and gains, for the cities of Mar del Plata (38°00'S, 57°33' W) and Bahía Blanca (38°44' S, 62°16'W) the number of heat losses is far greater than the potential gains.

5. Conclusions

In this work we made a comparative analysis of different buildings and their energy performance. We could observe the relationship between the heat loads required to heat up/cool a building and its volume. To conclude with a hypothesis we can state that the smaller buildings show less concern for energy savings; these buildings involve heating systems that do not impact significantly on the initial cost. In turn, in buildings with a volume greater than three thousand cubic meters we observed that as the size of the building increases, the heat load decreases.

The analysis of individual cases and the comparison with the previous sample showed the growth in energy demand for heating in recent years, which implies an increase in energy inefficiency.

At the same time, we studied the energy behavior of a building for different cities with diverse weather charac-

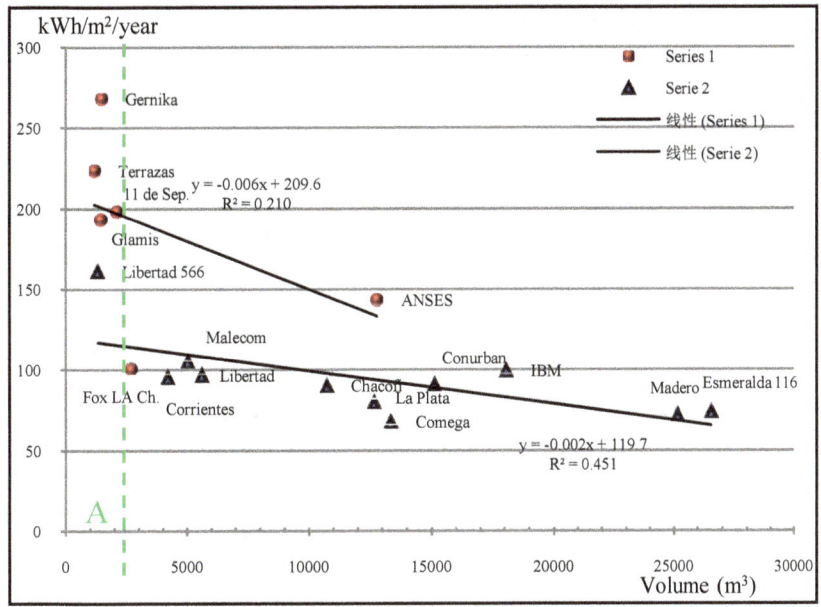

Figure 9. Heat load per square meter of the different analyzed buildings. Comparison with previous work.

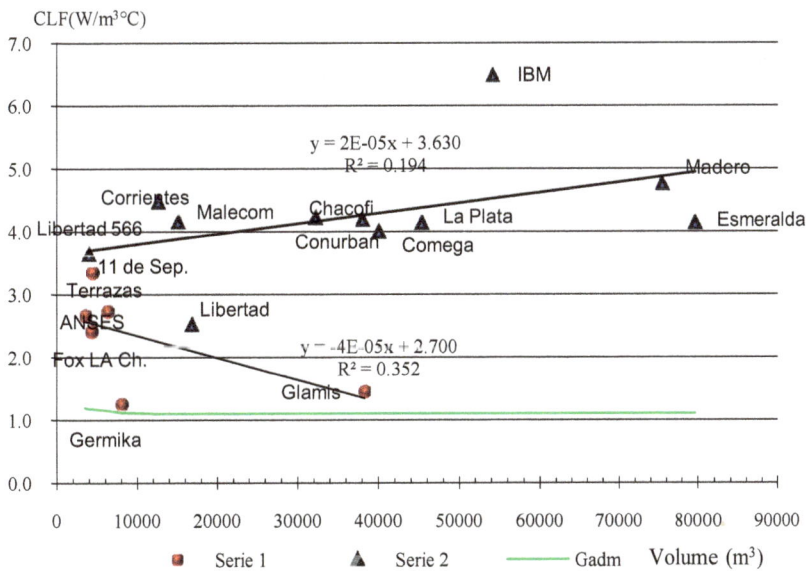

Figure 10. Relationship between the volume to be air-conditioned and U value. Comparison with previous work.

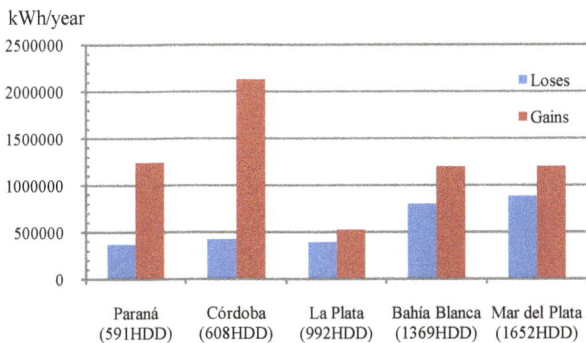

Figure 11. Heat losses and heat gains of Gernika building (Figure 6) in different cities of Argentina.

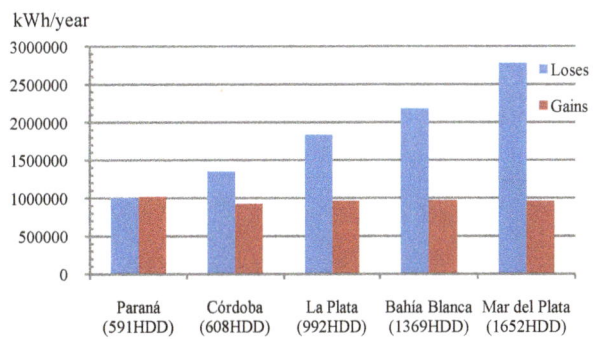

Figure 12. Heat losses and heat gains of ANSES building (Figure 4) in different cities of Argentina.

teristics, but all within the temperate/mesothermal climate (Group C) [13]. It was interesting to observe the different results on heat losses and gain relationship which was obtained from the alleged location of the same instance in different cities of Argentina.

Additionally, the difference between buildings' *UL* and admissible *UL*, determined by IRAM Standard 11,604, leads us to think about the poor quality of buildings as regards energy efficiency. It should be noted that in all cases the *UL* value is well above the admissible values by the rules.

With respect to indicators of energy efficiency in buildings, *UL* coefficient is appropriate to characterize the thermal quality of buildings and enable comparisons, while the annual heat load per area unit can display the behavior they will have on the site they are deployed. IRAM standards provide admissible heat load values on cooling but there is still to incorporate this indicator for heating.

6. Ackonwledgements

This work was supported by Agencia Nacional de Promoción Científica y Tecnológica (ANPCyT), Consejo Nacional de Investigaciones Científicas y Técnicas (CONICET) and Universidad Nacional de La Plata (UNLP), Argentina. It was developed in LAyHS-FAU-UNLP. It is framed within one of the main research lines developed in the laboratory, aimed at energy efficiency in building for urban areas. It is part of the objectives of PICT 06#956 "Energy efficiency in building for metropolitan areas" and of the project accredited by UNLP "Energy Efficiency and Sustainability for the materialization of Buildings in the Context of Climate Change Adaptation."

REFERENCES

[1]   IPCC, "Tercer Informe de Evaluación del Intergovernmental Panel on Climate Change: Cambio Climático," Geneva, 2001.

[2]   A. Vergara and J. De Las Rivas, Territorios Inteligentes. La Ciudad Sostenible," Fundación Metrópoli, Madrid, 2004.

[3]   L. Perez-Lomnard, *et al.*, "A Review on Buildings Energy Consumption Information," *Journal Energy and Buildings*, Vol. 40, No. 3, 2008, pp. 394-398.

[4]   B. Edwards, "Guía Básica de la Sostenibilidad," Gustavo Gili, 2008.

[5]   Secretaría de Energía de la Nación, "Informe de Auditoría de Gestión del Programa de Políticas Energéticas," Ministerio de Planificación Federal Inversión Pública y Servicios, Buenos Aires, 2008.

[6]   J. Kneifel, "Life-cycle Carbon and Cost Analysis of Energy Efficiency Measures in New Commercial Buildings," *Journal Energy and Buildings*, Vol. 42, No. 3, 2010, pp. 333-340.

[7]   J. F. Liernur, "Voz 'Torre'," Diccionario de Arquitectura en la Argentina, Editorial Clarín, Buenos Aires, 2004.

[8]   E. Rosenfeld and J. Czajkowski, "Catálogo de Tipologías de Viviendas Urbanas en el Área Metropolitana de Buenos Aires," Su Funcionamiento Energético y Bioclimático, IDEHAB-FAU-UNLP, La Plata, 1992.

[9]   IRAM 11659-2, "Aislamiento Térmico de Edificios. Verificación de sus Condiciones Higrotérmicas. Ahorro de Energía en Refrigeración. Parte 2: Viviendas," Instituto Argentino de Normalización y Certificación, Buenos Aires, 2007.

[10]   IRAM 11604, "Aislamiento Térmico de Edificios. Verificación de sus Condiciones Higrotérmicas. Ahorro de Energía en Calefacción. Coeficiente Volumétrico G de Pérdidas de Calor. Cálculo y valores Límites," Instituto Argentino de Normalización y Certificación, Buenos Aires, 1990 (Under Review; 2013).

[11]   C. Corredera and J. Czajkowski, "Evolución en el Diseño de Torres de Oficinas en la Argentina Desde un Enfoque Ambiental," ENCAC, Curitiba, 2003.

[12]   J. Czajkowski, *et al.*, "EnergoCAD. Sistema Informatizado Para el Diseño Bioclimático de Alternativas Edilicias," *15a Reunión de ASADES*, Catamarca, 1992.

[13]   M. C. Peel, *et al.*, "Updated World Map of the Köppen-Geiger Climate Classification," Hydrology and Earth System Sciences, Vol. 11, No. 5, 2007, pp. 1633-1644. http://www.hydrol-earth-syst-sci.net/11/1633/2007/hess-11-1633-2007.html

23

Wood Pellet Co-Firing for Electric Generation Source of Income for Forest Based Low Income Communities in Alabama

Ellene Kebede[1*], Gbenga Ojumu[2], Edinam Adozsii[1]

[1]Department of Agriculture and Environment Science,
Tuskegee University, Tuskegee, USA
[2]Department of Agriculture, Nutrition & Human Ecology,
Prairie View A&M University, Prairie View, USA

ABSTRACT

Alabama imports coal from other states to generate electricity. This paper assessed the direct and indirect economic impacts of wood pellet production to be co-fired with coal for power generation in Alabama. Four sizes of wood pellet plants and regional input-output models were used for the analysis. The results showed that the economic impact increases with the size of the plant. Wood pellet production will have a multiplier effect on the economy especially, forest-related services, retail stores, the health service industry, and tax revenue for the government. Domestic wood pellet production can reduce the use of imported coal, allow the use of local woody biomass, and create economic activities in Alabama's rural communities. Policies that support the production of wood pellet will serve to encourage the use of wood for power generation and support the rural economies.

Keywords: Wood Pellet; Electricity; Co-Firing; Coal; Input-Output; Forest Industry

1. Introduction

Renewable energy is widely recognized as a substitute for fossil fuels that can reduce the United States' dependence on foreign petroleum and enhance the domestic economy [1]. To date, emphasis has been on producing biofuels from field crops such as corn, sorghum, and oilseeds. Recently, however, advanced biofuels derived from nonfood feed stocks such as switch grass, agricultural residue, and woody biomass have received growing attention and are considered to be the future of the biofuels industry [2]. Regulations grouped under the Renewable Portfolio Standard (RPS) are also designed to increase the production of energy from renewable energy sources. The policy, a result of legislation passed in 1978 under the umbrella of the Public Utility Regulatory Policies Act, mandated increased energy production from renewable resources. The regulations introduced guidelines that a minimum percentage of electricity supply tobe produced from renewable energy sources. Producers with a certified renewable energy generator earn certificates for every unit of electricity they produce [3].

The renewable energy certificate is an incentive for electricity producers to use renewable feedstocks in their power generation operations. A good example is the European Union 2020 Energy policy, which is committed to reaching 20% share of renewable energy sources by 2020 [4]. There is a wider use of co-firing for power generation in Europe to substitute for coal. Imported wood pellets are mainly used for co-firing. Canada was previously the main source of supplier, but currently, the US-based wood pellet industry is gaining a major share. The newest plants in the southeast Georgia, Florida, and Alabama are designed for export markets. The largest wood pellet plant in the world is located in the state of Georgia, USA... Production is exported mainly to The Netherlands and the United Kingdom [5]. As of 2011 a new export-based wood pellet plant is also under construction in Alabama. Initially, the plant produces 250,000 metric tons of wood pellets per year, and a plant capable of producing 500,000 metric tons per year at full capacity is under construction in Aliceville, Pickens County in Alabama. This plant will start deliveries in 2012 [6].

*Corresponding author.

Literature shows that woody biomass can be used for biofuel as liquid transportation fuel and as non-liquid source to generate heat or electricity [7-9]. Wood pellets are used to generate residential heating and commercial power. Residential use in Europe is concentrated mainly in Sweden and Austria and to a lesser extent in Spain and Portugal [10]. The residential wood pellet fuel industry in North America was created in the early 1980s in response to the energy crisis. Currently, almost one million tons of wood pellets are sold each year to heat nearly 500,000 pellet stoves and fireplace in homes in the United States and Canada. Consumption is greatest in the Pacific Northwest and Northeastern states, where wood pellets are manufactured from sawmill and wood product residues and where heating energy requirements are significant [11].

Using wood pellets has the potential to reduce the use of fossil fuels and also attract new business opportunities for investors to consider processing in the rural timber-based communities. States in the Southern US could play a dominant role in the woody biomass industry for generating power. The South is dominated by private forest ownership, and 61% of the wood residues in the US come from the South [12]. Forest residues and excess mill residues, as well as urban residues, agricultural residues, and dedicated energy crops are assumed to be grown to support energy facilities [13]. Using woody biomass for bioenergy production will create a market for nontraditional sources of fuel such as logging residues, small diameter trees, and thinning residues, which can also be used as feedstocks [14,15]. An assessment by [16] of the potential impact of a new bioenergy sector examined using three sources of new energy demands for the South: export, cellulosic ethanol, and biomass electricity. They concluded that because of the established supply chain, relatively low cost and abundant supply of wood, and the consistency of wood's material characteristics, it is reasonable to expect that renewable energy markets would select wood as a preferred biomass feed stock.

Biomass for generating electricity is in its infancy, and economic analysis of biomass feed stock is limited. It is known, however, that co-firing with coal in producing electricity has proven to be technically feasible and cost effective [17]. Alabama Power is the major supplier of electricity in Alabama, and imported coal from other states is used to produce about 85% of the state's electricity [18]. The company has future plans to substitute renewable sources for fossils fuel, mainly coal. In co-firing, a percentage of biomass is introduced as fuel into an existing coal-fired boiler, often directly blended with the coal itself. Co-firing coal with switch-grass has been tried, and the electricity produced during the tests has been made available for sale to customers through a renewable pricing program [19,20]. Co-firing green pine chips with coal was also tested successfully, with one of the findings being that ampere, the current flow of the mill, was related to the percentage of dry wood in the fuel mix [21].

Forestry is an important sector in Alabama. Only nine of 68 counties in the State of Alabama are less than one-half forested with the lowest concentration in the North and the highest in the West Central and Southeast [22]. About 95% of forest land in Alabama is privately owned, and the area of timber land has increased by 5% in the 20-year span of 1997 to 2007 [12]. Private forests are composed of 78% nonindustrial private forest (NIPF) and 16% forest industry. The industrial forest has declined by 16%, whereas the NIPF increased by 12% between 1987 and 2002. This indicates a transition from industrial to nonindustrial timber and non-timber uses of forested land. The pulp and paper industry has declined; accordingly, so too has the utilization of forest and forest-processing residues [23]. With the decline of the pulp and paper industry, the utilization of forest and forest-processing residues could provide opportunities not only to reduce fossil fuel consumption but also to create and sustain employment and income thus contributing to local economies. Past studies have shown the potential effect of woody biomass for cellulosic ethanol production [24]. Cellulosic ethanol is not produced commercially in Alabama, but co-firing of coal with woody biomass has been tested. Given these various factors, the purpose of the present paper is to assess the direct and indirect socioeconomic impacts of small scale wood pellet production for domestic co-firing on forest landowners and rural communities.

2. Wood Pellet in the South

The Southeast and South Central US are timber-producing states and consist of more than half of the recoverable logging residues in the USA... Georgia, Alabama, and Mississippi are among the top three states for logging residues from growing stock. As such, the region, these states would be favorable places for commercial development of biomass fueled power generating plants and reducing carbon emissions from coal-generated electricity [25].

Generating electricity through co-firing biomass with coal reduces the out flow of pollutant gases compared with coal alone. An existing power plant facility can blend biomass (up to 5%) with coal or inject biomass separately (up to 20%) into the boiler [26]. The Southern Company has partnered with the USDA Forest Service, National Forests in Alabama, Forest Southern Research Station, Auburn University, Forest Products Development, and the CAWACO Resource Conservation and Development Council to test co-firing green wood chips in a boiler. Subsequently, green wood chips were co-fired successfully in blends with coal between 8% and 15% wood by weight. With 10% co-firing, boiler efficiency was about the same as coal alone, whereas a slight reduction was observed inefficiency with 15% wood [21].

Wild fire is a burning problem in many parts of the United States, and studies showed that thinning treatment will reduced wild fire and improve forest health. Alabama has a prescribed burning program to burn fallen branches and trees, low-quality wood, dried grasses, and the like that contribute to wild fire and affect forest health and productivity [23]. Forest thinning could generate feed stock for co-firing and wood pellets for residential and business space heating fuel [27,28]. It is estimated that combined bio power use by the industrial sector and electric utilities will meet about 4% ofenergy demand in 2010 and 5% in 2020 [13].

Alabama ranks third in the nation for forest and primary mill residues, which come mostly from the West and South regions of Alabama. The lumber market has lost ground since 1995 due to non-wood substitutes, and the paper mill industry, which is concentrated in southern Alabama, has also declined because recycled materials increased to 38% of the total fiber need by 1998 [26, 29].The availability of wood biomass makes Alabama attractive for producing biomass-based biofuels and bioenergy. In addition, biomass as a feedstock has a positive externality by lowering greenhouse gas emissions. If CO_2, as a social cost is incorporated in economic evaluations of generating electricity, logging residues will become a competitive fuel source [25,30].

The woody biofuels markets can create additional revenues to non-industrial private forest landowners and other economic agents that can stimulate employment which could contribute to rural development and benefit local communities [8]. These developments are also expected to contribute to the diversification of local economies and rural communities, in particular those that traditionally depend on timber production [31,32]. A national study using input-output and Policy System Analysis (POLSYS) model estimated the amount of ethanol that can be produced from cellulosic feedstock and the cumulative gain in new jobs, taxes, and reduced petroleum imports [33]. An input-output and CGE model based assessment of the economic impacts of wood biomass as bioenergy feedstock in Florida showed an increase in gross state product, employment, and a slight decrease in gasoline use [34].

A study by [35] estimated the benefits of using logging residues to generate electricity in East Texas and showed that their use reduces site preparation cost. The input-output model result showed that the logging residue use and electricity generation together would have a ripple effect on employment and output. Although biomass-based power generation has a relatively high initial investment, the benefits of using local feedstock in the long run will trickledown to the local economy compared to the use of coal for generating power [32]. Research has also shown that the high moisture of the green wood chips and coal mixtures resulted in low mill temperatures and caused a 5% reduction than its rated maximum power when co-firing [21]. Low moisture content and long storage time are the two advantages of wood pellet. Taken together, the low moisture content and consistent texture make wood pellet a better feedstock for power generation.

3. The Model

An input-output (I-O) model was employed to assess the economic impact of wood pellet for power generation in Alabama. I-O models trace commodity flows from producers to intermediates and finally consumers. Industries produce goods and services to meet final demand and purchase raw materials from producers. Producers, in turn, purchase goods and services from other industries. The total industry purchases of commodities, services, value-added, and imports are ultimately equal to the value of the commodities produced. I-O models also provide multipliers that estimate the relationship between the initial effect of a change in final demand and the total effects of that change [36,37]. An I-O model can be written in the matrix form as follows:

$$X = AX + Y \qquad (1)$$

$$X = (I - A)^{-1} Y \qquad (2)$$

where X is the vector of total output; A is the matrix of technical coefficients (a_{ij}), the amount of output of sector i consumed by sector j); Y is the vector of final demand. Equation (1) wasrearranged to provide Equation (2). The matrix $(I - A)$ is the Leontief matrix and $(I - A)^{-1}$ the Leontief inverse is a matrix of multipliers.

A multiplier for an industry is expressed as a ratio of direct, indirect, and induced effects, and is used to estimate the impacts on output throughout the economy. A Type I multiplier is direct plus indirect effects divided by direct effects. A Type II multiplier is direct plus indirect plus induced impact divided by direct impacts. A Type II multiplier tends to provide a higher estimate than Type I. Type II multipliers are used in the present study.

The multiplier is a coefficient that relates a change in output, employment, and value added as a consequence of change in final demand. The employment multiplier measures the total employment in all sectors in the economy attributable to the job created directly by the sector under consideration. The output multiplier of a sector measures the total production in all sectors of the economy that is necessary in meet the demand of the sector under consideration.

In the present study, a regional I-O model that included eight contiguous counties in the South and West regions of Alabama was developed to assess the economic impact on households(value-added employment compensation, proprietor income, other property income) and government (indirect business taxes) and the regional economy. The counties included in the model were: Pickens, Sumter, Greene, Hale, Marengo, Perry, Dallas, and Wilcox. These counties have the highest timberland in Alabama. Greene and Hale counties have 53,000 to 67,000 acres each under timber, and the other six counties have 67,000 to 105,000 acres of timberland each [22].

Table 1 shows the per capita personal income as a percent of state average and the unemployment rate of the counties included in the model. There is slight increase in share of per capita personal income between 2005 and 2010, but these counties have the lowest per capita personal income in the State of Alabama. They also experience the highest unemployment rate which ranges between 11% and 22% compared to state average of 9.5% in 2010.

The data for earnings by industry indicates that the government, at the state and federal levels, was the main source of income, accounted for 20% to 35% of the total earnings by industry.

This was followed by manufacturing (10% - 25%), health care and social assistance (15%), and retail trade (7% - 8%). Forestry and logging, which accounted for less than 1% was reported in four of the eight counties, and none of the counties reported agriculture and forestry support services in 2010. The paper industry was also important in the state during the 1970s and 1980s, but the counties included in the study, except Sumter County, never supported paper manufacturing [38].

Table 1. Per capita personal income as percent and unemployment rate.

County	Per capita personal income[1]		Unemployment rate[2]
	2005	2010	2010
Dallas	82	84	17.3
Greene	93	95	16.9
Hale	83	90	12.1
Marengo	91	95	12.4
Perry	76	74	16.5
Pickens	79	86	11.3
Sumter	68	73	14.2
Wilcox	66	75	21.7

[1][39]. [2][40].

The data for these eight counties were obtained from the 2009 IMPLAN Alabama economic data set [41]. Two IMPLAN sectors were selectedf or the analysis: forestry, forest products, and timber tract production (sector 15) and commercial logging (sector 16). It is also assumed that 25% and 75% of the feedstock originates from sectors 15 and 16, respectively [42].

4. Model Assumptions

The study assumes that the demand for wood pellet for co-firing is in place and pine chips are used as a raw material for producing wood pellets. The demand for wood chips was based on the following three assumptions: 1) raw material will be obtained within a 100-mile radius, which also covers the counties in the model; 2) pine chips have about 40% moisture content [43]. Based on the literature, co-firing is efficient with 15% wood [21], the pelleting process reduces the moisture content by about 25%, from 40% to 15%; and 3%) the plants will operate 16 hours per day, a 67% operational rate for 365 days.

$$C_d = T_h \times P_s \qquad (3)$$

$$R_c = C_d \times W_p \qquad (4)$$

where Cd is the annual wood chips demand; T_h is the tons of pine chips per hour; P_s is the plant size; F_c is the cost of raw material; and W_p the price of pine chips [44]. The price includes transportation costs within the 100-mile radius.

Plant size affects the efficiency and feasibility of a plant. The efficiency of producing pellets increases with size, and larger pellet producers are often more profitable than smaller producers [45]. Capital investment costs per ton decrease with an increase incapacity, and pellet mills are cost effective when they produce more than 10 tons per hour (t/h) of pellets [46,47]. This study shows the economic impact of four different plant sizes expressed in tons of wood pellet per year: 10 t/h or 50,000 tons per year; 20 t/h or 100,000 tons per year; and 40 t/h or 200,000 tons per year plants and the current export-based production with approximately 95 t/h 500,000 tons per year. The estimated annual wood chip cost for different plant sizes were imported to the regional input/output model.

5. Results and Discussions

The South and West regions of Alabama have a large forested area and are experiencing a higher level of unemployment accompanied with lowest per capita personal income in the State and can benefit from the establishment of woody biomass processing plants like wood pellets. As indicated by the input-output results, the 10 industries that will benefit from the wood pellet production are the main suppliers and related support services. The top three sectors that accounted for 70% of the employment and income are: commercial logging (sector 16); forestry, forest products and timber tract production (sector 15), and support activities for agriculture and forestry (sector 18). In addition, the food and beverage services sector will benefit from the increase in demand and income in the economy. The other sectors that gain from the wood pellet production are: private household operations; nursing and residential care facilities; retail stores for food and beverages; wholesale trade businesses; and health services.

The ripple effect is associated with the demand from these industries to supply services required by the wood pellet industry. This is captured in the Type II employment and output multipliers. **Table 2** compares the multipliers of sector 15 and 16 with the paper manufacturing sector (sector 105). The employment multiplier is what every job created in the sector will create in other sectors of the economy. Sector 15 had a larger employment multiplier (4.533) than sector 16 (1.39), generating more overall jobs for each job created in the sector. A job created in the forest/forest related sector will create 4.533 jobs in the economy, whereas the logging sector will create 1.39 jobs in the economy for each job created in the sector. The output multiplier for the sectors, therefore, is not significantly different.

The multipliers apply to any size plants, but the total effect will vary with the plant size.

The results of the economic impact of the four plant sizes analyzed are provided in **Tables 3-6**. Based on past studies, the increase in plant size will enhance cost effectiveness, and for this analysis an increase in the plant size increased the total impact on the regional economy. The increase in plant size from10 t/h to 40 t/h increased labor income, value added, and output by300%, and increased

Table 2. Type II employment and output multipliers.

Sector	Employment Multiplier	Output Multiplier
Commercial Logging	1.390	1.499
Forest Products and Timber Tract Production	4.533	1.655

Table 3. The results of the economic impact of 10 tons per hour wood pellet plant.

Type of Impact	Direct Effect	Indirect Effect	Induced Effect	Total Effect	Indirect + Induce/ total Effect
Employment	19	11.4	5.9	36.3	0.48
Labor Income (M $)	1.04	0.39	0.17	1.61	0.35
Value Added (M $)	1.41	0.51	0.35	2.27	0.38
Output (M $)	3.33	0.96	0.59	4.87	0.32

Table 4. The results of the economic impact of 20 tons per hour wood pellet plant.

Type of Impact	Direct Effect	Indirect Effect	Induced Effect	Total Effect	Indirect + Induce/total Effect
Employment	38	22.8	11.7	72.51	0.48
Labor Income (M $)	2.08	0.79	0.34	3.22	0.35
Value Added (M $)	2.82	1.02	0.71	4.55	0.38
Output (M $)	6.66	1.91	1.18	9.74	0.32

Table 5. The results of the economic impact of 40 tons per hour wood pellet plant.

Type of Impact	Direct Effect	Indirect Effect	Induced Effect	Total Effect	Indirect + Induce/total Effect
Employment	75.9	45.6	23.5	145	0.48
Labor Income (M $)	4.17	1.58	0.69	6.43	0.35
Value Added (M $)	5.64	2.04	1.41	9.09	0.38
Output (M $)	13.31	3.82	2.35	19.49	0.32

Table 6. The results of the economic impact of 95 tons per hour wood pellet plant.

Type of Impact	Direct Effect	Indirect Effect	Induced Effect	Total Effect	Indirect + Induce/total Effect
Employment	158	95	49	302	0.48
Labor Income (M $)	86.72	32.76	14.34	133.83	0.35
Value Added (M $)	117.37	42.34	29.40	189.12	0.38
Output (M $)	276.87	79.45	48.91	405.33	0.32

to 800% when the plant size increases to 95 t/h. Most of the employment was created in the commercial logging and forestry-related sectors. These sectors had an important indirect and induced impact on the economy especially in the 10 major sectors. The share of the indirect and induced to total effect showed that 48% of the employment, 35% of the labor income, 38% of the value added, and 32% of the total output resulted from the indirect and induced effects.

Distribution of value added showed that employment compensation (wages and salaries) accounted for 57%; other property type income (rental) accounted for 18%; proprietor income accounted for 16%; and indirect business taxes to the government accounted for 10% of the value added. The logging industry uses heavy machinery and equipment and the higher compensation could be associated the skilled manpower employed by the logging industry.

Notably, the region has the highest forest cover where forestry logging is less than 1% of income generated in the economy. Establishing a wood pellet plant could stimulate the forest industry and commercial logging, which could increase the income earned from forestry.

Furthermore, it could be an incentive to the establishment of the forest-related services sector that is not currently making a significant contribution to the regional economy.

6. Conclusion

Woody biomass is a major resource that could be used as a substitute for coal in generating electricity in Alabama. Wood pellet is not used widely for power generation in the United States, especially in the South. However, the State of Georgia has one of the largest wood pellet plants in the world, and Alabama has one wood pellet plant that produces products for export. The present study estimated the socioeconomic impacts of small-scale wood pellet plants for co-firing in power generating plants in the south and west regions of Alabama. Alabama Power Company, the major electricity supplier in the state, has a coal-based plant in Greene County with a generating capacity of 1,220,000 kW [20]. Wood pellet plants in the counties studied will be within a good proximity to the power generation plant. The company has successfully tested co-firing coal with green wood, and the results showed that wood can be co-fired up to 15%, but moisture content affects the ampere, the current production. Wood pellet has the added advantage of low moisture and a consistent texture to mitigate the loss of current output. The present paper assumed demand levels for wood pellet and assessed the economic impact of wood pellet for co-firing for generating power. The study tested four sizes of wood pellet plants and showed that the impact increases with the increase in plant size. Most of the employment, value added, and output will be generated in the commercial logging sector and forestry and forest-production tracts sector. These sectors will create demand for skilled manpower related to logging, equipment handlers, and transportation as well as provide income to the owners of forested land. The high employment multiplier showed that using wood pellets for co-firing will generate additional employment in the service sectors. The increase in demand for wood will encourage the use of forest residues and other biomass that have not been used to date that could generate income to property owners. The economic impact of the current large size plant for export is larger than the small-scale plants, but because it is export-oriented, its impact on reducing coal import and carbon emission in the state is none. The present study has shown that small-scale wood pellet plants can play a triple role in the economy, enhance the eco-nomic activity of the region, reduce the use of imported coal, and reduce CO_2 emissions. The use of wood biomass might be expensive, but studies [25,30] have shown that if the social cost of CO_2 emissions is considered, woody biomass can be competitive for producing electricity. Given the current 10% co-firing [21], which is regarded as efficient, the use of wood pellet will reduce coal import and carbon emission and generate economic activity in the region. In conclusion, the use of woody biomass for generating power will have a long-term economic impact on the community and the region. These benefits to the region and the community could be the basis for government support for developing the wood pellet sector.

7. Acknowledgements

The paper was part of a study funded by USDA Office of the Chief Economist/Energy Policy and New Uses.

REFERENCES

[1] B. Antizar-Ladislao and J. L. Turrion-Gomez, "Second-Generation Biofuels and LocalBioenergy Systems," *Biofuels Bioproducts and Biorefining*, Vol. 2, No. 5, 2008, pp. 455-469.

[2] USDA, "A Regional Roadmap to Meeting the Biofuels Goals of the Renewable Fuel Standard by 2022," USDA Biofuels Strategic Production Report, 2012. http://www.usda.gov/documents/USDA_Biofuels_Report_6232010

[3] K. S. Cory and B. G. Swezey, "Renewable Portfolio Standards in the States: Balancing Goals and Implementation Strategies," National Renewable Energy Laboratory, Technical ReportNREL/TP-670-41409, 2007.

[4] European Commission, "Energy 2020, a Strategy for Competitive, Sustainable and Secure Energy," Publications Office of the European Union, European Union Luxembourg, 2011.

[5] Renewable Energy World, "World's Largest Wood Pellet Plant to Feed REW Europe Power Plants," 2012. http://www.renewableenergyworld.com

[6] Westervelt Company, "Aliceville Selected for Fuel Pellet Production Facility," 2012. http://westervelt.com/westervelt-newsroom/pressreleases

[7] R. C. Brown, "Biorenewable Resources: Engineering New Products from Agriculture," Blackwell Publishing, Iowa State Press, Ames, 2003.

[8] A. P. C. Faaij and J. Domac, "Emerging International Bio-Energy Markets and Opportunities for Socio-Economic Development," *Energy for Sustainable Development*, Vol. 1, No. X, 2006, pp. 7-19.

[9] B. Jackson, R. Schroeder and S. Ashton, "Pre-Processing and Drying Woody Biomass. Sustainable Forestry for Bioenergy and Bio-Based Products," Fact Sheet 2007, pp. 141-144.

[10] R. Jannasch, R. Samson, A. de Maio, T. Adams and C. H. Lem, "Changing the Energy Climate: Clean and Green Heat from Grass Biofuel Pellets," Energy Probe, 2001. www.energyprobe.org

[11] H. Spelter and D. Toth, "North America's Wood Pellet Sector, "United States Department of Agriculture, Forest Service," Forest Products Laboratory Research Paper FPL-RP-656, 2009, pp. 5-23.

[12] W. B. Smith, P. D. Miles, C. H. Perry and S. A. Pugh, "Forest Resource of the United States, 2007," General Technical Report-WO-78, United States Department of Agriculture Forest Service, 2012.

[13] R. D. Perlack, L. L. Wright, A. F. Turhollow, R. L. Graham, B. J. Stokes and D. C. Erbach, "Biomass as Feedstock For a Bioenergy and BioproductsIndustry: The Technical Feasibility of a Billion-Ton Annual Supply," US Department of Agriculture, DOE/GO-102005-2135, ORNL/TM, 2012.

[14] E. M. White, "Woody Biomass for Bioenergy and Biofuels in the United States," United States Department of Agriculture, Forest Service, Technical Report PNW-GTR-825, 2012.

[15] B. L. Polagye, K. T. Hodgsonb and P. C. Maltea, "An Economic Analysis of Bio-Energy Options Using Thinning from Overstocked Forests," Biomass and Energy Biomass, Vol. 31, No. 2-3, 2007, pp. 105-125.

[16] G. Comatas and J. Shumaker, Cross Reference Wear et al. 2007, 2009.

[17] S. Nienow, K. McNamara, A.Gillespie and A. Preckel, "A Model for the Economic Evaluation of Plantation Biomass Production for Co-Firing with Coal in Electricity Production," Agricultural and Resource Economics Review, Vol. 1, No. 28, 1999, pp. 106-118.

[18] Energy Information Administration, "Alabama, Overview and Analysis 2009," 2012. http://www.eia.gov/state/state-energy-profiles.cfm?sid=AL

[19] Southern Company, "Southern Company and Renewable Energy, Overview 2007," 2012. http://southerncompany.com/planetpower/pdfs/renewable_energy.pdf

[20] Alabama Power, "Power Generating Plants," 2012. http://www.alabamapower.com/about/plants.asp

[21] D. Boylan, K. Roberts, B. Zemo and T. Johnson, "Phase 2 Co-Firing Testing of Wood Chips at Alabama Power's Plant Gadsden 2008," 2012. www.cawaco.org/biomass/Phase2report.pdf

[22] University of Alabama, "Total Woodland Areas by County 2007," 2012. http://alabamamaps.ua.edu/

[23] B. W. Smith, P. D. M. Patrick, J. S. Vissage and S. A. Pugh, "Forest Resources of the United States, 2002," A Technical Document Supporting the USDA Forest Service an Update of the RPA Assessment, North Central Research Station, Forest Service—US Department of Agriculture, 2004.

[24] C. Bailey, J. F. Conner, J. F. Dyer and L. Teeter, "Assessing the Rural Development Potential of Lignocellu-

losic Biofuels in Alabama," Biomass and Bioenergy, Vol. 35, No. 4, 2011, pp. 1407-1417.

[25] J. Gan and C. T. Smith, "A Comparative Analysis of Woody Biomass and Coal for Electricity Generation under Various CO_2 Emission Reductions and Taxes," Biomass and Bioenergy, Vol. 30, No. 4, 2006, pp. 296-303.

[26] D. N. Wear, D. R. Carter and J. Prestemon, "The US South's Timber Sector in 2005. A Prospective Analysis of Recent Change," USDA Forest Service Southern Research Station, Asheville, 2007.

[27] K. E. Skog and R. J. Barbour, "Estimating Woody Biomass Supply from Thinning Treatments to Reduce Fire Hazard in the US West," Proceedings RMRS-P-41, USDA Forest Service, 2006, pp. 657-672.

[28] D. G. Neary and E. J. Z. Elaine, "Forest Bioenergy System to Reduce the Hazard of Wildfires: White Mountains, Arizona," Biomass and Bioenergy, Vol. 31, No. 9, 2007, pp. 638-645.

[29] A. Millbrandt, "A Geographic Perspective on the Current Biomass Resource Availability in the United States," National Renewable Energy Laboratory, Technical Report NREL/TP-560-39181, 2005, pp. 18-27.

[30] M. Gronowska, S. Joshi. and H. L. MacLean, "A Review of US and Canadian Biomass Supply Studies," BioResources, Vol. 4, No.1, 2009, pp. 341-369.

[31] J. Bliss and C. Bailey, "Pulp, Paper, and Poverty: Forest-Based Rural Developmentin Alabama, 1950-2000," In: R. Lee and D. Field, Eds., Communities and Forests: Where People Meet the Land, Oregon State University Press, Corvallis, 2005, pp. 138-158.

[32] J. Domac, J. Richard and S. Risovic, "Socio-Economic Drivers in Implementing Bioenergy Projects," Biomass and Bioenergy, Vol. 28, No. 2, 2005, pp. 97-106.

[33] D. G. De La Torre Ugarte, B. C. English and K. Jensen, "Sixty Billion Gallons by 2030: Economic and Agricultural Imparts of Ethanol and Biodiesel Expansion," American Journal of Agricultural Economics, Vol. 89, No. 5, 2007, pp. 1290-1295.

[34] A. W. Hodges, J. S. Thomas and M. Rahmani, "Economic Impacts of Expanded Woody Biomass Utilization on the Bioenergy and Forest Products Industries in Florida," Institute of Food and Agricultural Sciences Food, Gainesville, 2012. www.fred.ifas.ufl.edu/economic-impact-analysis/pdf/Woody-Biomass-Utilization.pdf

[35] J. B. Gan and T. C. Smith, "Co-Benefits of Utilizing Logging Residues for Bioenergy Production: The Case for East Texas, USA," Biomass and Bioenergy, Vol. 31, No. 9, 2007, pp. 623-630.

[36] R. E. Miller and P. D. Blair, "Input-Output Analysis Foundation and Extension," Prentice Hall, Inc., Upper Saddle River, 1985.

[37] J. D. G. Hewing, "Regional Input-Output Analysis," SAGE Publications, Beverly Hills, 1985.

[38] Bureau of Economic Analysis, "Personal Income and Earning by Industry," 2012.
http://www.bea.gov/regional/index.htm

[39] University of Alabama, "Income Poverty and Employment," 2012. http://cber.cba.ua.edu/

[40] Alabama Department of Industrial Relations, "Unemployment Statistics," 2012.
http://www2.dir.state.al.us/LAUS/default.aspx

[41] Minnesota IMPLAN Group Inc., "Alabama Economic Data," Hudson, 2009.

[42] University of Alabama, "Assessment of Wood-Based Syngas Potential for Use in Combined Cycle Power Plants in Alabama," Draft Prepared by University Center for Economic Development, University of Alabama, Tuscaloosa, 2012.

[43] M. D. Gibson, "Moisture Content and Specific Gravity of the Four Major Southern Pines under the Same Age and Site Conditions," *Wood and Fiber Science*, Vol. 18, No. 3, 1986, pp. 428-435.

[44] T. Mart-South, "A Quarterly Report of the Market Conditions for Timber Products of the US South," 2012.
http://www.timbermart-south.com/pdf/2Q2011news.pdf

[45] A. Wolf, A. Vidlund and E. Andersson, "Energy-Efficient Pellet Production in the ForestIndustry—A Study of Obstacles and Success Factors," *Biomass and Bioenergy*, Vol. 30, No. 1, 2006, pp. 38-45.

[46] S. Mani, S. Bi, X. Sokhansanj and A. Turhollow, "Economics of Producing Fuel Pellets from Biomass," *Applied Engineering in Agriculture*, Vol. 22, No. 3, 2006, pp. 421-426.

[47] P. Porter, J. Barry, R. Samson and M. Doudlah, "Growing Wisconsin Energy: A Native Grass Pellet Bio-Heat Roadmap for Wisconsin," Agrecol, 2012.
http://datcp.wi.gov/uploads/Business/pdf/22061Agrecol.pdf

Permissions

The contributors of this book come from diverse backgrounds, making this book a truly international effort. This book will bring forth new frontiers with its revolutionizing research information and detailed analysis of the nascent developments around the world.

We would like to thank all the contributing authors for lending their expertise to make the book truly unique. They have played a crucial role in the development of this book. Without their invaluable contributions this book wouldn't have been possible. They have made vital efforts to compile up to date information on the varied aspects of this subject to make this book a valuable addition to the collection of many professionals and students.

This book was conceptualized with the vision of imparting up-to-date information and advanced data in this field. To ensure the same, a matchless editorial board was set up. Every individual on the board went through rigorous rounds of assessment to prove their worth. After which they invested a large part of their time researching and compiling the most relevant data for our readers. Conferences and sessions were held from time to time between the editorial board and the contributing authors to present the data in the most comprehensible form. The editorial team has worked tirelessly to provide valuable and valid information to help people across the globe.

Every chapter published in this book has been scrutinized by our experts. Their significance has been extensively debated. The topics covered herein carry significant findings which will fuel the growth of the discipline. They may even be implemented as practical applications or may be referred to as a beginning point for another development. Chapters in this book were first published by Scientific Research Publishing Inc.; hereby published with permission under the Creative Commons Attribution License or equivalent.

The editorial board has been involved in producing this book since its inception. They have spent rigorous hours researching and exploring the diverse topics which have resulted in the successful publishing of this book. They have passed on their knowledge of decades through this book. To expedite this challenging task, the publisher supported the team at every step. A small team of assistant editors was also appointed to further simplify the editing procedure and attain best results for the readers.

Our editorial team has been hand-picked from every corner of the world. Their multi-ethnicity adds dynamic inputs to the discussions which result in innovative outcomes. These outcomes are then further discussed with the researchers and contributors who give their valuable feedback and opinion regarding the same. The feedback is then collaborated with the researches and they are edited in a comprehensive manner to aid the understanding of the subject.

Apart from the editorial board, the designing team has also invested a significant amount of their time in understanding the subject and creating the most relevant covers. They scrutinized every image to scout for the most suitable representation of the subject and create an appropriate cover for the book.

The publishing team has been involved in this book since its early stages. They were actively engaged in every process, be it collecting the data, connecting with the contributors or procuring relevant information. The team has been an ardent support to the editorial, designing and production team. Their endless efforts to recruit the best for this project, has resulted in the accomplishment of this book. They are a veteran in the field of academics and their pool of knowledge is as vast as their experience in printing. Their expertise and guidance has proved useful at every step. Their uncompromising quality standards have made this book an exceptional effort. Their encouragement from time to time has been an inspiration for everyone.

The publisher and the editorial board hope that this book will prove to be a valuable piece of knowledge for researchers, students, practitioners and scholars across the globe.

List of Contributors

Rajesh Kumar and Arun Agarwala
Instrument Design Development Centre, Indian Institute of Technology Delhi, New Delhi, India

Avijit Choudhury
Agm-Energy Services, Enfragy Solutions India Pvt. Ltd., New Delhi, India

Fabio Correa Leite, Decio Cicone Jr., Luiz Claudio Ribeiro Galvão and Miguel Edgar Morales Udaeta
Energy Group of Electric Energy and Automation Engineering Department, Polytechnic School, University of São Paulo, São Paulo, Brazil

Doug Hargreaves
Queensland University of Technology, Brisbane, Australia

Vishal Sardeshpande
A.T.E. Enterprises Private Limited, 2 Shreenivas Classic, Pune, India

Vijay Patil and Cherat Murali
Birla Accucast Limited, MIDC Waluj Industrial Area, Aurangabad, India

Xunmin Ou, Xiliang Zhang, Xu Zhang and Qian Zhang
Institute of Energy, Environment and Economy, Tsinghua University, Beijing, China
China Automotive Energy Research Center, Tsinghua University, Beijing, China

Xiaoyi He
Institute of Energy, Environment and Economy, Tsinghua University, Beijing, China

Nand Kishore Gupta
M.P. Housing & infrastructure Development Board, Bhopal, India

Anil Kumar Sharma and Anupama Sharma
Maulana Azad National Institute of Technology, Bhopal, India

Talakonukula Ramesh and Ravi Prakash
Department of Mechanical Engineering, Motilal Nehru National Institute of Technology, Allahabad, India

Karunesh Kumar Shukla
Department of Applied Mechanics, Motilal Nehru National Institute of Technology, Allahabad, India

Satoru Okamoto
Department of Mathematics and Computer Science, Shimane University, Matsue, Japan

Habib Reyhani Farashah, Seyed Ahmad Tabatabaeifar, Ali Rajabipour and Paria Sefeedpari
Department of Agricultural Machinery Engineering, Faculty of Agricultural Engineering and Technology, University of Tehran, Karaj, Iran

Jarman T. Jarman
Ministry of Interior, Riyadh, KSA

Essam E. Khalil and Elsayed Khalaf
Faculty of Engineering, Cairo University, Giza, Egypt

Muhammad Ery Wijaya and Tetsuo Tezuka
Department of Socio-Environmental Energy Science, Graduate School of Energy Science, Kyoto University, Kyoto, Japan

Andrea Giachetta, Katia Perini and Adriano Magliocco
Department of Architectural Sciences, University of Genoa, Genoa, Italy

Desley Vine, Laurie Buys and Peter Morris
Science & Engineering Faculty, Queensland University of Technology, Brisbane, Australia

Michael E. A. Warwick
Department of Chemistry, University College London, Christopher Ingold Laboratories, London, UK
UCL Energy Institute, Central House, London, UK

Ian Ridley
School of Property, Construction and Project Management, RMIT University, Melbourne, Australia

Russell Binions
School of Engineering and Materials Science, Queen Mary University of London, London, UK

Mohammad Malekizadeh and M. F. M. Zain
Department of Architecture, Universiti of Kebangsaan Malaysia (UKM), Bangi, Malaysia

Amiruddin Ismail
Sustainable Urban Transport Research Centre (SUTRA), Faculty of Engineering and Built Environment, Universiti Kebangsaan Malaysia (UKM), Bangi, Malaysia

Ahmad Hami
Department of Landscape Architecture, Universiti of Tabriz, Tabriz, Iran

Mala Abba-Aji
Department of Mechanical Engineering, University of Maiduguri, Maiduguri, Nigeria

Vincent Ogwagwu
Federal University of Technology, Minna, Nigeria

Bukar Umar Musa
Department of Electrical and Electronics Engineering, University of Maiduguri, Maiduguri, Nigeria

Mohamed Darwish
Qatar Environment and Energy Research Institute, Doha, Qatar

Nelson Fumo and Roy Crawford
The University of Texas at Tyler, Tyler, USA

María Belén Salvetti, Jorge Czajkowski and Analía Fernanda Gómez
Laboratorio de Arquitectura y Hábitat Sustentable (LAyHS), Facultad de Arquitectura y Urbanismo, Universidad Nacional de La Plata, La Plata, Argentina

Ellene Kebede and Edinam Adozsii
Department of Agriculture and Environment Science, Tuskegee University, Tuskegee, USA

Gbenga Ojumu
Department of Agriculture, Nutrition & Human Ecology, Prairie View A&M University, Prairie View, USA

www.ingramcontent.com/pod-product-compliance
Lightning Source LLC
Chambersburg PA
CBHW050452200326

41458CB00014B/5153